■ Encyclopedia of Invasive Species

■ Encyclopedia of Invasive Species

From Africanized Honey Bees to Zebra Mussels

Volume 1: Animals

Susan L. Woodward and Joyce A. Quinn

GREENWOOD

AN IMPRINT OF ABC-CLIO, LLC
Santa Barbara, California • Denver, Colorado • Oxford, England

Copyright 2011 by ABC-CLIO, LLC

Library of Congress Cataloging-in-Publication Data

Woodward, Susan L., 1944 Jan. 20–
  Encyclopedia of invasive species : from africanized honey bees to zebra mussels / Susan L. Woodward and
Joyce A. Quinn.
     p. cm.
  Includes bibliographical references and index.
  ISBN 978–0–313–38220–8 (cloth : alk. paper) — ISBN 978–0–313–38221–5 (ebook)
1. Introduced organisms—Encyclopedias. I. Quinn, Joyce Ann. II. Title.
QH353.W66   2011
578.6′2—dc23   2011026543

ISBN: 978–0–313–38220–8
EISBN: 978–0–313–38221–5

15 14 13 12 11   1 2 3 4 5

This book is also available on the World Wide Web as an eBook.
Visit www.abc-clio.com for details.

Greenwood
An Imprint of ABC-CLIO, LLC

ABC-CLIO, LLC
130 Cremona Drive, P.O. Box 1911
Santa Barbara, California 93116-1911

This book is printed on acid-free paper ∞

Manufactured in the United States of America

■ Contents

Aquatic Plants

Forbs

Graminoids

Vines

Tables and Lists about Invasive Plants

■ General Introduction: Invasive Species—Concepts and Issues

Just What Is an Invasive Species?

Many different words are used to refer to those species that cause great concern among land managers, ecologists, and ordinary folk dealing with the consequences of organisms that have been transported from their places of origin and released to go wild in the waters, forests, grasslands, and deserts of the United States. Some are synonyms, others are not. Some terms may have subtle value-laden connotations, while others are attempts at scientific objectivity. "Plant people" use different words than "animal people." There is as yet no consensus among scientists on what are the best terms to use, so the literature remains inconsistent and definitions sometimes vague.

Several terms merely imply populations existing outside their native range and can be used more or less interchangeably: *alien, exotic, nonnative*, and *nonindigenous*. The term "alien" may be viewed as pejorative, causing unwarranted bias—and unnecessary actions—against all foreign species. Nonetheless, the official legal definition of "alien" was put forth by President Clinton's Executive Order 13112 (February 3, 1999), which states that an alien species is, "with respect to a particular ecosystem, any species, including its seeds, eggs, spores, or other biological material capable of propagating that species, that is not native to that ecosystem." This means that a species native to one part of the United States can be alien to another region of the country, as is the case with so-called native transplants or "domestic exotics" such as the American bullfrog (*Lithobates catesbeianus*) in western waterways, smooth cordgrass (*Spartina alterniflora*) in western salt marshes, or the House Finch (*Carpodacus mexicanus*) in eastern states.

What constitutes a native species is open to question. Native species are generally considered "natural" and in their place of origin, i.e., the place where they belong. In the United States, native species are often those considered to have been here when Europeans first colonized the North American continent in the sixteenth century. This qualification, however, is based on two questionable concepts: (1) that North American environments were pristine before the arrival of Europeans in the fifteenth century, and (2) that only habitats not influenced by human activity are natural. Realizing that aboriginal peoples may have had profound impacts on North American ecosystems through their use of fire, their hunting, their trade, and their agriculture, some argue that the date dividing pristine from disturbed environments should be pushed back a few tens of thousands of years to when humans—the most invasive species of all—first set foot on the continent.

Biogeographers tend to define *native range* as that geographic region where a species evolved to its contemporary form and to which it was long restricted by natural barriers. In the history of life, overcoming such barriers is a common occurrence, as is the removal of barriers. Species constantly disperse from their place of origin as their populations grow, as they evolve, as continents move and oceans shrink, as drainage systems pirate headwater streams from other river systems, and so forth. Long-distance migrant species populated distant oceanic islands and led to the exchange of plants and animals among continental

About Range Maps

Range maps are usually constructed from records of species occurrence compiled in museum and herbaria collections. Each verified specimen is represented by a dot on a map, and usually the outermost dots are connected to form the range boundary. Adjustments are often made based on knowledge of the specific environmental conditions of the area and of the habitat requirements and ecological tolerances of the species in question. A given species thus does not occur everywhere in its mapped range, but only where conditions are suitable within the designated region.

Information on the distribution of many species is often a matter of presence or absence in a particular area. Much of the information available for invasive plants, especially, comes as records of occurrence (or absence) in a political unit such as an entire state. Maps based on this information necessarily provide highly generalized distribution patterns. Larger political units, such as Texas or China, mean even more generalization. Absence of a species according to the maps may be because it is not yet recognized or recorded at a particular location.

In order to present the best possible depiction of a species' range the authors have consulted a variety of sources and synthesized often-conflicting information. Native ranges may be especially problematic because biological surveys are not advanced in many parts of the world and reliable range maps did not exist. In many cases the place of origin of a species that is invasive in the United States is simply not known or at best uncertain. Such instances are indicated by question marks on the maps. Plants, animals, fungi, and microorganisms have been carried around the world with humans for millennia, obscuring the history of their origins. Some have undergone subsequent evolution that complicates the picture further. Increasingly, genetic studies are confirming earlier scientific conclusions or shedding new light on the past travels of species and pinpointing their starting points with increased accuracy.

Maps, like the species themselves, are dynamic. As new sightings and collections are made, the range maps may change. The reader wanting the most up-to-date information or more detailed information than can be derived from the small maps in this work is encouraged to seek out recent publications and local experts in the field.

When comparing maps from different sources, the reader is cautioned that the appearance of distribution areas will vary when different map projections—methods of converting the spherical globe into a flat map—are used. Distortions in size, shape, or cardinal direction (N, S, E, W) are inevitable. The projections chosen for this volume (Eckert III for the world and Albers Equal-Area Conic for the United States) display fairly accurately true area or size relationships with minimal shape distortion. Areas closer to polar regions, as indicated by Greenland, are increasingly more distorted in shape and direction.

landmasses many times in the geological past. Species are still on the move, actively and passively, in the same manner as in the past. Those that arrive "on their own," such as the Cattle Egret (*Bubulcus ibis*), are referred to as adventive species. The vast majority of species today, however, are transported, deliberately or accidentally, by humans. Such newcomers to a region are known as introduced species.

Not all nonnative species are invasive; in fact, most are not. Legally, that term should be reserved for "an alien species whose introduction does or is likely to cause economic or environmental harm or harm to human health" (Executive Order 13112 of February 3, 1999). A legally defined term that predates "invasive" is nuisance species, which appeared in the Nonindigenous Aquatic Nuisance Prevention and Control Act of 1990 (Public Law 101-636) and continues in use. An aquatic nuisance species was defined as a nonindigenous species that "threatens the diversity or abundance of native species or the ecological stability of infested waters, or commercial, agricultural or recreational activities dependent upon such waters." It is the same thing as an invasive species.

Some scientists dislike having the definition of invasive species dependent on negative impacts and prefer a more neutral definition that focuses on the population growth and range expansion of certain nonindigenous species. They would argue that impacts can be somewhat subjective and vague, and that range expansion or spread is the main process that distinguishes some nonnative species from others. For example, many a nonindigenous species arrives, reproduces, and establishes a population that is self-sustaining over several generations and perhaps for a very long time in natural and seminatural habitats at or near the point of entry; but it never spreads beyond the immediate areas of its introduction—i.e., it never become "invasive." Such alien plants and animals may be categorized as established. Where they have become regular, functioning parts of a recipient ecosystem, even if far from the point of entry, they are considered naturalized.

Some invasive plants are identified as "noxious" species, a category defined in the 1974 Federal Noxious Weed Act as "any living stage, such as seeds and reproductive parts, of any parasitic or other kind of plant, which is of foreign origin, is new to or not widely prevalent in the United States, and can directly or indirectly injure crops, other useful plants, livestock, or poultry or other interests of agriculture, including irrigation, or navigation, or the fish or wildlife resources of the United States or the public health." Noxious plants are so designated by federal and state laws. For animals, the term injurious is used to describe those "species, including offspring and eggs, determined to be injurious to the health and welfare of humans, the interests of agriculture, horticulture or forestry, and the welfare and survival of wildlife resources of the U.S." (Branch of Invasive Species, U.S. Fish and Wildlife Service). Under the Lacey Act (see Appendix B), the designation of such species is made by the U.S. Department of the Interior; and the importation and interstate transport of wild mammals, wild birds, fish, mollusks, crustaceans, amphibians, and reptiles so designated are regulated by the secretary of the interior. (Currently, no amphibians are on the list of injurious species, but in September 2009, the Defenders of Wildlife petitioned Secretary Salazar to place all amphibians on the list unless certified free of the chytrid frog fungus [*Batrachochytrium dendrobatidis*].)

The Size of the Problem

According to many resource managers and ecologists, invasive species and potential invaders together constitute one of the major environmental threats facing the United States in the twenty-first century. Preventing or controlling the establishment of nonindigenous species is complicated by a number of uncertainties, including how many and which of the nonnative species will become abundant and widespread and inflict significant impacts on natural ecosystems; national, regional, and/or state economies; or public health. According to one estimate, some 50,000 species of plant, animal, fungi, and microbe have been transported to the United States. Of these, perhaps 6,500 have established viable, self-sustaining populations and 500 or so have become invasive in the sense both of demonstrating rapid population growth and range expansion and of causing harm or altering natural ecosystems. Actual numbers are difficult to obtain and are more reliable for some classes of organisms than others. Attention focused earlier and with greater federal funding on agricultural pests and nonnative aquatic species than on others, so more data are available for those groups. Inconsistent use of terms often further muddies the waters of how many species are invasive as opposed to simply introduced or naturalized. The figures that follow should only be considered indicators of the size of the problem and not definitive, accurate measures.

Invasive Plants

Plants in the United States consist of approximately 18,000 native species and 5,000 non-indigenous species. As is true for most if not all types of invasive species, the greatest numbers occur in Florida and Hawai'i. Florida is home to more than 900 naturalized plant species; Hawai'i reports 946. However, only some are invasive. Most plants that have become invasive were intentionally brought to the United States as ornamentals or for uses as varied as erosion control or herbal medicine. Others were accidental introductions, primarily as contaminants in crop seeds or hay or carried in ship ballast. Many nonnative plants alter natural ecosystems, displacing native species, many of which are rare or endangered, but impacts are far-reaching and varied. Nonnative plants are estimated to comprise 73 percent of the weeds of cultivated farmland and 45 percent of pasture weeds. Some plants may be easily contained in drier western climates where limited water prevents their spread, but may be unstoppable in the rainier eastern states. Others may be invasive in grasslands or deserts, but not in ecosystems where competition from other plants keeps them under control.

Overall, 138 trees and shrubs are considered invasive in the United States. Some of the most notorious trees are salt cedar or tamarisk (*Tamarix* spp.), Australian melaleuca (*Melaleuca quinquenervia*), and Brazilian peppertree (*Schinus terebinthifolius*). Tamarisk depletes surface and groundwater supplies and makes soils so salty that other plants are unable to grow. Melaleuca infestations are not only threatening the Everglades' unique ecosystem, but the tree is also detrimental to human health and causes increased incidence of fire. Brazilian peppertree is ranked with melaleuca as one of the worst invaders of the Everglades. In Hawai'i, unless checked, velvet tree (*Miconia calvescens*) could devastate native forests and alter hydrology as it has in Tahiti.

Shrubs can be equally devastating. Tropical soda apple (*Solanum viarum*), from South America and now invasive in the southeastern United States, is a competitive plant in agricultural fields, both pasture and cropland. It also harbors insects or pathogens that severely damage and reduce yields of food crops. Gorse (*Ulex europaeus*), a large bushy shrub with many attractive yellow flowers, forms thorny thickets that impede passage and degrade the quality of recreational activities. The oily foliage and deep accumulation of dry matter cause fires to occur more frequently and to burn hotter. The common garden plant lantana (*Lantana camara*) is also a fire hazard in southeastern states, where it invades citrus groves. Its leaves and unripe fruit are toxic to animals and can cause death.

More than 500 introduced forbs and graminoids are considered noxious weeds. Forbs may be as small as the 12 in. (30 cm) tall fig buttercup (*Ficaria verna*) or grow as large as giant hogweed (*Heracleum mantegazzianum*), which can be 25 ft. (7.6 m) tall with leaves

The Problem with Common Names and Why Scientific Names Are Surer Bets

The common names of plants and animals are not always reliable indicators of what species is being discussed. Different names may be used by people in different parts of the country. For example, in the eastern United States the leafy weed known in most places as Japanese knotweed is called Japanese bamboo, although it is not even a relative of true bamboos, which are grasses. The misnomer points to another difficulty: common names frequently do not reflect the true identify of a plant or animal. Among the many tiny invasive animals causing problems in the Great Lakes is the spiny water flea, not an insect like real fleas but a crustacean.

Scientific names may at first be strange tongue-twisters because they have been latinized, but they have been approved by international committees and linked to one species and one species only. Often the name provides insights into the nature of the organism or, the place where it was first discovered. Scientific names are revised from time to time as scientists come to new understandings about a species' evolution and relation to other species, but while in current use they leave no doubt what plant, animal, fungus, or microorganism is being discussed. Each scientific name has two parts. The first word refers to the genus into which biologists have classified the species and the second identifies the particular member of the genus. Both words must be used to correctly designate a particular species.

In the Encyclopedia's species accounts, common names appear as the title of each entry, and entries are arranged within major categories alphabetically by common name. More frequently used alternate names, especially those of regional prominence, are given as appropriate. The scientific name—and some others that have been used in older literature (synonyms)—follow. A reader seeking more information should search according to the scientific name, or at least verify that the scientific name is the same, to be sure he or she is gathering data about the same organism as the one described in a particular entry.

as wide as 5 ft. (1.5 m). Fig buttercup displaces plants that are important sources of pollen and nectar for pollinating insects, threatening their existence. Contact with the sap of giant hogweed, a plant recently invasive in New England and the upper Midwest, causes a person's skin to be susceptible to severe sunburns. Several thistles and thistle-like knapweeds degrade range and pasture land, and some, especially yellow star-thistle (*Centaurea solstitialis*), are very poisonous to livestock. Prickly Russian thistle (*Salsola tragus*), more commonly recognized as tumbleweed in the arid western states, looks more like a shrub than a forb. As an annual, however, it lives only one year, leaving behind the woody skeleton of its branches that rolls away in the wind, distributing seeds.

Many alien grasses alter the fire regime of rangeland or desert scrub, destroying native plant species that are not fire adapted, and in turn displacing the animals dependent on that ecosystem. These include the infamous cheatgrass (*Bromus tectorum*) that plagues western grazing lands, lowering the nutrient value of rangeland and injuring livestock with their sharp awns. The feathery plumes of fountain grasses (*Pennisetum* spp.) and Pampas grass (*Cortaderia selloana*) that make them desirable as ornamental plants in the garden often make it difficult to convince horticulturalists of the harm those plants can do when they escape into the natural landscape, where they displace native plants and create fire hazards. Dense stands of giant reed (*Arundo donax*), with vast intertwined root systems, clog stream channels, altering hydrology and water flow.

Several vines grow so efficiently that they completely cover the ground and overtake shrubs and even tall trees. Probably the best known is kudzu (*Pueraria montana*), which blankets acres of land in southeastern states, not only smothering vegetation but breaking trees and utility lines with the weight of the vines and foliage. Alien swallow-worts (*Cynanchum* spp.) endanger monarch butterflies because eggs laid on the plants, in the absence of the butterfly's preferred host, do not survive. Japanese dodder (*Cuscuta japonica*), a parasite just beginning to gain a foothold in California, poses such a significant threat, not only to natural landscapes but also to crops, that it could affect international trade of crop seeds.

Among the 160 or so aquatic plants that are nonindigenous, the best known and most troublesome invasives are probably waterhyacinth (*Eichhornia crassipes*) and hydrilla (*Hydrilla verticillata*). Both species grow into dense mats that alter natural ecosystems, downgrade water quality, impede water transportation, and limit recreational activities.

Invasive Mammals

Introduced mammals include all domesticated livestock; some of their feral descendents that run free in natural habitats are among the world's worst invaders. In the continental United States, feral pigs (*Sus scrofa*) are a particular scourge; on Hawai'i and islands in general, feral goats (*Capra hircus*) are other voracious destroyers of native plants and the habitats supporting endemic birds and other animals. Other mammals that were never domesticated but nonetheless transported to the United States by humans, either deliberately or accidentally, include such invasive species as the black or roof rat (*Rattus rattus*), Norway rat (*Rattus norvegicus*), nutria (*Myocastor coypus*), and the Indian mongoose (*Herpestes javanicus*).

Introduced birds are some of the better-studied species as well as the animals perhaps most familiar to the American public. The urban Rock Pigeon (*Columba livia*) is actually a feral bird first brought to the country in colonial times as a food animal. Two other common avian pests in cities and towns today were deliberately released to "enhance" the wild bird fauna of the United States: the House Sparrow (*Passer domesticus*) and the European Starling (*Sturnus vulgaris*). The most abundant bird in Hawai'i, the Japanese White-eye (*Zosterops japonicus*), came from eastern Asia.

Reptiles introduced from other continents include the huge Burmese python (*Python molurus bivittatus*)—now rapidly increasing its numbers in the Everglades and spreading beyond the park boundaries to invade other parts of Florida—as well as the brown anole (*Norops sagrei*), a small lizard that appears to be spreading out of Florida. Among invasive amphibians is the annoying little Cuban treefrog (*Osteopilus septentrionalis*), which gets into peoples' houses, but the most ecologically damaging is a native frog from the eastern United States introduced into western waters, the American bullfrog (*Lithobates catesbeianus*). As further indication of how aquatic ecosystems have been affected by so-called native transplants like the bullfrog, almost two-thirds of the fishes found in American drainages were introduced by people from other U.S. waterways, sometimes in the same state and sometimes from across the continent. Examples of such native invasive fishes are the alewife (*Alosa pseudoharangus*) in the Great Lakes, the western mosquitofish (*Gambusia affinis*) in states west of its native range limits in Texas, and rainbow trout (*Oncorhynchus mykiss*), native to Pacific drainages but now stocked throughout the United States and indeed much of the world. Alien invasive fish include various Asian carps and such strange creatures as walking catfish (*Clarius batrachus*) and the northern snakehead (*Channa argus*).

Invertebrates, fungi, and microorganisms pose an even greater challenge when estimating numbers of invasive species. We do not even know how many native species exist in most taxonomic groups. According to one estimate, some 4,500 arthropods have been introduced to the United States, more than 2,500 of them to Hawai'i. The hemlock woolly adelgid (*Adelges tsugae*) has decimated hemlocks in natural forests and opened forest streams to sunlight, while the Asian longhorned beetle (*Anoplophora glabripennis*) threatens shade and ornamental trees in New York and Chicago but could become a forest pest on a par with the invasive gypsy moth (*Lymantria dispar*) if it invades eastern broadleaf deciduous forests. Some invading insects, such as the Formosan subterranean termite (*Coptotermes formosanus*), cause devastating structural damage to buildings; while others, such as the common bed bug (*Cimex lecturius*) and brown marmorated stink bug (*Halyomorpha halys*), are simply serious annoyances inside man-made structures. Among the more potentially damaging insect invaders are those that could disrupt crop pollination by negatively affecting European honey bees—themselves an introduced (domesticated) species. Africanized honey bees (*Apis mellifera scutella*) hybridize with European honey bees and take over their hives. Two introduced arachnids, the honeybee tracheal mite (*Acarapis woodi*) and the varroa mite (*Varroa destructor*) parasitize bees and lead to the demise of bee colonies.

Snails, clams, and mussels are among other significant invertebrate invaders. Zebra mussels (*Dreissena polymorpha*) can transform freshwater communities and clog water intake and distribution pipes. Their rapid spread through Great Lakes and into the Mississippi River drainage in the 1980s was a major stimulus to the development of interest in exotic species in general in the United States.

Invasive fungi are largely associated with diseases affecting trees, such as Dutch elm disease and sudden oak death, but a fungus is also responsible for a disease infecting frogs and toads in many parts of the United States. One of the newest invaders, the fungus *Geomyces destructans*, is implicated in bat white-nose syndrome, a condition currently ravaging bat colonies in the eastern United States.

The smallest of invasive species, the microorganisms, are represented by only three entries in this encyclopedia. Humans have been transporting protists, bacteria, and viruses as long as human migrations have taken place. The human diseases brought by early settlers from overseas—smallpox, influenza, measles, to name a few—decimated Native American populations. More recently, HIV and new strains of influenza have run rampant through the U.S. population. Emerging infectious diseases such as dengue fever or ebola may be just around the corner. These are more appropriately dealt with in a book on epidemiology than one on invasive species. The three organisms chosen for inclusion here have close ties to natural habitats and infect wild animals (as well as, in some cases, humans). Avian malaria threatens rare, endemic birds in Hawai'i. The bacterium *Borrelia burgdorferi* has a complicated life cycle involving two mammalian hosts and expresses itself as Lyme disease in humans. The West Nile virus infects birds, horses, and humans. It is now controlled in horses by vaccination; its long-term impact on wild bird populations remains to be seen.

To put the above information in some perspective, it should be noted that about 200,000 species are believed to inhabit the United States. About 91,000 have been described, leaving a large number of plants, animals, fungi, and microbes yet to be discovered. As of the latest count, plants account for nearly 19,000 of the native species, and vertebrates for about 3,000. (According to one count, there are 1,154 native fishes, 295 amphibians, 311, reptiles, 784 birds, and 428 mammals.)

Native aquatic species—mussels, fishes, salamanders, and turtles—are quite diverse by global standards, as are habitats for both terrestrial and aquatic life. Nearly one-third of our species are considered at risk, including almost 70 percent of freshwater mussels and more than 50 percent of native crayfishes. Natural vegetation has been removed or greatly altered in more than 60 percent of the land area in the lower 48 states. Experts generally agree that destruction of habitat is the greatest threat to our native species, but that the introduction of nonnative species is the second greatest cause of decline and disappearance of native plants and animals.

The Invasion Process

For a nonnative species to gain a foothold and become an abundant and widespread inhabitant in an area beyond its native range, several steps or stages are required. Each step involves overcoming some sort of barrier. Species are normally held in their native ranges by geographic barriers to dispersal, such as an ocean, a different and inhospitable (for them) climate region, a drainage divide, a mountain range, distance, wind or ocean current direction, or some other natural feature of the planet. The first stage, the transport stage, requires getting through or over the unfavorable conditions imposed by what is normally a geographic barrier and entering a new site. Plants and animals have done this successfully on their own over millions of years, either by chance or because changes to the environment weaken or remove the barriers. Thus, species have colonized oceanic islands and moved from one continent to another. When humans began to migrate and then to engage in trade, they accelerated the process by either deliberately or unintentionally providing plants, animals, and microorganisms safe transport to new areas in their provisions, packing materials, ships, and other vehicles. Some organisms became desirable commodities in their own right as exemplified today by the pet trade and horticultural industry.

Getting to a new site is only the first step. Once a species has arrived or "been introduced," it must be able to reproduce and establish a self-sustaining population before it can be in a position to become invasive. Surmounting limitations imposed by small founding population sizes and environmental barriers in the new location are the challenges of the second or establishment stage. Typically, only a small number of individuals of a given species occur in the new area, a factor that by itself makes them vulnerable to extinction. The so-called Allee effect depresses reproductive rates since the few individuals present may be sparsely distributed and have difficulty finding mates, or the sex ratio may be skewed toward one gender or the other, both limiting the number of offspring that can be produced during the first few generations. In many instances, a new species persists in an area only because it is repeatedly introduced and not because the species is reproducing at the site.

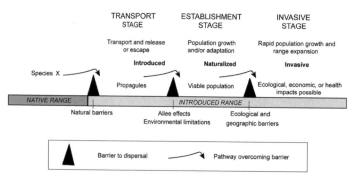

THE INVASION PROCESS

The three stages of the invasion process, showing population status and dispersal patterns in each.

Since the new environment will not be identical to that of the native range, the species may need time to adapt to a new climate, substrate, ecological community, or other aspect of the receiving habitat before its success is assured. If a population does not evolve, it remains highly localized or goes extinct. Even when the demographic and environmental barriers are overcome, it can take 10–25 years for a new species to become numerous enough for people to detect its presence. Success in the establishment phase means reproduction occurs regularly and not only sustains the population, but allows for population growth. The species nonetheless remains near the initial point of introduction at this stage.

When the number of individuals increases, the tendency is for a species to expand its distribution area—that is, to spread. Eventually, it will have to overcome local geographic barriers to dispersal. As it spreads into new regions in its adopted home, it enters the invasive stage and begins to produce viable offspring at some distance from the original place of introduction. During the invasive stage, a species continues to increase in abundance and expand its range. Typically, it will first occupy sites disturbed by human activities (such as farmland and settlements) and then be found in seminatural and natural ecosystems. If the new species is perceived to harm or change native ecosystems, it may be deemed officially an "invasive species." Some ecologists prefer that infliction of ecological and economic harm on the recipient region be considered an additional, fourth stage in the invasion process. Species can spread without causing harm, as has happened with the Cattle Egret.

It is frequently stated that, on average, about 10 percent of species actually move from one stage to the next. This means that one in ten of those transported to a new area actually survives and will be found outside of captivity or other controls (i.e., is introduced). One in ten of those will establish self-replacing populations (i.e., become established), and one in ten of established species will spread and become invasive. The result is that very few species arriving in areas they previously have not occupied pass through all the barriers and become invasive. Most species fail to gain even a temporary foothold, let alone invade new regions. Studies have shown that this so-called "tens rule" describes the history of nonindigenous terrestrial vertebrates, fishes, insects, mollusks, and plant pathogens in the continental United States fairly well, but fails to accurately describe nonnative birds in Hawai'i, where more than half of all birds known to have been introduced have become established. It should be noted that the original work that proposed the tens rule focused on plants introduced into the United Kingdom; 10 percent was the average of a range running from 5 to 20 percent. Research examining only those vertebrates coming into the United States from Europe determined that 25 percent became invasive. At this point, it remains unknown how many actually can be expected to succeed, or what makes a species a successful invader, or what allows a native ecosystem to resist, accommodate, or succumb to invasive alien species.

For species that do become "invasive," the process proceeds in a typical pattern. Initially, only a few individuals are spotted in a new area. There is a long lag time before, suddenly, what seemed like an innocuous new member of an ecological community explodes in numbers (*irrupts*) and becomes a major nuisance or pest and may even begin to transform native ecosystems. Very often, at least in non-island situations, after the initial irruption, the nonindigenous population declines; and the native system accommodates the newcomer. When this happens, the exotic becomes a naturalized member of the community.

Recognition of stages in the invasion process helps resource managers design control and mitigation programs. The easiest and most cost-effective control scheme is to intercept arrivals and prevent establishment and/or spread in the first place. This involves detecting incoming species through inspection and quarantines. Constant monitoring is necessary to measure the success or failure of such measures. Once a population is growing and spreading, management becomes problematic and eradication nearly impossible.

Pathways of Introduction

The means by which species are introduced into new areas are known as pathways. These are generally separated into deliberate or intentional pathways and accidental or unintentional pathways (Table 1 and Table 2). Escapes from confinement or captivity may be thought of as a third, hybrid pathway in which the initial importation was intentional, but the release of free-living individuals or populations was an accident. Generally speaking, most plants and vertebrates have been introduced deliberately; most invertebrates and microorganisms arrived by accident.

Plants naturally disperse by sending forth spores and seeds and, for a few, seedlings or parts capable of sprouting vegetatively. These propagules are transported by wind, ocean and stream currents, or animals to places beyond the parent plant and sometimes beyond the range limits of the species. If they land in new territory, they may or may not survive and establish a colony. Humans may have first changed the dispersal process for some plants (and animals) by creating disturbed habitats around campsites and unintentionally carrying seeds on their bodies or in their digestive tracts—just as other animals do. They also harvested and stored seeds and deliberately transported them as they moved from site to site. Some plants adapted to the new dispersing agent and became "camp-followers," showing up unbidden at each new settlement. *Cannabis sativa* (marijuana) was such a plant in the Old World. The precursors of many crops may have acted similarly. Mammals such as the dog and pig may have also joined the retinue of people on the move along with arthropods such as cockroaches, lice, silverfish, and other such species that came to live in close association with human beings.

Once domesticated crops and livestock were available, people deliberately transported them to far-off places. Seeds and cuttings were traded first among neighboring villages and later to distant shores. With rafts and sailing ships, people helped terrestrial and freshwater species overcome the geographic barrier of the sea. Plants and animals were carried intentionally and accidentally around the world, especially by Europeans and Polynesians. As seagoing technology improved, more distant parts of the planet became connected. Sailing ships were pretty much confined to the natural routes determined by wind and current. But sea-lanes were cut across these old routes when steamships came of age. Seaports were the first point of entry for many species new to an area. Accidental travelers were hidden in the dry or solid ballast of ships under sail. Others, such as shipworms, burrowed in the wooden hulls. During the colonial period, the flow of goods ran largely from the Americas to Europe. Ships from England, France, Spain, and Holland sailed without full loads and took on needed ballast at their home ports. The ballast was offloaded at the port of destination, and if nonnative species had survived the journey, they might become established near wharves and piers and at ballast dumping grounds. Botanists looking for "new" species hunted in these locations. Seeds and insects also stowed away in the hay and straw carried aboard ships as packing material or as fodder and bedding for livestock. Rats climbed the mooring lines at one port to sail to and disembark at another. Everything was on the move.

From seaports, a few new arrivals were able to move inland along the canals built to tie the port city to its hinterland. Purple loosestrife (*Lythrum salicaria*), for example, a plant that came to dominate marshes in New England, was first collected from wetlands near the Erie and Delaware-Ruritan canals. Railroads penetrated even farther inland; their right-of-ways

Table 1. Examples of Invasive Species That Have Been Introduced by Intentional Pathways[a]

Aesthetic amenities, sentiment, or nostalgia	European Starling (*Sturnus vulgaris*)
	House Sparrow (*Passer domesticus*)
	Ornamentals, including but not limited toLantana (*Lantana camara*)
	Multiflora rose (*Rosa multiflora*)
	Strawberry guava (*Psidium cattleianum*)
	Wisterias (*Wisteria sinensis, W. floribunda*)
Aquarium trade	Common salvinia (*Salvinia minima*)
	Hydrilla (*Hydrilla verticillata*)
	Lionfish (*Pterois volitans*)
Bait bucket releases	Nightcrawler (*Lumbricus terrestris*)
	Rusty crayfish (*Orconectes rusticus*)
Biological controls	Common myna (*Acridotheres tristis*)
	Grass carp (*Ctenopharyngodon idella*)
	Mosquitofish (*Gambusia affinis, G. holbrooki*)
	Multicolored Asian lady beetle (*Harmonia axyridis*)
	Indian mongoose (*Herpestes javanicus*)
Bioterrorism (potential)	Infectious diseases
Botanical gardens	Bush honeysuckles (*Lonicera* spp.)
	Japanese barberry (*Berberis thunbergii*)
	Velvet tree (*Miconia calvescens*)
Domestic use (dyes, fish poisons)	Common mullien (*Verbascum thapsus*)
	Dyer's woad (*Isatis tinctoria*)
	Yellow toadflax (*Linaria vulgaris*)
Erosion control/bank stabilization	Australian pine (*Casuarina equisetifolia*)
	Giant reed (*Arundo donax*)
	Ice plant (*Carpobrotus edulis*)
	Japanese knotweed (*Fallopia japonica*)
	Kudzu (*Pueraria montana*)
Food and beverage	Asian clam (*Corbicula fluminea*)
	Chinese mitten crab (*Eriochor sinensis*)[b]
	Chinese mystery snail (*Cipangopaludina chinensis malleata*)
	Fire tree (*Morella faya*)
	Golden apple snail (*Pomacea canaliculata*)
	Himalayan blackberry (*Rubus armeniacus*)
	Northern snakehead (*Channa argus*)
Forage fish	Alewife (*Alosa pseudoharengus*)
Furbearers	Nutria (*Myocastor coypus*)
Livestock abandonment	Feral goat (*Capra hircus*)
	Feral horse (*Equus caballus*)
	Feral pig (*Sus scrofa*)
Livestock forage or fodder	Buffelgrass (*Pennisteum ciliare*)
	Cogon grass (*Imperata cylindrica*)
	Gorse (*Ulex europaeus*)
Medicinal herbs or seasonings	Common St. Johnswort (*Hypericum perforatum*)
	Garlic mustard (*Alliara petiolata*)
	Giant hogweed (*Heracleum mantegazzianum*)
	Japanese hops (*Humulus japonicus*)

(Continued)

Table 1. (Continued)

Packing material	Japanese stilt grass (*Microstegium vimineum*)
	Princess tree (*Paulownia tomentosa*)
	Smooth cordgrass (*Spartina alterniflora*)
Pet abandonment	Burmese python (*Python molurus bivittatus*)
	Feral cat (*Felis silvestris catus*)
	Nile monitor (*Varanus niloticus*)
Pet trade	Burmese python (*Python molurus bivittatus*)
	Giant African snail (*Achatina fulica*)
	Nile monitor (*Varanus niloticus*)
Research	African clawed frog (*Xenopus laevis*)
	Gypsy moth (*Lymantria dispar*)
Sport fishing	Brown trout (*Salmo trutta*)
	Rainbow trout (*Oncorhynchus mykiss*)
Timber/reforestation/firewood	Carrotwood (*Cupaniopsis anacardioides*)
	Melaleuca (*Melaleuca quinquenervia*)
Wildlife habitat or food	Chinese lespedeza (*Lespedeza cuneata*)
	Russian olive (*Elaeagnus angustifolia*)
Windbreaks/fencerows	Melaleuca (*Melaleuca quinquenervia*)
	Russian olive (*Elaeagnus angustifolia*)
	Tamarisk (*Tamarisk chinensis, T. ramosissima*)

[a]See table "Pathways of Introduction for Plants," in Volume 2, for a more complete listing of plant pathways.
[b]Probable means of introduction.

became avenues of expansion for weedy plants, while other organisms rode the rails as hitchhikers in cargo and packing materials to almost all parts of the country. Airplanes now reach the most isolated places, so nowhere is immune to the introduction of nonnative species.

Sometimes military traffic is implicated in the transport of unwanted species to American shores. Perhaps the most notorious example is the arrival of the brown tree snake (*Boiga irregularis*) on the U.S. territory of Guam after World War II. But a noxious weed, Canada thistle (*Cirsium arvense*), had also moved on Union military steamships during the American Civil War, reaching its southern limits at the Virginia town of Remington. More recently, the Argentine ant (*Linepithema humile*) arrived in Hawai'i on military ships during World War II, and the Formosan subterranean termite (*Coptotermes formosanus*) reached Houston, Texas, on ships returning from the Pacific theater.

Shipping remains an important pathway for the entry of new species. Fouling organisms such as lacy crust bryozoan (*Membraniphora membranacea*), chain tunicates (*Botrylloides violaceus*), and colonial tunicates (*Didemnum vexillum*) made their way to U.S. waters in this manner. Today, seawater is used as ballast, and both larval and adult aquatic organisms, plants as well as animals, have been transported from one port across an ocean to another port, where the ballast water is discharged. The zebra mussel (*Dreissena polymorpha*) and the spiny water flea (*Bythotrephes longimanus*) both presumably made their way to the Great Lakes in this manner. Canal construction opened the Great Lakes to invasion by such fish as sea lamprey (*Petromyzon marinus*) and round goby (*Neogobius melanostomus*). The shipment of used tires on cargo ships provided shelter and breeding sites for Asian tiger mosquitoes (*Aedes albopictus*), allowing their entry into the United States.

Table 2. Examples of Invasive Species That Have Been Introduced by Unintentional Pathways[a]

Ballast and bilge water discharge	Asian green mussel (*Perna viridis*)
	Common reed (*Phragmites australis* spp. *australis*)
	Eurasian watermilfoil (*Myriophyllum spicatum*)
	Purple loosestrife (*Lythrum salicara*)
	Quagga mussel (*Dreissena rostriformes bugensis*)[b]
	Smooth cordgrass (*Spartina alterniflora*)
	Spiny water flea (*Bythotrephes longimanus*)
	Zebra mussel (*Dreissena polymorpha*)
Crop seed/plant contaminants	Canadian thistle (*Cirsium arvense*)
	Cheatgrass (*Bromus tectorum*)
	Prickly Russian thistle (*Salsola tragus*)
	Quackgrass (*Elymus repens*)
	Yellow starthistle (*Centaurea solsititialis*)
Dry (solid) ballast	Earthworms
	Common periwinkle (*Littorina littorea*)
Escapes from aquaculture	American bullfrog (*Lithobates catesebianus*)
	Bighead carp (*Hypophthalmichthys nobilis*)
	Giant salvinia (*Salvinia molesta*)
	Grass carp (*Ctenopharyngodon idella*)
	Water chestnut (*Trapa natans*)
	Waterhyacinth (*Eichhornia crassipes*)
Escapes from fur farms	Nutria (*Myocastor coypus*)
Escapes from gardens	Chocolate vine (*Akebia quinata*) and most others that originated as ornamental plants
Escapes from pet owners	Feral cat (*Felis silvestris catus*)
	Monk Parakeet (*Myopsitta monachus*)
	Mute Swan (*Cygnus olor*)
Escapes from research labs	Africanized honey bee (*Apis mellifera scutellata*)
	Gypsy moth (*Lymanatria dispar*)
Fish stock contaminants	African clawed frog (*Xenopus laevis*)
	New Zealand mud snail (*Potamopygrus antipodarum*)[b]
Hull fouling	Chain tunicate (*Botrylloides violaceus*)
	Colonial tunicate (*Didemnum vexillum*)
	Lacy crust bryozoan (*Membranipora membranacea*)
Ocean currents	Asiatic colubrina (*Colubrina asiatica*)
	West Indian marsh grass (*Hymenachne amplexicaullis*)
Removal of natural barriers once separating bodies of water (e.g., canal construction)	Round goby (*Neogobius melanostomus*)
	Sea lamprey (*Petromyzon marinus*)
	Zebra mussel (*Dreissena polymorpha*)
Stowaways in cargo	Asian tiger mosquito (*Aedes albopictus*)
Stowaways in clothing, bedding, and luggage	Common bed bug (*Cimex lectularis*)
Stowaways in equipment, including military	Canada thistle (*Cirsium arvense*)[a]
	Formosan subterranean termite (*Coptotermes formosanus*)
	Argentine ant (*Linepithema humile*)
Stowaways in grain shipments	Argentine ant (*Linepithema humile*)
Stowaways in packing materials	Asian longhorned beetle (*Anoplophora glabripennis*)
	Brown marmorated stink bug (*Halyomorpha halys*)

(Continued)

Table 2. (Continued)

Stowaways in wood products, including firewood	Dutch elm disease fungus (*Ophiostoma novo-ulmi*)
	Emerald ash-borer (*Agrilus planipennis*)
Stowaways on live plants, bulbs, or root balls	Glassy-winged sharpshooter (*Homalodisca vitripennis*)
	Hemlock wooly adelgid (*Adelges tsugae*)
	Coqui (*Eleutherodactylus coqui*)
	Chestnut blight fungus (*Cryphonectria parasitica*)
	Earthworms
	Japanese beetle (*Popillia japonica*)
	Red imported fire ant (*Solenopsis invicta*)
Stowaways on ships	Black rat (*Rattus rattus*)
	Cuban treefrog (*Osteopilus batrachus*)
	Norway rat (*Rattus norvegicus*)

[a]See table "Pathways of Introduction for Plants," in Volume 2, for a more complete listing of plant pathways.
[b]Probable means of introduction.

Most fungal pathogens arrive with their natural host plants. In the case of the chestnut blight fungus (*Cryphonectria parasitica*), live plant material was involved. Dutch elm disease arrived on veneer logs. Fungi may also be inadvertently transported in soil and in root balls. Insects, too, infest plants, including wood products, seed supplies, and grain shipments.

Terrestrial vertebrates are mostly imported and transported deliberately. Amphibians and reptiles are often valued as bait and pets and used for biomedical research. They may also hitchhike on plants; the coqui (*Eleutherodactylus coqui*), which has invaded Hawai'i, is a case in point. Nonnative birds have been released as game animals, but most come in as pets and either are released or escape captivity. Many of the more troublesome invasive mammals were deliberately imported as livestock and then deliberately released to the wild (e.g., goats and pigs). Some, such as the Indian mongoose (*Herpestes javanicus*), were deliberately released as potential biological controls for agricultural pests. Terrestrial invertebrates have been introduced as aquarium novelties, food, pets or ornamentals, and biological control agents.

Freshwater aquatic vertebrates (primarily fish) are often deliberately stocked in ponds and streams for sport fishing or as future food sources. Some (e.g., alewives) were released to serve as forage fish for larger game fishes. Invasive aquatic amphibians and reptiles tend to result from pet releases, bait bucket releases, or biological control efforts. Contamination of fish stock by the African clawed frog (*Xenopus laevis*) led to the establishment of that amphibian in parts of California, but American bullfrogs in western waters stem from releases for food and escapes from aquaculture facilities.

Impacts

Invasive species are associated with a variety of ecological, economic, public health, and aesthetic impacts. Sometimes, whether these are positive or negative lies in the eye of the beholder. Usually "impact" is assumed to be negative, and known or potentially negative impacts are the reasons why invasive species are receiving so much attention in the twenty-first century. Indeed, the official definition of "invasive" in the United States includes reference to the harm a species can or does do. Yet it should be remembered that many

Table 3. Potential Ecological Impacts of Invasive Species[a]

Genetic	Change in genetic information in a native species through introgression
	Hybridizing with native species to produce offspring (or new species) that outcompete parents
Individual	Changes in foraging, pollination, or reproductive behavior
Population	Reductions in population size; niche shifts; local extinction
Community	Changes in species composition and interactions
Ecosystem	Changes in nutrient cycles and disturbance regimes

[a]See table, "Impacts of Invasive Plants," in Volume 2, for the complete listing of impacts of invasive plant species.

species are initially introduced because someone foresees a benefit, be it a beautiful blossom, a challenging game animal, a fascinating pet, or a way to control a pest or ameliorate an environmental problem.

Ecological Impacts in Natural and Semi-natural Ecosystems

As new species spread into the wild and semi-wild habitats of the United States, they have the potential to affect life in our forests, grasslands, and deserts at all biological/ecological levels (Table 3). Native organisms may respond as individuals to a new predator or competitor by altering their behavior. A case in point is the avoidance of some rodents to areas infested by the red imported fire ant. As a result, the mice may forage in less protected areas and become more vulnerable to predation by owls.

Individual organisms can also be affected by introduced pathogens, predators, or competitors for limited resources to the extent that the impact becomes evident at the population level, when increased mortality rates threaten the survival of the entire population of a given organism. Avian malaria threatens to decimate several endemic honeycreepers in Hawai'i. West Nile virus caused significant population declines in songbirds such as crows and chickadees when it first spread through eastern states. In the Great Lakes, the round goby (*Neogobius melanostomus*) displaces native fish such as the mottled sculpin (*Cottus bairdi*) from its customary spawning grounds and competes with it and other fish for food. Certainly the most notorious recent invader is the predatory brown tree snake, which arrived in Guam sometime between the end of World War II and 1952 and, in the next 20 years, caused the extinction of 10 species of native forest birds and decimated lizards, causing the local extinction of 4 species. In addition, the brown tree snake is implicated in the loss of two species of bat from Guam. While the example of the brown tree snake is unusually dramatic and illustrative of what can happen when a new predator is introduced to an island previously lacking predators, it does serve as a cautionary tale of how wrong things can go.

The zebra mussel (*Dreissena polymorpha*), through its rapid population growth and ability to grow on the shells of native unionid mussels, can physically overwhelm the host and reduce its access to nutrients. Among plants, spotted knapweed (*Centaurea stoebe*), a perennial forb, is an introduced competitor species. It produces an allelopathic chemical that depresses the germination or growth of native plants, such as the endemic Mt. Sapphire rockcress (*Arabis fecunda*), and thereby preserves a greater share of light, water, and nutrients for the invader.

Native populations may also be affected at the genetic level through hybridization and introgression. Hybridization involves the cross-breeding of members of two species. When viable offspring are produced, they may exhibit hybrid vigor and grow faster or larger than either parent and reproduce more quickly than either. If sterile offspring are produced, the parent species have wasted their gametes, a practice that may be costly if their numbers are already low. The rainbow trout (*Oncorhynchus mykiss*), a native transplant to western waters, produces fertile offspring when it mates with the California golden trout (*O. mykiss aguabonita*) and the threatened Paiute cutthroat trout (*O. clarki seleniris*). The hybrid offspring of rainbow trout and golden trout can backcross with both parent species and contaminate the gene pool of golden trout by introducing genes of rainbow trout, a process known as introgression. Through this means the native genotype can disappear. Smooth cordgrass (*Spartina alterniflora*), another native transplant to California, readily hybridizes with the native California cordgrass (*Spartina foliosa*). The first-generation hybrids have higher growth rates and greater reproductive success than either parent. The hybrids also tolerate a broader range of salinity and invade open mudflats, changing the physical environments in an estuary.

Loss of one or more species, as well as the addition of new species, has repercussions for an entire ecological community. Mutualistic relationships such as predator/prey and pollinator/host can be disrupted. Niche shifts can occur as new species are accommodated. Where the brown anole is present, the native green anole (*Anolis carolinensis*), which customarily seeks prey on the ground or lower regions of tree trunks, forages higher in the tree canopy. Chestnut blight fungus (*Cryphonectria parasitica*) essentially eliminated the American chestnut from the tree layer of eastern forests. The relative abundance of other trees in the forests changed as white oak (*Quercus alba*), chestnut oak (*Q. prinus*), and red oak (*Q. rubra*) increased in the absence of chestnut. The red imported fire ant has changed the composition of ant communities in areas it has invaded, which in turn may have reduced the dispersal of seeds and thereby affected the plant community as well. Competition for nutrients, water, shelter, or breeding grounds most affects those natives already threatened with extirpation from other forces. Among the more than 950 species listed as endangered or threatened in the United States, some 400, or about 42 percent, are believed to be at risk in part due to the impacts of invasive species. Interestingly, despite the demise of some species, at the local level, species richness usually increases, as invaders more than make up for the loss of natives.

When abiotic as well as biotic elements are affected, impacts are occurring at the ecosystem level. Two important ecosystem-level changes involve nutrient cycles and disturbance regimes. By filtering such huge amounts of water, dense populations of zebra mussel increase the amount of nitrogen and phosphorus in the water column and reduce the amount of carbon, which migrates down to the bottom-dwellers in the mussels' pseudofeces. On Hawai'i, a plant alters the nitrogen cycle, with ecosystem-wide consequences. The fire tree (*Morella faya*) fixes nitrogen from the atmosphere, which allows it—unlike any native trees—to live on the nitrogen-poor volcanic substrates of the islands. Fire trees add nitrogen to the soil and provide suitable conditions for a variety of other plants to colonize the area, thereby giving rise to a whole new community of plants.

On western rangelands, the dominance of cheatgrass has altered the disturbance regime from one where fires occurred on average once every 6–10 years to a burn cycle of 3–5 years. Such frequent fires eliminate shrubs, forbs, and native grasses and produce a monoculture of cheatgrass, an annual grass with low forage value.

At a regional or landscape scale, the mixing of species from all over the globe leads to the homogenization of the world's biota and a loss of global biodiversity. While locally, the total number of species may increase, the same species tend to be added everywhere. At the same time, geographically restricted and unique species tend to be disappearing. This impoverishment of the variety of life on earth is viewed with alarm, for homogenization is occurring at all levels—genetic, species, community, and ecosystem—potentially interfering with ecosystem functioning, ecosystem services, and the ability to evolve and adapt to changing environmental conditions. Furthermore, the world becomes a less interesting place as a great sameness spreads across not only our human-made townscapes and cityscapes, but the natural world as well.

The impacts of new invasive species are often not as much of a problem as initially feared. A case in point is the Monk Parakeet (*Myiopsitta monachus*), which had the potential of becoming a major agricultural pest based on the habits of its extinct relative, the Carolina Parakeet (*Conuropsis carolinensis*). However, the Monk Parakeet has not spread beyond urban areas, where many people are pleased to have these colorful birds visit their feeders. Unfortunately, it is still impossible to predict which introduced species will become invasive and which invasive species will become major ecological problems.

Economic Impacts of Invasive Species

Estimates of the costs of invasive species to the United States are commonly reported at well over $100 billion a year, but it is impossible to know the exact figure, which is likely much higher (Pimental et al. 2000; Pimental et al. 2004). It is difficult to put a dollar value on ecological damages or to separate the financial impacts of the combination of factors affecting agriculture, forestry, fisheries, industry, land values, and human well-being, not to mention the price of control and measures directed at preventing introductions, managing invasive species, and implementing remediation measures to repair the damage.

In agriculture, direct damage to crops and pastures affect yields and the quality of the product, which in turn can have repercussions on market value. Weeds, insects, and pathogens cost agriculture about $25 billion each year in lost production and another $3 billion for pesticides to control them (Pimental et al. 2000). Measures set up to protect plant and animal life and human health in the United States can become trade barriers and violate World Trade Organization (WTO) agreements, causing costly boycotts among international buyers. The inspections, quarantines, monitoring, and response to introduced insects and pathogens are all expensive but necessary to protect our food supply as well as our farmers. Forestry suffers financial loss not only when trees sicken and die, but also when quarantines prevent sales of wood products. Some 9 percent of lumber, pulpwood, and other forest products are thought to be lost to insect damage at an annual cost of $7 billion. The "Slow the Spread Program" for gypsy moth costs the federal government $8–10 million a year, with additional funds provided by affected states (Tobin 2008). Similar programs are needed for the emerald ash borer (*Agrilus planipennis*) and other invasive insects. The green crab (*Carcinus maenas*) is assumed to have caused the demise of the soft shell clam (*Mya arenaria*) fishery in New England in the 1950s and is also implicated in declines of the commercially important northern quahog (*Mercenaria mercenaria*), a scallop (*Argopecten irradians*), and other shellfish with annual harvests worth $44 million in 2000 (Perry 2000). Fouling damage by Asian clams (*Corbicula fluminea*) in the United States reportedly causes $1 billion a year. The damage by zebra mussels and control of those mollusks at raw water-using and electricity-generating industries in the Great Lakes region are estimated to cost more than $100 million a year (Wisconsin Department of Natural Resources, 2004). Other Great Lakes invaders affect the recreation industry in the region, where losses from sport fishing alone reach an estimated $200 million a year. In Florida, hydrilla clogs waterways and costs $14.5 million a year to control, but there are still financial impacts on the recreation and tourism industries (Pimental et al. 2000; Pimental et al. 2004).

Public Health and Well-Being Impacts

The most obvious impacts to human health are new pathogens. Global epidemics are expected to become more common with ever-increasing international travel and the globalization of world trade. Such pathogens, past and future, are not covered in this encyclopedia, but they have followed or will follow the same pathways as other invasive species. For some, their virulence depends on finding reservoirs and transmitters among both native and non-native animals, as in the case of the West Nile virus and the bacterium that causes Lyme disease. A warming climate and invasive mosquitos likely mean the establishment of (currently) tropical diseases like dengue fever in the near future; but the ubiquity and rapidity of air travel opens the U.S. population to all sorts of emerging infectious diseases.

A host of simply annoying species have invaded the United States. Asian multicolored lady beetles and brown marmorated stink bugs are two recent examples, while the common bed bug represents a very old traveler now experiencing a resurgence. For folks living in parts of Hawai'i, the noisy coqui can be added to the list, and for people in Florida, the Cuban treefrog lurking in the toilet fits the bill. Such annoyances can be expensive. The din of calling coquis can lower property values. Bug-bitten customers sue hotels and landlords, adding to the costs incurred in trying to eradicate bed bugs.

Invasion Science: A Brief History

The seminal work in the scientific study of biological invasions is generally considered to be Charles S. Elton's 1958 book, *The Ecology of Invasions by Animals and Plants*, which called attention to some of the most damaging of invaders in various parts of the world. Among featured invasive organisms were the chestnut blight fungus, which had ravaged forests in the eastern United States, and the sea lamprey, which had rapidly spread through the Great Lakes after the opening of the Welland Canal and decimated native fisheries. He described the impact of the American smooth cordgrass on the tidal flats in his native England, where it had hybridized with another cordgrass (*Spartina maritima*) to create a new species now known as common cordgrass (*S. anglica*). The hybrid was replacing both parent species and spreading quickly in wetlands on both sides of the English Channel. (Today, smooth cordgrass is causing similar problems in California.) He also noted how the European Starling had expanded its range across the United States and Canada in little more than 60 years after the first pairs had begun to breed in New York's Central Park. Elton's compendium obviously drew on earlier work by others, including classic papers by M. T. Cooke (1928) on the spread of the starling between 1891 and 1926 and by J. C. Phillips (1928) on the history of the spread of nonnative birds across the continent. Probably the earliest records of European plants growing in North America were published by the English traveler John Josselyn in 1672, when he catalogued plants observed in New England. Nonetheless, Elton's work was especially influential, and he became the "Father of Invasion Ecology." He had sounded the alarm about the disruptions to native ecosystems that can or could occur from successful invaders and called for renewed efforts in the conservation of native biodiversity, in studying the effects of invasive species, and in learning how best to control them and increase the resistance of both natural and human-dominated ecosystems to invasion.

For those with a more "objective," less value-driven perspective on invasive species, Joseph Grinnell's 1919 paper, "The English House Sparrow Has Arrived in Death Valley: An Experiment in Nature," is a harbinger of the study of invasives as a way to understand basic biogeographical, ecological, or evolutionary processes. Three papers by Richard Johnston and Robert Selander (1964, 1973, 1973) demonstrated that evolution in nonnative populations could be fast. They focused on the House Sparrow, which in a little more than 100 generations had diverged into distinct regional morphs based on size and plumage, displaying well-known biogeographic patterns described by Bergmann's Rule and Gloger's Rule, respectively. In 1964, H. G. Baker and G. L. Stebbins convened a conference on "The Genetics of Colonizing Species" and published contributed papers the following year. The 1960s also saw publication of the paradigm-changing book *The Theory of Island Biogeography* by Robert MacArthur and E. O. Wilson, which modeled how new species fit into existing communities, envisioning an equilibrium number of species on an island that meant that a new colonizer would cause the extinction of a previously established species. Experimental work by Daniel Simberloff and others tested this hypothesis, which dominated ecological research and thought for a couple of decades.

On the conservation/management side of the invasive species issue, George Laycock's 1966 *Alien Animals* set the tone. Laycock condemned the negative consequences and the human foibles associated with transplanting animals to new locations and strongly advocated against future introductions. Laycock helped bring exotic animals in general to the attention of the public and to resource managers, but may also have biased opinions against all nonnative species. Nonetheless, he stimulated research on nonnatives at a time when scientists were focusing their efforts on supposedly pristine or natural ecological systems. By the 1970s, many scientific journals began to accept articles on invasive species, although some range and wildlife management journals still refused manuscripts on nonnatives. Invading plants and animals increasingly became important research subjects in efforts to discover and understand ecological and evolutionary patterns and processes.

International interest in invading species as modifiers of natural communities with the potential to drive or at least exacerbate system-wide changes was growing at the same time research was developing in the United States. In the early 1980s, the Scientific Committee on Problems of the Environment (SCOPE), part of the International Council of Scientific Unions (ICSU), launched a program on the "Ecology of Biological Invasions." The program sponsored scientific meetings in Great Britain, South Africa, Australia, and the Netherlands, as well as a major symposium in Asilomar, California, and a final convention to synthesize results in Hawai'i in 1986. This global initiative resulted in 15 volumes, including *Ecology of Biological Invasions of North America and Hawaii*, edited by H. A. Mooney and J. A. Drake (1986). The program addressed two questions that remain at the core of invasion ecology today: what makes a good invader, and what determines whether an ecosystem is prone to, or resistant to, invasion.

Invasion ecology had gained a separate space within ecology by the beginning of the twenty-first century, as attested to by a spate of recent textbooks and other works devoted to invasion ecology or invasion biology (e.g., Cadotte et al. 2006; Lockwood et al. 2007; Ruiz et al. 2003; and Sax et al. 2005). Increasingly, invasive species were also being seen as a component of global environmental change, expected to both affect and be affected by climate change, altered nutrient cycles and disturbance regimes, and changing land-use patterns. Another project of SCOPE, *Invasive Species in a Changing World* (Mooney and Hobbs 2000), highlighted some of these problems and also brought attention to the implications of social views, monetary costs, and the global economy to the management and control of invasive species.

The Human Factor

Species that are invasive today depended on human choices, values, activities, and systems for their initial introduction as well as establishment and further spread. As outlined above, almost all arrived in their receiving habitats as a consequence of their association with humans. (The Cattle Egret is a major exception and, according to some definitions, may not really be "invasive.") They have followed people from place to place either surreptitiously or as desired members of the human entourage. Most gained their first footholds in or near human habitation or sites of commerce and then found suitable habitats in areas where native plant and animal communities had been disturbed or destroyed by forest clearing, cultivation, domestic livestock grazing, or urban development. In continental situations in particular, their successful entry into wild and semi-wild ecological communities is more likely when some native elements have already been removed or weakened by, for example, predator control, fire suppression, selective grazing, trampling, or a changing climate. Invasive species reflect us: our history of colonization, exploitation, and trade; our changing technology; our human curiosity and aesthetics. Even our love of life and freedom come into play when we release domesticated and pet animals to the wilds.

How one views invasive species and deals with them also reflects human values. Individual rights, property rights, animal rights, the right to make a living, all influence whether or not we take measures to prevent introductions, eradicate populations before they begin to spread, or manage species that are beyond the stage where eradication is possible. Some of the largest offenders today in terms of bringing new species into the country are the pet trade and horticultural/nursery trade. Inspection and quarantine are expensive, and it is difficult to prove cost savings from species that do not make it through our ports of entry and to gain public support for such measures. Conflicting viewpoints between conservationists and animal rights groups can make management programs unwieldy at best. But who is right? It depends on what Americans want and value as the natural heritage they pass to future generations. Informed decision making is paramount. People should at the very least know the risks and trade-offs. In the entries contained in the two volumes of this encyclopedia, we try to present a sample of species, pathways, and impacts—actual and potential—so the reader can be better able to make these decisions. Many problems arise because we simply do not understand or are unaware of the consequences of our actions. The species accounts that follow will help the reader identify invasive species and learn some of the ways each of us can slow their spread or best manage them as fellow inhabitants of our land. The choice to do something or nothing is ours.

References

Baker, H. G., and G. L. Stebbins, eds. *The Genetics of Colonizing Species*. New York: Academic Press, 1965.

Benson, A. J. "Documentation over a Century of Aquatic Introduction in the United States." In *Nonindigenous Freshwater Organisms: Vectors, Biology, and Impacts*, edited by R. Claudi and J. H. Leach, 1–31. Boca Raton, FL: Lewis Publishers, 1999. Cited in Lockwood, Julie L., Martha F. Hoopes, and Michael P. Marchetti. *Invasion Ecology*. Oxford: Blackwell Publishing, 2007.

Cadotte, Marc William, Sean M. McMahon, and Tadashi Fukami, eds. *Conceptual Ecology and Invasion Biology: Reciprocal Approaches to Nature*. *Invading Nature*, vol. 1. Dordrecht: Springer, 2006.

Cooke, M. T. "The Spread of the European Starling in North America (to 1928)." *Circular of the U.S. Department of Agriculture* 40: 1–9, 1928.

Davis, M. A. "Invasion Biology 1958–2005: The Pursuit of Science and Conservation." In *Conceptual Ecology and Invasion Biology: Reciprocal Approaches to Nature*, edited by Marc William Cadotte, Sean M. McMahon, and Tadashi Fukami, 35–64. *Invading Nature*, vol. 1. Dordrecht: Springer, 2006.

Elton, Charles S. *The Ecology of Invasions by Animals and Plants*. London: Chapman and Hall, 1958.

Evans, Edward A. "Economic Dimensions of Invasive Species," *Choices*, 2003. http://www.choicesmagazine.org/2003-2/2003-2-02.htm.

"Executive Order 13112 of February 3, 1999. Invasive Species." *Federal Register* 64, no. 25 (February 8, 1999). http://www.invasivespeciesinfo.gov/laws/execorder.shtml.

"Exotic, Invasive, Alien, Nonindigenous, or Nuisance Species: No Matter What You Call Them, They're a Growing Problem." Great Lakes Environmental Research Laboratory, NOAA, 2007. http://www.glerl.noaa.gov/pubs/brochures/invasive/ansprimer.pdf.

Federal Noxious Weed Act of 1974. Public Law 93-629. Sections 2801–2814, enacted January 3, 1975. (Superseded by the 2000 Plant Protection Act, except for Sec. 2814.)

Fritts, Thomas H., and Dawn Leasman-Tanner. "The Brown Treesnake on Guam: How the Arrival of One Invasive Species Damaged the Ecology, Commerce, Electrical Systems, and Human Health on Guam: A Comprehensive Information Source." U.S. Geological Survey, Fort Collins Science Center, 2001. http://www.fort.usgs.gov/resources/education/bts/bts_home.asp.

Grinnell, J. "The English House Sparrow has Arrived in Death Valley: An Experiment in Nature." *American Naturalist* 53: 468–472, 1919.

Huennecke, L. "SCOPE Program in Biological Invasions: A Status Report." *Conservation Biology* 2: 8–10, 1988.

"Injurious Wildlife." Branch of Invasive Species, U.S. Fish and Wildlife Service. http://www.fws.gov/fisheries/ans/ANSInjurious.cfm.

"Invasiveness in Exotic Plants: Immigration and Naturalization in an Ecological Continuum," In *Conceptual Ecology and Invasion Biology: Reciprocal Approaches to Nature*, edited by Marc William Cadotte, Sean M. McMahon, and Tadashi Fukami, 65–106. *Invading Nature*, vol. 1. Dordrecht: Springer, 2006.

Johnston, R. F., and R. K. Selander. "Evolution of the House Sparrow. I. Intrapopulation Variation in North America." *The Condor* 69: 217–258, 1967.

Johnston, R. F., and R. K. Selander. "Evolution of the House Sparrow. II. Adaptive Differentiation in North American Populations."*Evolution* 25: 1–28, 1971.

Johnston, R. F., and R. K. Selander. "Evolution of the House Sparrow. III. Variation in Size and Sexual Dimorphism in Europe and North and South America." *American Naturalist* 107: 373–390, 1973.

Johnston, R. F., and R. K. Selander. "House Sparrows: Rapid Evolution of Races in North America." *Science* 144: 548–550, 1964.

Josselyn, John. *New England's Rarities, discovered in Birds, Beasts, Fishes, Serppents, and Plants of that Country*, 1672. Reprint, Boston: William Veazie, 1865. Republished, Bedford, MA: Applewood Books, n.d.

Laycock, George. *The Alien Animals. The Story of Imported Wildlife*. New York: Ballantine Books, 1966.

Lockwood, Julie, Martha F. Hoopes, and Michael P. Marchetti. "An Introduction to Invasion Ecology," *Invasion Ecology*, 1–17, 2007.

Lockwood, Julie L., Martha F. Hoopes, and Michael P. Marchetti. "Ecological Impacts of Invasive Species." Chapter 9 of *Invasion Ecology*. Oxford: Blackwell Publishing, 2007.

Lockwood, Julie L., Martha F. Hoopes, and Michael P. Marchetti. *Invasion Ecology*. New York: Blackwell Publishing, 2007.

Mooney, H. A., and J. A. Drake, eds. *Ecology of Biological Invasions of North America and Hawaii*. New York: Springer-Verlag, 1986.

Mooney, H. A., and R. J. Hobbs, eds. *Invasive Species in a Changing World*. Washington, DC: Island Press, 2000.

Murphy, Helen T., Jeremy VanDerWal, Lesley Lovett-Doust, and Jon Lovett-Doust. "Invasiveness in Exotic Plants: Immigration and Naturalization in an Ecological Continuum," In *Conceptual Ecology and Invasion Biology: Reciprocal Approaches to Nature*, edited by Marc William Cadotte, Sean M. McMahon, and Tadashi Fukami, 65–106. *Invading Nature*, vol. 1. Dordrecht: Springer, 2006.

"Nonindigenous Aquatic Nuisance Prevention and Control Act of 1990." As Amended through P.L. 106–580, December 29, 2000. http://www.anstaskforce.gov/Documents/nanpca90.pdf.

Osborn, Liz. "Number of Native Species in the United States." Current News Nexus. Research News and Science Facts, 2010. http://www.currentresults.com/Environment-Facts/Plants-Animals/number-of-native-species-in-united-states.php.

Perry, Harriet. "*Carcinus maenas*." USGS Nonindigenous Aquatic Species Database, Gainesville, FL, 2008. http://nas.er.usgs.gov/queries/factsheet.aspx?SpeciesID=190.

Phillips, J. C. "Wild Birds Introduced or Transplanted in North America." *Technical Bulletin* of the U.S. Department of Agriculture, 1928.

Pimentel, D., L. Lach, R. Zuniga, and D. Morrison. "Environmental and Economic Costs of Non-Indigenous Species in the United States." *Bioscience* 50(1): 53–65, 2000.

Pimentel, David, Rodolfo Zuniga, and Doug Morrison. "Update on the Environmental and Economic Costs Associated with Alien-Invasive Species in the United States," 2004. Available online at http://ipm.ifas.ufl.edu/pdf/EconomicCosts_invasives.pdf.

Richardson, David M., Petr Pyšek, Marcel Rejmánek, Micahel G. Barbour, F. Dane Panetta, and Carol J. West. "Naturalization and Invasion of Alien Plants: Concepts and Definitions." *Diversity and Distributions* 6: 93–107, 2000.

Ruiz, Gregory M., and James T. Carlton. *Invasive Species: Vector and Management Strategies.* Washington, DC: Island Press, 2003.

Sax, Dov F., John J. Stachowicz, and Steven D. Gaines. *Species Invasions. Insights into Ecology, Evolution, and Biogeography.* Sunderland, MA: Sinauer Associates, Inc., 2005.

Simberloff, D. "A Rising Tide of Species and Literature: A Review of Some Recent Books on Biological Invasions." *Bioscience* 54: 247–54, 2004.

Stein, Bruce A., Lynn S. Kutner, and Jonathan S. Adams. *Precious Heritage: The Status of Biodiversity in the United States.* New York: Oxford University Press, 2000.

Tobin, Patrick C. "Cost Analysis and Biological Ramifications for Implementing the Gypsy Moth Slow the Spread Program." USDA, Forest Service, Northern Research Station, General Technical Report NRS-37, 2008. http://nrs.fs.fed.us/pubs/9238

Williamson, Mark, and Alastair Fitter. "The Varying Success of Invaders." *Ecology* 77(6): 1661–66, 1996.

Wisconsin Department of Natural Resources. "Zebra Mussels (*Dreissena polymorpha*)," n.d. http://dnr.wi.gov/invasives/fact/zebra.htm.

■ Preface

Invasive species have gained our attention in different ways. Susan Woodward, who wrote about invasive microorganisms, fungi, and animals in Volume 1, had her interest in invasive animals first sparked as a student of biogeography in the 1970s. Birds such as the European Starling and House Sparrow were featured in textbooks to demonstrate how animals spread in an environment that was new to them or how quickly they evolved adaptations to varying local conditions across a whole continent. Graeme Caughley's work on irruptions of red deer in New Zealand was new, and the modeling of invasions and management of exotics in their infancy. As a doctoral student at UCLA, she studied feral burros along the lower Colorado River, viewing them as an example of humans "changing the face of the earth" (the buzzwords of those days) by transporting domesticated and wild animals around the world. Under contract to the Bureau of Land Management (BLM) at that time, she collected baseline data on population dynamics, diet, home range size, and other aspects of burro behavior and ecology that would help that agency devise policies and practices for the animals' management. With an applied aspect to her work, she straddled what has become two main perspectives on invasive species in general: an academic interest in the science of invasions and a management interest in preventing arrivals, eradicating or controlling the spread of those species that were able to establish populations, and managing those whose numbers and distribution were for all practical purposes already beyond eradication.

Joyce Quinn, a biogeographer whose research dealt with distribution of plants and their relationships with the natural environment, such as climate and soils, wrote the invasive plants section in Volume 2. In spite of her experience, she found that researching and writing this book led her to learn more. Some plants that she had thought were an integral part of the "natural" landscape, such as common mullein, are actually alien plants that had become naturalized and are now widespread in the United States. She notes:

> Several years ago, an uninvited plant sprouted unexpectedly in my yard. I tried for years to get rid of it, dutifully pulling off the little sprouts as they emerged. After four or five years, I gave up and decided to let the plant grow. In a couple of years, it became an attractive tree about 10 ft. (3 m) tall. It had smooth speckled bark, long lacy compound leaves, and clusters of small purple star-shaped flowers. As I was doing research for this *Encyclopedia*, I discovered that my new plant was a chinaberry tree. While attractive, it had few redeeming qualities so I decided to eliminate it. A friend helped me saw the trunk, about 8 in. (20 cm) in diameter, slightly below ground level. I immediately poured undiluted glyphosate on the freshly cut stump, thinking that was the end of it. I paid no attention until four months later when I saw a 6 in. (15 cm) sprout! I sprayed it with herbicides, but another sprout soon emerged. I sprayed again, but at the time of this writing, I still do not know if I have managed to kill the invader. I fear that it will be an on-going process. If I have had such trouble with just one alien invasive plant, the challenges that land managers, conservationists, and agriculturalists have in battling invasive species seem insurmountable.

Scope

The purpose of the *Encyclopedia of Invasive Species* is to provide an introduction to the species, issues, and management options involved with invasive animals, fungi, microorganisms, and plants. The number of plants and animals introduced into the United States is staggering. Only a relatively few establish self-sustaining populations, and very few of these actually become invasive (in the scientific sense of greatly and rapidly expanding their range in the United States). Still, there are hundreds of invasive species—too many to be included in a reference book of this sort. For many species, much remains to be learned, and it is premature to develop full entries for them, but this still leaves many to choose from. In selecting the 168 species for inclusion in the *Encyclopedia*, we have tried to offer a wide spectrum of invasive species that includes some present in the United States from colonial times, and some that have just been detected; some that completed their spread across the country long ago, and others that are in the midst of rapid population growth and range expansion. We also wanted to include some species that are found throughout the country, and some that are limited to a region or single state.

For animals, we aimed to include representatives from all major classes of vertebrates and a good variety of invertebrates. The reader will find common, well-known invaders and others that may be a surprise. We also wanted to showcase a few fungi, especially those that have been major transformers of urban, suburban, and natural forests, and at least acknowledge the presence of invading microorganisms with a tiny sample of those threatening the health of native animals and, in some cases, humans as well. Finally, we wished to have a geographically broad selection of invasive animals, with all 50 states and Puerto Rico having some members of their nonindigenous fauna represented. Florida, Hawai'i, and California have the largest numbers of officially recognized invasive species. Residents of these states will undoubtedly find nonnative organisms causing significant impact in natural and artificial ecosystems missing from our accounts. This was necessary in order to include some organisms limited to other states.

For plants, we also tried to include a little bit of everything. Volume 2 addresses a variety of growth forms, ranging from aquatic plants to trees and vines, and all regions of the United States. Some plants are widespread throughout the country, while others are localized. Many plants were deliberately brought to the United States as ornamentals or for some useful characteristic, while others were accidentally introduced. The length of entries dedicated to each species is variable. The taxonomic relationships of some plants and similar species are not always clearly defined. Some plants, for example, hybridize so freely that it becomes difficult to distinguish different species. A few accounts of invasive plants treat two or more related species in the same entry because their effects and management are similar. As with animals, we could always find "one more" species that should be included, but it was not possible to include all. A wealth of information from various sources can be accessed by the reader wanting to know more. The General Bibliography at the end of Volume 2 has a list of recommended resources, including websites.

The *Encyclopedia* is specifically meant for high school and college students, but addresses many of the informational needs of the curious naturalist, horticulturalist, or any homeowner or environmentally concerned citizen who is interested in the origins and consequences of invasive plants and animals.

Although some invasive species have been part of the landscape of the United States for literally hundreds of years, the wide-reaching effects of most are only beginning to become realized. Some invasive species are detrimental to native ecosystems and threaten biodiversity, others are more economically damaging to crops and livestock, and a few pose a danger to human health. Invasive species are a major part of current global environmental change. Experts consider them the second-greatest threat to native species after habitat destruction and fragmentation. The control and interdiction of invasive species coupled with the damage some incur on crops, pastures, livestock, native ecosystems, and human health and well-being costs billions of dollars each year. The invasive species problem is dynamic—as those in the Mid-Atlantic states weathering their first onslaught of the brown marmorated stink bug know well—and endlessly fascinating.

How to Use the *Encyclopedia*

Volume 1 begins with an introduction to inform the reader of the nature and scope of issues related to invasive species in the United States. Separate sections deal with the terminology related to invasive species, the invasion process from an ecological point of view, the pathways by which nonnative species have been and continue to be introduced to the United States, some of the ecological and economic impacts of invasives, and a brief outline of the history of modern invasion science. A final section of the introduction describes the human factors that determine what species come in, where they succeed, and if and how they are managed.

The introduction is followed by 88 entries describing microorganisms, fungi, invertebrates, and vertebrates. Entries are arranged alphabetically within major taxonomic groups. The species described represent the large number introduced and invasive in the continental United States, Hawai'i, and Puerto Rico.

Each entry in both volumes includes the following elements, unless noted otherwise:

Native Range

Distribution in the United States

Description

Related or Similar Species

Introduction History

Habitat

Diet (animals only)

Life History (animals, fungi, and microorganisms only)

Reproduction and Dispersal (plants only)

Impacts

Management

Selected References

Additionally, each entry in both volumes is accompanied by at least one photograph and maps that show the original and invasive range of the species in question according to the best information available. Often range maps are, by necessity, approximate. This is especially true for organisms not native to the United States or Europe, where biological surveys are more complete than on other continents.

In Volume 1, the entries are followed by a list of state-by-state occurrences of invasive animals, fungi, and microorganisms; a glossary; and an index to both volumes.

Volume 2 begins with a brief overview of invasive plants in the United States, which loosely follows the organization within species accounts, describing in general the scope of the invasive plant problem, including the ways, both intentional and accidental, that plants were brought into the country; and some of the effects invasive plants have on native plant and animal species, natural ecosystems, agricultural or fishing industries, recreational activities, or human health. The ways in which invasive plants reproduce and expand their range is summarized, as is information on management and prevention of invasive plants species. A sidebar on herbicides accompanies the overview.

The 80 entries on invasive plants are arranged by growth form categories: aquatics, forbs, graminoids, shrubs, trees, and vines. Photographs of each species show different parts of the plant. Interesting facets of a plant's use or history or of strategies attempted for its control are related in sidebars.

Several supplementary lists follow the invasive plant entries to provide background information, and various tables summarize plant data in different, easily accessible ways, including a table of common and scientific names of both plants and animals briefly mentioned in the text of Volume 2, and a list of organizations concerned with invasive plants in the United States. Two tables of noxious or invasive plants, one organized by state and the other by species, as well as a table of species listed by type of impact are also available.

Volume 2 concludes the set with these appendices to the *Encyclopedia*: a list of American species that are invasive in other parts of the world; a list of federal laws related to the prevention and management of invasive species; international agreements and conventions dealing with invasive species; and the IUCN/SCC Invasive Species Specialist Group's list of 100 of the "World's Worst Invasive Alien Species," with an indication of those covered in the *Encyclopedia*. The glossary, a selected bibliography of classic and contemporary writings and online information sources, and the index to the set complete Volume 2.

It is our hope that our efforts will stimulate thought and make the natural world more accessible to the general public. Informed readers can help make the decisions that will curtail the spread of species that have only recently arrived, prevent the arrival of yet others, and manage those that are currently invasive

Acknowledgments

Both authors thank the photographers who graciously allowed their photos to be used, often donating them, or sometimes providing them at a reduced fee. They deserve our special thanks for giving life to the species descriptions. Bugwood.org and its associated personnel at the Center for Invasive Species and Ecosystem Health, University of Georgia, deserves special mention as a clearinghouse for providing informational sources and photographs. Joyce Quinn prepared the excellent maps for the species accounts.

Each author is most appreciative of the other's contributions to the development of the project, the overall organization of the volumes, and constructive critiques of text and illustrations throughout the manuscript preparation process. We complement each other and work well as a team. We acknowledge Kevin Downing, originally of Greenwood Press and now serving the broader ABC-CLIO community as editorial operations manager, who initiated the proposal for the *Encyclopedia* set, and David Paige of ABC-CLIO, who guided us through subsequent discussions and organizational details. Anne Thompson, development editor, later offered guidance in the specifics of the manuscript, and Erin Ryan helped with the specifications for the illustrations. We thank all four for creating a positive and flexible working environment and offering valuable suggestions all along the way.

■ Alphabetical List of Invasive Microorganisms, Fungi, and Animal Entries

Entries in the encyclopedia are arranged by categories. Following are the entries in Volume 1 in alphabetic order.

African Clawed Frog (*Xenopus laevis*)

Africanized Honey Bee (*Apis mellifera scutellata*)

Alewife (*Alosa pseudoharengus*)

American Bullfrog (*Lithobates catesbeianus*)

Argentine Ant (*Linepithema humile*)

Asian Clam (*Corbicula fluminea*)

Asian Green Mussel (*Perna viridis*)

Asian Longhorned Beetle (*Anoplophora glabripennis*)

Asian Swamp Eel (*Monopterus albus*)

Asian Tiger Mosquito (*Aedes albopictus*)

Australian Spotted Jellyfish (*Phyllorhiza punctata*)

Avian Malaria (*Plasmodium relictum capistranoae*)

Bat White-Nose Syndrome Fungus (*Geomyces destructans*)

Bighead Carp (*Hypophthalmichthys nobilis*)

Black Rat (*Rattus rattus*)

Brown Anole (*Norops[=Anolis] sagrei*)

Brown Marmorated Stink Bug (*Halyomorpha halys*)

Brown Trout (*Salmo trutta*)

Burmese Python (*Python molurus bivittatus*)

Cattle Egret (*Bubulcus ibis*)

Chain Tunicate (*Botrylloides violaceus*)

Chestnut Blight Fungus (*Cryphonectria parasitica*)

Chinese Mitten Crab (*Eriocheir sinensis*)

Chinese Mystery Snail (*Cipangopaludina chinensis malleata*)

Chytrid Frog Fungus (*Batrachochytrium dendrobatidis*)

Colonial Tunicate (*Didemnum vexillum*)

Common Bed Bug (*Cimex lectularius*)

Common Myna (*Acridotheres tristis*)

Common Periwinkle (*Littorina littorea*)

■ Microorganisms

■ | Avian Malaria

Scientific name: *Plasmodium relictum capistranoae*
Family: Plasmodiidae

Native Range. Eurasia. The exact place of origin is unknown, but genetic evidence strongly suggests it arose somewhere in Eurasia, and it is known to infect Eurasian birds such as House Sparrows (*Passer domesticus*) and Common Mynas (*Acridotheres tristis*), both of which have been introduced to Hawai'i.

Distribution in the United States. Hawai'i. Avian malaria is most common at elevations of 3,000–5,000 ft. (900–1,500 m) on the moist windward sides of all the main islands.

Description. This parasitic protozoan requires microscopic examination of blood and tissues for identification. The clinical symptoms of malaria in Hawai'i's endemic honeycreepers (Drepaniidae) include weight loss, lethargy, lack of appetite, and high death rates. Infected birds will have a prominent sternum or breast bone (keel). Necropsies reveal enlarged, chocolate-brown or black livers and spleens and thin, watery blood; up to 50 percent of circulating red blood cells are infected by the microorganism.

Related or Similar Species. Other *Plasmodium* species are responsible for malaria in reptiles, birds (including poultry), and mammals, including humans. Different subspecies of *P. relictum* are known to infect birds in other parts of the world.

Introduction History. The introduction of avian malaria to Hawai'i required the arrival of both the parasite and a vector. The mosquito vector was unintentionally brought to Hawai'i in 1826. It seems to have reached Maui first and may have come in water barrels aboard a ship sailing from Mexico. The hundreds of exotic birds released in the islands likely led to the accidental introduction of *Plasmodium relictum* in the early twentieth century. Species such as the House Sparrow and Common Myna, from Europe and India, respectively, are the most probable sources. With no prior exposure to the parasite, native birds quickly fell victim to avian malaria.

Habitat. Avian malaria occurs where its vector species live. In Hawai'i, the vector is the southern house mosquito (*Culex quinquefasciatus*), itself an introduced species. This mosquito is most common at elevations below 5,000 ft. (1,500 m), and avian malaria is most prevalent among native birds in moist lowland forests. Cool temperatures prevent the development of mosquito larvae, and temperatures below 55°F (13°C) restrict the development of malarial parasites in adult mosquitoes. Recent studies, however, suggest that the mosquito—and hence the disease—is beginning to occur at higher elevations in the islands. This may be a product of warmer summer temperatures.

The disease is found in dry habitats if water is seasonally available. Mosquito breeding sites occur in standing water, including ditches, stock ponds, the artificial containers commonly found in human settlements, tree fern cavities, pools in intermittent streams, and wallows made by feral pigs. Cavities in lava flows also trap water and create corridors for mosquitoes to move between forest fragments.

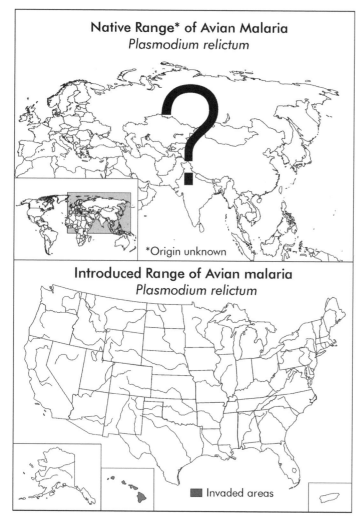

Top: The place of origin of the protozoan that causes avian malaria is unknown, but somewhere in Eurasia is likely. Bottom: Avian malaria currently only affects birds in Hawai'i.

Life History. P. *relictum* undergoes both sexual and asexual reproduction at different stages of the life cycle in both its bird and mosquito hosts. When a mosquito bites an infected bird, it will ingest gametocytes, the male and female reproductive cells of the protozoan. In the gut of the mosquito, the gametocytes produce encysted cells that develop into motile spores (sporozoites) which move to the salivary gland and are injected into a bird when the mosquito again feeds on blood. Thus the sporozoites are the infectious stage of the protozoan. In the bird, sporozoites attack red blood cells and develop within them, replicating their nuclei and other organelles. The resulting multinucleated cells, called schizonts, grow and cause the red blood cell that they have infected to burst. The schizonts then break apart and release single-nucleus daughter cells (sporozoites) into the bloodstream. (These daughter cells produce toxins that, in human malaria, cause the characteristic chills and fever.)

Impacts. Avian malaria infects passerine (perching) birds, which in Hawai'i include the Hawaiian Crow (*Corvus hawaiiensis*) and honeycreepers, a group of 57 birds (half of which are now extinct), all believed to have evolved from a single finch-like ancestor and all endemic to the islands. The introduction of avian malaria to the Hawaiian Islands has been implicated in a modern wave of population reductions, range restrictions, and possibly extinction of a number of Hawai'i's rare birds. Only native bird species, having evolved during a long period of isolation, seem to be deleteriously affected by the parasite. For some of them, such as the 'Iiwi (*Vestiaria coccinea*) and the Maui 'Alauahio (*Paroreomyza montana*), infection by the malaria parasite is almost always fatal. Other honeycreepers, such as the 'Omao (*Myadestes obscurus*), show greater resistance to the disease. Lowland populations of Hawai'i 'Amakihi and O'ahu 'Amakihi (*Hemiagnathus virens* complex) in the Puna area of the Big Island appear to have evolved some resistance to avian malaria since coming into contact with the parasite.

Most surviving honeycreepers today live at elevations above 5,000 ft. (1,500 m), where both *Culex* mosquitoes and the malaria protozoan have difficulty reproducing. The

Akiapola'au (*Hemignathus munroi*), for example, once occupied forests as low as 1,600 ft. (500 m) on the island of Hawai'i, but the remnant populations of this endangered species are now found only in high-elevation forests. On Maui, the six remaining endemic passerines (four are extinct) inhabit a narrow strip of high-elevation rainforest on the slopes of Haleakalā Volcano.

Efforts are underway to reestablish some of Hawai'i's threatened and endangered honeycreepers in protected patches of high-elevation forest. If mosquitoes continue to expand their distribution into higher and higher elevations as climate changes, these efforts could be doomed. Development of resistance to *Plasmodium relictum* seems to be occurring in some populations and may help some bird species survive; however, resistant birds still harbor the parasite and can transmit it to mosquitoes. They serve as reservoirs for the disease, further threatening vulnerable species.

Management. Control measures are directed at the vector, the southern house mosquito. The most effective way of preventing the disease from spreading is to eliminate or reduce populations of the mosquito by removing man-made habitats of standing water where the mosquito breeds or applying Bt (*Bacillus thuringiensis israelensis*) to impoundments such as horse watering troughs that cannot be removed. This attack on mosquitoes has also targeted feral pigs (see Vertebrates, Mammals, Feral Pig), which dig in the forest floor in search of food and leave behind depressions that catch rainwater and harbor mosquito larvae. The pigs also push over tree ferns and eat the starchy pith, creating water-holding cavities for mosquitoes to breed in.

Selected References

Atkinson, Carter T. "Ecology and Diagnosis of Introduced Avian Malaria in Hawaiian Forest Birds." USGS FS 2005-3151. Pacific Island Ecosystems Research Center, U.S. Geological Survey, 2005. http://biology.usgs.gov/pierc/Native_Birds/Avian_malaria.pdf.

Atkinson, Carter T., and Dennis A. LaPointe. "*Plasmodium relictum* (Micro-organism)." ISSG Global Invasive Species Database, 2005. http://www.invasivespecies.net/database/species/ecology.asp?si=39&fr=1&sts.

Beadell, Jon S., Farah Ishtiaq, Rita Covas, Martim Melo, Ben H. Warren, Carter T. Atkinson, Staffan Bensch, Gary R. Graves, Yadvendradev V. Jhala, Mike A. Peirce, Asad R. Rahmani, Dina M. Fonseca, and Robert C. Fleischer. "Global Phylogeographic Limits of Hawaii's Avian Malaria." *Proceedings of the Royal Society B: Biological Science*, 272(1504): 2935–44. Published online August 22, 2006. http://www.ncbi.nlm.nih.gov/pmc/articles/PMC1639517/. doi:10.1098/rspb.2006.3671.

LaPointe, Dennis A. "Feral Pigs, Introduced Mosquitoes, and the Decline of Hawai'i's Native Birds." USGS FS 2006-3029. Pacific Island Ecosystems Research Center, U.S. Geological Survey, 2006. http://biology.usgs.gov/pierc/Fact_Sheets/Pigs_and_mosquitoes.pdf.

■|Lyme Disease Bacterium

Scientific name: *Borrelia burgdorferi*
Phylum: Spirocheates

Native Range. North America and Europe. *B. burgdorferi sensu stricto* occurs in the United States, while two other strains—*B. burdorferi garinii* and *B. burgdorferi afzelli*—are found in Europe.

Distribution in the United States. Maine to Virginia; Great Lakes region; northern California.

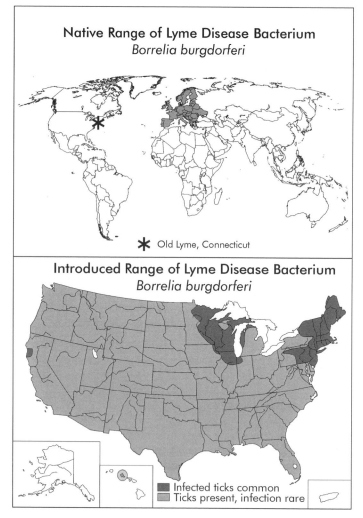

Native Range of Lyme Disease Bacterium
Borrelia burgdorferi

✱ Old Lyme, Connecticut

Introduced Range of Lyme Disease Bacterium
Borrelia burgdorferi

Infected ticks common
Ticks present, infection rare

Top: The spirochete responsible for Lyme disease is known from both North America and Europe, but the disease was first reported in Europe, making that continent the more probable place of origin. Bottom: *Borrelia burgdorferi* infects ticks throughout the lower 48 states, but it is most prevalent in the Northeast, western Great Lakes region, and parts of northern California. (Adapted from American Lyme Disease Foundation, "U.S. Maps and Statistics," http://www.aldf.com/usmap .shtml.)

Description. B. burgdorferi has the characteristic corkscrew appearance of spirochetes in general. Its clinical manifestation as Lyme disease in humans is marked first by a diagnostic bull's-eye rash that usually surrounds a tick bite and expands with time (erythema migrans). The rash may last for 2–3 weeks and be accompanied by flu-like symptoms. Untreated, stage 2 symptoms develop, including intermittent inflammatory arthritis, facial palsy, and extreme fatigue and malaise. In stage 3 of the disease, inflammation of the brain (encephalitis) and spinal chord (myelitis) and weakening of the lower limbs (paraparesis) occur; symptoms of fibromylagia can develop.

The microscopic spirochete is transmitted to humans when they are bitten by infected ticks. In the eastern and midwestern United States, the black-legged or deer tick (*Ioxodes scapularis*) is the vector; in the Pacific Northwest, the western black-legged tick (*I. pacificus*) is involved. Adult deer ticks are teardrop-shaped and 0.1 in. (3 mm) long, about the size of an apple seed. Females, the only ones that seek blood meals, have black heads and dorsal shields and dark red abdomens. In males, the hard shield or scutum covers the entire back, and the head and whole body are black. The nymphs of these tiny arachnids are about the size of a poppy seed, have black heads and translucent bodies; they are most apt to transmit Lyme disease to people because they are so small that they may not be noticed and removed. Western black-legged ticks are very similar in appearance.

Introduction History. Lyme disease was first recognized in 1975 by Dr. Allen Steele of Yale University, when about 50 children in the town of Lyme, Connecticut, developed rashes and joint pain, swelling, and inflammation similar to arthritis. Indeed, the new disease was originally called "Lyme arthritis." It was soon determined that the symptoms were similar to a tick-borne infection that had been known in Europe since at least 1883, when it was reported by the German physician Alfred Buchwald. It was not until 1982, however, that Dr. Willy Burgdorfer discovered the causative agent of the disease in black-legged ticks. The spirochete was named in his honor.

Since 1975, the incidence of the disease has increased and spread in the United States, where it is now the most common tick-borne illness. More than 20,000 new cases are reported each year. Ten states account for more than 90 percent of occurrences: Delaware, Maryland, Massachusetts, Minnesota, New Jersey, New York, Pennsylvania, Rhode Island, and Wisconsin. The increasing prevalence of the disease is due to a variety of factors, including better identification and record keeping; encroachment of residential areas into tick habitats; exploding deer and tick populations associated with suburban sprawl and forest fragmentation, as well as reforestation in the northern United States; and the range expansion of black-legged ticks into newly available habitats.

Habitat. The tick vectors of Lyme spirochetes inhabit the understory of moist, deciduous forests and open grassy areas with tall vegetation. It is often found along paths and roadsides.

Diet. Blacklegged ticks are external parasites on warm-blooded animals. They require a blood meal in order to molt and develop to the next stage of life. Field mice, especially the white-footed mouse (*Peromyscus leucopus*), serve as important hosts for larvae. Nymphs and adults feed on deer, dogs, and humans.

Life History. The bacterium *B. burgdorferi* circulates between ticks and a variety of vertebrates, each species affecting the survival of the spirochete because of varying competence as host species and different rates of infestation by ticks. Ticks can obtain the spirochete as larvae. Hatching during the summer, a larval tick waits on the ground for a small mammal or bird to come into contact with it, whereupon it attaches to the passing animal and begins feeding, sucking blood for a few days. If the host, most commonly a white-footed mouse, is infected with the pathogen, the larval tick will likely also become infected. (Mice serve as "reservoir hosts": they easily acquire the spirochete and sustain viable bacteria in their blood, allowing it to increase in numbers. They accept tick larvae again and again, and

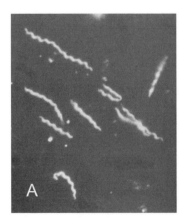

A. The corkscrew shape of the *Borrelia* spirochete. (Centers for Disease Control and Prevention.)
B. Black-legged or deer tick (*Ioxodes scapularis*), the vector for Lyme disease in eastern and central states. (Hardin MD/University of Iowa and CDC, http://www.lib.uiowa.edu/hardin%5Cmd/cdc/1669.html.) C. Western black-legged tick (*I. pacificus*), adult female. This arachnid is the vector for Lyme disease in western states. (Hardin MD/University of Iowa and CDC, http://www.lib.uiowa.edu/hardin/md/cdc/ticks5.html.)

thereby can transmit the bacteria to other larvae. Mice themselves show no signs of disease.) Once the larva has gorged itself, it will not feed again, but drops off its host and, in the fall, molts into a nymph. Nymphs remain inactive until early the following spring. When the nymphs become active, they wait in ambush on vegetation for a new host to brush past them. They attach to a deer, person, dog, or cat and feed for 4–5 days. If they were already infected with the Lyme spirochete as larvae, they may transmit it to the host. If they were free of infection, they may become infected during this time should the host already carry the pathogen. Peak activity for nymphs is spring and summer, and this is when humans are most apt to acquire Lyme disease. Once the nymph is fully engorged with blood, it drops off the host and molts into an adult. Adults are active throughout the fall and wait on grass or leaf tips about 3 ft. (1 m) above the ground to attach to deer or other larger mammals, including people. Adult deer ticks are most active in late fall, but humans are less apt to become infected at this time of year because the adults are large enough to be seen and picked off before the spirochete can be transmitted. Adult ticks without hosts become inactive when temperatures drop below 45°F (7°C) and typically find shelter in leaf litter on the forest floor. With warming temperatures in early spring, the adults seek a final blood meal, usually on deer, that will allow them to mate. Mating occurs on or off a host. The female lays some 3,000 eggs under leaf litter and then dies, completing the two-year life cycle.

Infection of humans by *B. burgdoferi* requires that a tick be attached for at least 24 hours and more likely for 2–3 days. Only small numbers of the spirochetes occur in a tick until it feeds. With the intake of blood, the bacteria multiply in the tick's gut. After 2–3 days, they migrate to the salivary glands, where they are injected into the host as the tick completes its feeding. This is why undetected nymphs usually spread the disease to humans. An estimated 85 percent of infected persons received the bacteria from nymphs in the spring; the other 15 percent obtained the infection from adults in autumn. Only about 1 percent of tick bites in an area where Lyme disease is prevalent result in infection.

Impacts. Some people are able to clear the infection without developing any symptoms. In others, the bacteria spread through the body and elicit inflammatory responses in the skin (erythma migrans) and joints, tendons, and bursae, especially in the knees, ankles, and wrists. These may be accompanied by fever and general malaise. Untreated, the conditions last a week or more and often reoccur for up to 10 years. The most common neurological condition associated with Lyme disease is facial paralysis, which may last for a couple of months. More serious manifestations of the disease include inflammation of the brain, a form of meningitis presenting as a headache, stiff neck, and sensitivity to light. It can be quite debilitating for a long period of time. Most cases are successfully treated with antibiotics.

Management. Control is aimed at preventing infection of humans. People should try to avoid tick habitats, particularly in spring when the nymphs are active. Removal of vegetation that is prime tick, mouse, and/or deer habitat—tall grass, brush, and dead leaves—from housing and work areas is also advisable. Precautions such as wearing light-colored clothing, long-sleeved shirts, hats, and closed shoes, plus tucking pant legs into socks or boots, can delay the attachment of ticks to skin and make them more visible. Insect repellent on clothing and skin (except on the face) also helps. After outdoor activity, a careful body inspection should be done, and any ticks should be removed with tweezers. Deer ticks and western black-legged ticks seek body folds such as armpits, groin, back of the neck, and back of the knee. Shower and wash clothes in hot water.

Vaccines to prevent Lyme disease are now available for dogs and cats.

Selected References

"Deer Tick Ecology." American Lyme Disease Foundation, Inc., 2006. http://www.aldf.com/deerTickEcology.shtml.

Meyerhoff, John O. "Lyme Disease." Medscape, 2009. http://emedicine.medscape.com/article/330178-overview.

Todar, Kenneth. "*Borrelia burgdorferi* and Lyme Disease." Todar's Online Textbook of Bacteriology, 2008. http://www.textbookofbacteriology.net/Lyme.html.

■ | West Nile Virus

Also known as: WNV
West Nile Virus
Family: Flaviviridae

Native Range. Uncertain. West Nile virus was first described in Uganda and then found in other parts of Africa as well as in Europe, Southwest and Central Asia, and Australia. Whether it was native or introduced to these regions has yet to be determined.

Distribution in the United States. West Nile virus has been reported in all of the 48 contiguous states, although outbreaks do not occur in every state every year.

Description. This small flavivirus consists of a positive-sense single strand of RNA (ribonucleic acid) containing between 11,000 and 12,000 nucleotides. The RNA is surrounded by a protein coat (nucleocapsid) that is encased in a lipid membrane. The complete structure or virion is spherical and measures 40–65 nm in diameter.

Related or Similar Species. Other flaviviruses cause human diseases, including St. Louis encephalitis, yellow fever, dengue fever, and Hepatitis C.

Introduction History. West Nile virus was first described from Uganda in 1937. Subsequently, it was discovered in many parts of the Old World and in Australia and found to be one of the most widespread flaviviruses in the world. Confirmation of the virus in New York City in 1999 marked its first appearance in the Americas. Early in the spring of 2000, it showed up again in mosquitoes and birds and quickly spread to other parts of the eastern United States. In 2000, it ranged from New Hampshire to North Carolina; by 2001, it had crossed the Mississippi River; and by 2004, it was in every state except Alaska and Hawai'i.

The source of the initial introduction is unknown. Genetic studies point to origins in the Mediterranean or Middle East. The strain that reached the United States seems to have evolved into a more virulent form through the mutations of amino acids in a single gene. How it reached the United States is a mystery. Once on the continent, the virus spread geographically in migrating birds.

Habitat. West Nile virus cycles between two hosts, birds and mosquitoes. Other vertebrates act as dead-end hosts; for though the virus may cause illness in them, it never reaches high-enough levels in the blood to be transmitted to mosquitoes.

The virus thrives in environments where mosquitoes, especially of the genus *Culex*, breed: wetlands, urban areas with artificial containers that collect rainwater, and wherever else standing water persists for several weeks or more. Recent epidemics are correlated with unusual hot, dry periods.

Life History. West Nile virus replicates best in certain bird species that become the major amplification hosts. The virus has been isolated in well over 100 different species. Some are extremely susceptible to infection and die, while others are very tolerant of infection and show no signs of impairment. The West Nile virus is transmitted from bird to bird by

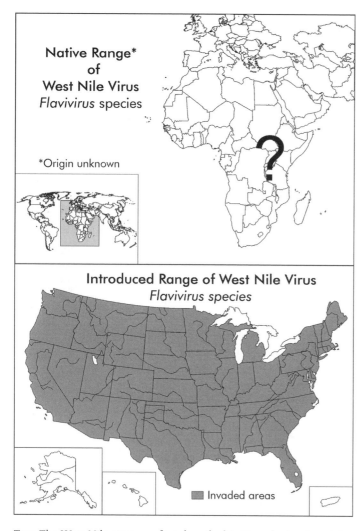

Native Range*
of
West Nile Virus
Flavivirus species

*Origin unknown

Introduced Range of West Nile Virus
Flavivirus species

■ Invaded areas

Top: The West Nile virus was first described in Uganda, but its origins remain unknown. Bottom: The West Nile virus has been reported in all of the lower 48 states.

mosquitoes, the most competent vectors in the United States being the southern house mosquito (*Culex quinquefasciatus*); the northern house mosquito (*C. pipiens*); the white-dotted mosquito (*C. restuans*); *C. salinarius*, a mosquito of salt marshes; and in the western United States, the encephalitis mosquito (*C. tarsalis*). These mosquitoes feed preferentially on birds, but also feed opportunistically on the blood of other vertebrates and can transmit the disease to humans, horses, and other species. Other mosquitoes such as *Culex nigripalpus*, a common mosquito in Florida; the Asian tiger mosquito, *Aedes albopictus*; the inland floodwater mosquito (*Aedes vexans*); and the eastern treehole mosquito (*Ochlerotatus triseriatus*) may also spread West Nile virus.

Some evidence suggests that squirrels (*Sciurus* spp.), eastern chipmunks (*Tamias striatus*), eastern cottontails (*Sylvilagus floridanus*), and alligators (*Alligator mississippiensis*) may build up sufficiently high levels of the virus in their blood to serve as reservoirs.

Impacts. During periods in 1999 and 2002–2003 when rates of human infection with West Nile virus were epidemic, high mortality rates were experienced among several common songbird species. Indeed, the widespread American Crow (*Corvus brachyrhynchos*) is so susceptible to fatal infection that dead individuals serve as indicators of the virus's presence in a given area. In some regions of the country, 45–100 percent of crows died during past outbreaks. Among other birds studied in the eastern United States, sharp population declines correlating with WNV epidemics occurred in the Blue Jay (*Cyanocitta cristata*), the Tufted Titmouse (*Baeolophus bicolor*), the American Robin (*Turdus migratorius*), the House Wren (*Troglodytes aedon*), the Black-capped and Carolina Chickadee (*Poecile atricapillus* and *P. carolinensis*, respectively) and the Eastern Bluebird (*Sialia sialis*). The Common Grackle (*Quiscalus quiscula*) was hard hit in Maryland. Others species showed great tolerance. These included Mourning Dove (*Zenaida macroura*), Northern Cardinal (*Cardinalis cardinalis*), Baltimore Oriole (*Icterus galbula*), Chipping Sparrow (*Spizella passerina*), and

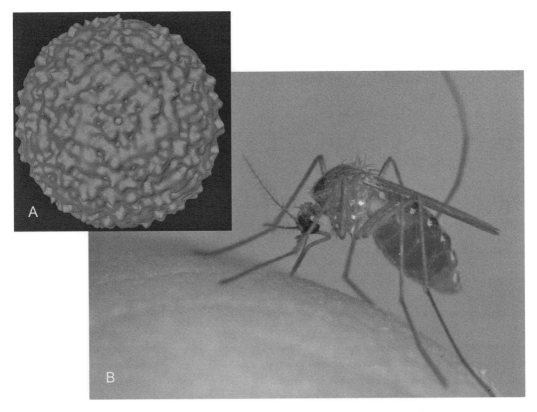

A. Image of West Nile virus particle produced by cryoelectron microscopy. (Purdue University Department of Biological Sciences.) B. The southern house mosquito transmits the West Nile virus from bird to bird. (U.S. Geological Survey.)

Catbird (*Dumetella carolinensis*). Populations of Blue Jays and House Wrens had rebounded by 2005, after the epidemic in humans had subsided, but populations of other birds remained low.

In the western states, Black-billed Magpies (*Pica hudsonia*), House Finches (*Carpodacus mexicanus*), and Greater Sage Grouse (*Centrocercus urophasianus*) have been severely affected.

The ecological impacts of rapid changes in bird populations is unknown, but shifts in species abundance and therefore possibly ecosystem functioning can be expected. Crows, for example, are important scavengers and are also predators of the nestlings of other birds and thereby control population sizes.

Many humans infected with the West Nile virus will develop no symptoms, but about 20 percent of infected people experience flu-like symptoms, including high fever, headache and body aches, general fatigue, rash, vomiting, and diarrhea. For relatively few—fewer than 1 percent of those known to be infected—the disease is fatal. The virus invades the nervous system and presents as encephalitis, meningitis, limb paralysis, or acute respiratory failure.

In horses, the symptoms of West Nile virus include weakness in the hindquarters, tremors, muscle rigidity, and paralysis. Between 1999 and 2002, the disease affected some 20,000 equines in the United States and was fatal in an estimated 38–57 percent of cases. Vaccination became available in 2002, but infection rates remained high.

Management. Control measures focus on eliminating breeding sites for the mosquitoes that are the vectors for the virus. This is primarily an urban and suburban problem and means removing containers such as old tires, drums, bottles, and cans in which standing water accumulates; repairing leaky pipes and outside faucets; unclogging gutters; and replacing water in birdbaths, watering troughs, and the like several times a week. People can minimize their risk of being bitten by mosquitoes by using insect repellent, wearing clothing that covers arms and legs, and minimizing outdoor activity at dusk and dawn when mosquitoes are most active. A vaccine is available for horses, but still under development for humans.

Selected References

LaDeau, Shannon L., A. Marm Kilpatrick, and Peter P. Marra. "West Nile Virus Emergence and Large-Scale Declines of North American Bird Populations." *Nature* 447: 710–13, 2007. doi:10.1038/nature05829.

National Biological Information Infrastructure (NBII) and IUCN/SSC Invasive Species Specialist Group (ISSG). "*West Nile Virus* (Micro-organism)." ISSG Global Invasive Species Database, 2006. http://www.issg.org/database/species/ecology.asp?si=304&fr=1&sts.

Weiss, Rick. "Bird Species Plummeted after West Nile: National Survey Finds Losses Among Many Songbirds." *Washington Post*, 2007. http://www.washingtonpost.com/wp-dyn/content/article/2007/05/16/AR2007051601032_pf.html.

"West Nile Virus." National Institute of Allergy and Infectious Diseases, 2009. http://www.niaid.nih.gov/topics/westnile/understanding/pages/what.aspx.

"West Nile Virus Detected in Arizona." Agency Directive, Arizona Game and Fish Department, 2003. http://www.azgfd.gov/w_c/diseases_west_nile.shtml.

■ Fungi

■ | Bat White-Nose Syndrome Fungus

Scientific name: *Geomyces destructans*
Phylum: Ascomycota
Family: Heliotiaceae

Native Range. Unknown. The fungus associated with bat white-nose syndrome may be native to Europe or may represent a soil fungus that has recently mutated in the northeastern United States to become a pathogen of hibernating bats.

Distribution in the United States. *Geomyces destructans* occurs from Vermont south to Virginia and west into Tennessee and Missouri. It probably also occurs in northwest Oklahoma.

Description. This fungus manifests itself as a white coating on the nose, ears, and wing membranes of hibernating bats. The dense covering of fine hyphae can be removed through grooming, but scars often remain. Under a microscope, the fungus is distinguished by asymmetrically curved conidia, the asexually produced spores. These conidia and the fact that the fungus has very low optimal temperatures for growth confirmed *G. destructans* was a new and separate species in the genus *Geomyces*.

Introduction History. Bat white-nose syndrome was first reported from Howe's Cave near Albany, New York, during the winter of 2006–2007. Three years later, its presence was confirmed in caves and mines in Connecticut, Massachusetts, New Jersey, Pennsylvania, Vermont, Virginia, and West Virginia. In February 2010, the disease was reported in Tennessee. It has also expanded into eastern Missouri, as well as eastern Canada. Its origins and potential for further spread remain unknown.

Habitat. *G. destructans* proliferates on many organic surfaces in the dark, cool, high-humidity environments of caves and underground mines, where it colonizes the skin of hibernating bats. It fails to thrive at temperatures in excess of 68°F (20°C). Thus this cold-loving fungus finds optimal growing conditions in the hibernacula used by nonmigratory bats that must over winter in a dormant state at latitudes above 40° N or at higher elevations in the southeastern United States.

Life History. Much is yet to be learned about this recently described fungus. It can grow on a variety of organic substances and seems to persist year round in caves and mines. It apparently becomes established in the skin tissue of bats when their body temperatures are lowered to 35°–50°F (2°–10°C) during torpor in the winter. The hyphae penetrate tissue by entering hair follicles and sebaceous glands. No inflammation or immune response seems to be elicited in the bat.

Although *G. destructans* produces what is known as white-nose syndrome, penetration of the wing membranes by the fungus may be the most detrimental aspect of the disease to bats. Current high mortality rates are one sign that it is a new pathogen that has not achieved a balance with its host. *G. destructans* is spreading rapidly from bat to bat and hibernacula to hibernacula. Spores easily attach to skin, hair, ropes, and clothing; it is not known for how long they remain viable outside their subterranean home or just how they disperse to other

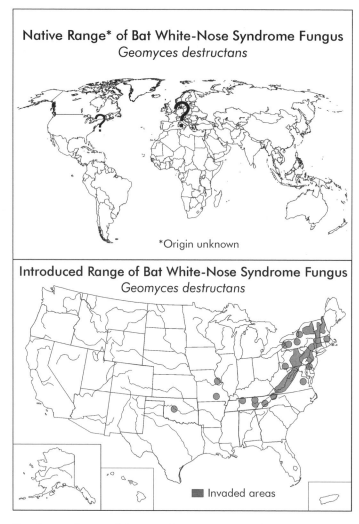

Native Range* of Bat White-Nose Syndrome Fungus
Geomyces destructans

*Origin unknown

Introduced Range of Bat White-Nose Syndrome Fungus
Geomyces destructans

■ Invaded areas

Top: It is not known if *Geomyces destructans* originated as a mutation in a soil fungus native to the northeastern United States or if it is native to Europe. Bottom: Areas in the United States where bat white-nose syndrome had been reported as of June 2010. (Adapted from map by Cal Butchkoski, Pennsylvania Game Commission.)

caves. Migrating bats and cavers likely disperse the fungus to new caves and underground mines.

Impacts. White-nose syndrome has killed some 500,000 insect-eating bats since it was first discovered in 2006. Dead bats are usually emaciated, but exactly how, or indeed if, the fungus kills them is still to be determined. In addition to the tell-tale white encrustations on ears and wing membranes and around the nose, affected bats display unusual winter behaviors. They tend to come out of torpor much more frequently than normal and fly around the cave or mine in which they have been hibernating or leave the hibernacula during the day to fly around outside. Bats are found closer to the cave entrance than normal, and large numbers of dead bats are clustered near or just outside the entrance. It may be that the fungus is such a skin irritant that it rouses the bats from hibernation. When they arouse from torpor and fly around, they use vital body fat reserves, which cannot be replenished because insects are not available for them to feed upon during the winter. Ultimately, the bats may starve to death.

Infection of the wing membranes may exacerbate the situation since the membranes play important physiological roles in regulating body temperature, blood pressure, water balance, and gas exchange. It may also be that infected bats disturb other nonaffected bats in the colony and rouse them from hibernation, too. Bats that have been sampled as they enter hibernacula in the fall seem to be healthy, strongly suggesting they acquire the pathogen in the winter cave.

Currently, six species of bats are known to be affected by white-nose syndrome. The most dramatic losses have been suffered by the little brown bat (*Myotis lucifugus*), which has undergone population declines of 93 percent in some caves. The Indiana bat (*Myotis sodalis*), a federally endangered species, has seen declines of 53 percent in certain hibernacula. Other bats affected include northern long-eared bat (*Myotis septentrionalis*), tri-colored bat

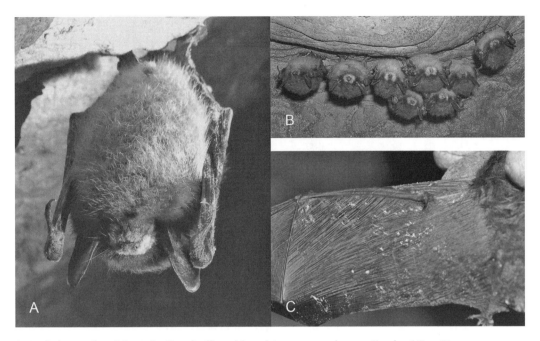

A. Little brown bat (*Myotis lucifugus*) affected by white-nose syndrome, Greeley Mine, Vermont. (Marvin Moriarty, U.S. Fish and Wildlife Service.) B. Hibernating bats with white-nose syndrome, New York. (Nancy Heaslip, New York Department of Environmental Conservation.) C. *Geomyces destructans* infection on wing membrane of bat. (Ryan von Linden/New York Department of Environmental Conservation.)

(*Perimyotis subflavus*; formerly known as Eastern pipistrelle [*Pipistrellus subflavus*]), eastern small-footed myotis (*Myotis leibii*) and big brown bat (*Eptesicus fuscus*). The fungus has recently been found in caves used by another endangered species, the Virginia big-eared bat (*Corynorhinus townsendii virginianus*), although white-nose syndrome has not (yet) been reported on this bat. In Tennessee, there is concern that it will affect the endangered gray bat (*Myotis grisescens*). Most of these bats reproduce slowly, having only a single young each year, so populations are not expected to rebound quickly from the devastating impacts of white-nose syndrome.

Such sudden and widespread death among hibernating bats was previously unknown. Lethal fungal skin infections are rare among mammals, but more common among "cold-blooded" vertebrates whose body temperatures hibernating bats approximate. Chytridiomycosis in amphibians (see Chytrid Frog Fungus below) may be an analogous condition.

Bats have an enormous capacity for consuming flying insects, including moths, mosquitoes, and many plant pests. Their loss would represent a major change in natural, suburban, and agricultural ecosystems in the eastern United States.

Management. While direct bat-to-bat transmission of the fungus is difficult if not impossible to prevent, it is likely that humans transfer *G. destructans* spores from cave to cave, and steps can be taken to reduce or halt that process. The U.S. Forest Service has closed caves and mines to recreational cavers in national forests in 33 states, and many states, caving clubs, and private owners have followed suit. The public should honor all cave closings in affected states and adjoining areas to control the spread of this deadly pathogen. Those entering caves for scientific purposes should carefully decontaminate all clothing and gear with household bleach or commercially available antibacterial cleaners. Ropes and harnesses should not be used at all or, if essential, should be dedicated to use in only a single cave. The U.S. Fish and Wildlife Service has issued containment and decontamination protocols and provides an online listing of cave closures at http://www.fws.gov/northeast/wnscavers.html.

Selected References

Gargas, A., M. T. Trest, M. Christensen, T. J. Volk, and D. S. Blehert. "*Geomyces destructans* sp. nov. Associated with Bat White-Nose Syndrome." *Mycotaxon* 108: 147–54, 2009. Available online at http://botit.botany.wisc.edu/toms_fungi/147gargas9-73.pdf.

"White Nose Syndrome: Could Cave Dwelling Bat Species in the Eastern US Become Endangered in Our Lifetime?" Bat Conservation and Management, 2010. http://www.batmanagement.com/wns/wns.html.

"White-Nose Syndrome Threatens the Survival of Hibernating Bats in North America." USGS Fort Collins Science Center, 2010. http://www.fort.usgs.gov/wns/.

■|Chestnut Blight Fungus

Also known as: Chestnut bark disease fungi
Scientific name: *Cryphonectria parasitica*
Synonym: *Endothia parasitica*
Class: Pyrenomycetes
Order: Diaporthales
Family: Valsaceae

Native Range. Mountainous areas of China and Japan.

Distribution in the United States. Eastern states throughout the range of the American chestnut (*Castanea dentata*), roughly from Maine west to Michigan and south to Georgia and Mississippi. It also occurs outside the tree's range wherever chestnuts have been planted.

Description. This fungus causes cankers, localized areas of dead tissue, on the trunks of American chestnut trees. On the surface of the tree, both swollen and sunken cankers form above the infection; the sunken type expands to girdle the tree. Often orange or yellow fruiting bodies the size of pin heads cover the canker. During moist weather, spores ooze out of the fruiting bodies, looking like tiny curled orange horns.

Early symptoms of infection include reddish-brown patches on the bark that later develop into cankers. The bark may crack, and yellow fruiting bodies appear on the surface. Leaves on affected branches turn brown, but do not fall from the tree for months. The entire tree above the infection will die as the canker grows, but the roots are not killed. An American chestnut will continue to resprout from the root collar for decades after the above-ground part of the tree has died.

Introduction History. The blight was first identified in the United States in 1904 at the Bronx Zoo in New York City. It may have entered the country as early as 1876, when Japanese chestnut trees (*Castanea crenata*) were first imported. By the end of the nineteenth century, the Japanese species was being offered for sale by most U.S. mail-order nurseries. In 1904, the blight was actually widespread north of Virginia. By 1926, it occurred throughout the natural range of the American chestnut. Its impact had been so devastating that in 1912, the U.S. Congress passed the Plant Quarantine Act, the first legal action to stem the flow of nonnative species into the United States, in an attempt to prevent future catastrophes. The law gave the federal government the authority to establish inspection stations and quarantine areas to intercept and prevent the spread of exotic pests and pathogens. Today, the Animal and Plant Inspection Service (APHIS), part of the U.S. Department of Agriculture, is in charge of these activities.

Habitat. In North America, *Cryphonectria parasitica* occurs almost exclusively in American chestnut trees. It infected mature trees in natural forests as well as those planted as ornamentals, and now attacks the saplings that still sprout from old stumps and root systems. The fungus does also attack Allegheny chinkapin (*Castanea pumila*) and bush chinkapin (*C. alnifolia*). Several oaks (*Quercus* spp.) also serve as hosts; but cankers usually remain small and superficial on them and do not kill the tree. Only the post oak (*Q. stellata*) seems to be seriously damaged by the fungus. Other trees on which chestnut blight has been reported include shagbark hickory (*Carya ovata*), red maple (*Acer rubrum*), and staghorn sumac (*Rhus typhina*).

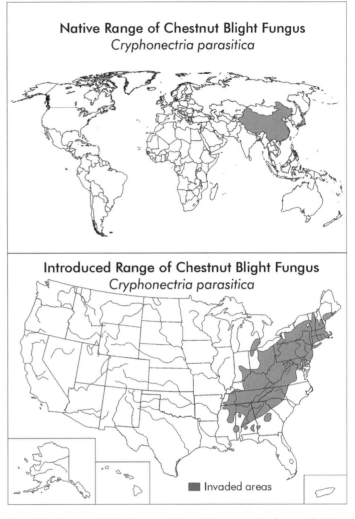

Top: The chestnut blight fungus originated in mountain forests of China and Japan. Bottom: The former range of the American chestnut (*Castanea dentata*), shown on the map, is a surrogate for the range of the blight fungus; although the blight also affects chestnuts planted beyond the tree's native range. (Adapted from Saucier 1973.)

Life History. The chestnut blight fungus produces small, sticky spores in the sexual fruiting bodies visible on the bark's surface. These so-called ascospores are forcibly ejected during warm rains and dispersed by the wind or on the feet of birds and insects. Nonmotile asexual spores called conidia are also produced. The fungal spores enter a tree through wounds or cracks, often at a branch crotch. The junctions of limbs and the trunk are particularly vulnerable because the movement of branches in the wind creates small ruptures in the bark in the crotches. The spores germinate in the inner bark and cambium, where masses of thread-like hyphae (mycelia) form into small brown fans. The cushion-like masses of solid mycelia (stromata) interfere with the flow of nutrients through the phloem and with the growth of the cambium. On American chestnut, stromata can reach densities of 8 per in^2 (50 per cm^2). First the leaves die, then the branches above the point

A. Developing cankers on young American chestnut tree. (USDA Forest Service-Region 8-Southern Archive, USDA Forest Service, Bugwood.org.) B. Leaves of the American chestnut. The once-mighty tree is now represented by root sprouts in the understory. (Linda Haugen, USDA Forest Service, Bugwood.org.) C. A row of dead American chestnuts after the blight struck. The split rail fence in the foreground is likely constructed of chestnut timber. (USDA Forest Service-Northeastern Area Archive, USDA Forest Service, Bugwood.org.)

of infection. The sunken cankers can expand enough to girdle the tree within a single growing season and kill the entire above-ground portion of the tree, or it may take several years. *C. parasitica* continues to live in dead wood, but is unable to enter the roots. Chestnuts continue to resprout for decades. While the sprouts may live for 5–10 years and attain heights of 25 ft. (8.0 m), they rarely live long enough to mature and flower or bear nuts before they, too, succumb to the blight.

Impacts. The chestnut blight fungus eliminated the American chestnut as a major tree in eastern forests. These trees were the largest in the broadleaf deciduous forests in the southern Appalachians, where they commonly grew to heights of 60–90 ft. (18–27 m) and diameters at breast height (DBH) of 3–5 ft. (1–1.5 m). Exceptional specimens were 120 ft. (35.6 m) tall and 7 ft. (2 m) in diameter. The crown of a single tree could be nearly 100 ft. (30 m) wide. Its sweet nuts were produced in great abundance every year and nourished deer, turkeys, squirrels, and other wildlife as well as people. The fruits that fell to the ground after the first frost were an important cash crop for many Appalachian families. The wood was slow to decay and was therefore used for building barns, fences, log cabins, furniture, and coffins. Split-rail fences made of chestnut still persist in woodlands grown up in abandoned pastureland. The bark and wood was rich in tannic acids and preferred over the bark of other trees for tanning leather. Large trees with their spreading branches and reliable nut harvests had been favored shade trees in towns and on farms in and beyond the native range of the chestnut.

Forest structure and composition changed dramatically with the loss of this major tree in the forest canopy. Various oaks (*Quercus* spp.), red maple (*Acer rubrum*), and hickories (*Carya* spp.) assumed dominance in the forest canopy.

By 1940, most large chestnut trees—an estimated 3.5 billion trees—had been killed. Lumber could be salvaged from dead trees for 10 years after the blight, and bark and wood was harvested for tannin until the late 1950s. Standing ghost trees became infested with small boring insects that left pin-sized holes in the wood. This so-called "wormy chestnut" is still valuable whenever it is discovered; it is used largely for making small craft items.

IUCN has nominated the chestnut blight fungus as among 100 of the world's worst invasive species.

Management. Control of the chestnut blight fungus has followed two routes, one trying to reduce the virulence of the fungus, and the other trying to increase the resistance of the tree. A strain of the fungus first identified in Italy was discovered that had been weakened by the presence of a virus. The spores of these so-called hypovirulent fungi can be inoculated into cankers on American chestnuts to slow their growth or, for a time at least, have the fungus produce only the swollen, nonlethal type of canker. A drawback of this procedure is that hypovirulent fungi spread much more slowly in nature than the virulent strains.

Cross-breeding resistant Asian chestnut species with American chestnut can impart a degree of resistance to the fungus in the hybrid generation. Several researchers are repeatedly back-breeding hybrids with American parents in an attempt to produce a tree with good forest form, high-quality timber, abundant and good-tasting nuts, and resistance to chestnut blight.

Some naturally resistant individual American chestnut trees survive both within and outside the original range of the tree. Attempts to graft scions from them to existing root stock are also part of restoration efforts.

Selected References

Anagnostakis, Sandra L. "Revitalization of the Majestic Chestnut: Chestnut Blight Disease." American Phytopathological Society, 2000. http://www.apsnet.org/publications/apsnetfeatures/Pages/ChestnutBlightDisease.aspx.

"Chestnut Blight." Missouri Botanical Garden, 2001–2010. http://mobot.org/gardeninghelp/plantfinder/IPM.asp?code=283&group=39&level=s.

National Biological Information Infrastructure (NBII) and IUCN/SSC Invasive Species Specialist Group (ISSG). "*Cryphonectria parasitica* (fungus)." ISSG Global Invasive Species Database, 2006. http://www.issg.org/database/species/ecology.asp?si=124&fr=1&sts.

Rellou, Julia. "Chestnut Blight Fungus (*Cryphonectria parasitica*)." Introduced Species Summary Project, Columbia University, 2002. http://www.columbia.edu/itc/cerc/danoff-burg/invasion_bio/inv_spp_summ/Cryphonectria_parasitica.htm.

Saucier, Joseph R. "American Chestnut ... an American Wood. (*Castanea dentata* (Marsh.) Borkh.)." FS-230, U.S. Department of Agriculture, Forest Service. Washington, DC: U.S. Government Printing Office, 1973. Available online at http://www.fpl.fs.fed.us/documnts/usda/amwood/230chest.pdf.

Treadwell, Judy C. "American Chestnut History." NCNatural.com, 1996. http://www.appalachianwoods.com/appalachianwoods/history_of_the_american_chestnut.htm.

■|Chytrid Frog Fungus

Also known as: Chytrid fungus, Bd
Scientific name: *Batrachochytrium dendrobatidis*
Order: Rhizophydiales
Family: Chytridiaceae

Native Range. Unknown. Africa may be the source of this fungus, because the earliest-known case of infection is on a museum specimen of the African clawed frog (*Xenopus laevis*) collected in 1938. The disease caused by this fungal agent, chytridiomycosis, was endemic to sub-Saharan Africa 23 years prior to its being found elsewhere in the world. However, recent genetic research suggests that chytrid frog fungus may have originated in Japan.

Distribution in the United States. Throughout.

Description. Chytrid frog fungus is an invisible external parasite of amphibians. Positive identification requires examination of tissue or water and sediment samples in a laboratory. The several symptoms of chytridiomycosis are not unique to this disease but can alert one to the possibility of infection by *Batrachochytrium dendrobatidis*. These include behavioral changes in the infected animal such as lethargy, inability to right itself if turned on its back, failure to flee from humans, and failure to seek shelter from the sun. Frogs and toads may sit in the water all the time. Redness can appear on the belly skin, and wet skin may slough off. The bodies of toads and frogs can become bloated from retention of fluids. Toe curling, head held in a tucked position, and other signs of mild paralysis may occur. Sudden mass die-offs occur in infected populations. No single symptom applies to all affected species, making definitive field identification impossible.

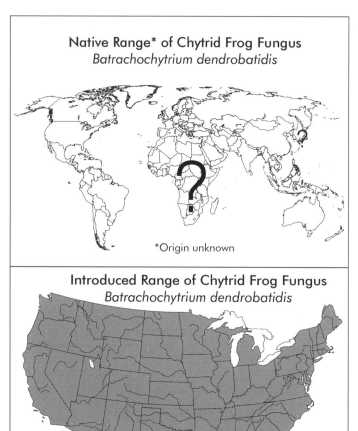

Native Range* of Chytrid Frog Fungus
Batrachochytrium dendrobatidis

*Origin unknown

Introduced Range of Chytrid Frog Fungus
Batrachochytrium dendrobatidis

■ Invaded areas

Top: Africa has been suspected as the place of origin for the chytrid frog fungus, but genetic evidence now points to Japan as the possible source area. Bottom: The chytrid frog fungus is found throughout the United States.

Related or Similar Species. *Batrachochytrium dendrobatidis* is the only chytrid fungus known to infect living vertebrates. Other species decompose cellulose and chitin.

Introduction History. It is unknown when or how the chytrid frog fungus entered the United States. Two hypotheses are being tested. The first poses that the fungus probably originated in Africa and only began to spread rapidly around the world during the twentieth century. The African clawed frog, a known host, was exported in large numbers for medical research beginning in the 1940s and again in the 1960s and 1970s, when its use in human pregnancy tests became popular. The first documented lethal outbreaks of chytridiomycosis did not occur until 1998, when reports came simultaneously from Australia and Central America. The fungus was not scientifically described and named until 1999. Today, it has been identified on over 350 species of amphibians on all continents except Antarctica.

A second hypothesis states that the chytrid frog fungus has been widely distributed, but not recognized, around the world for a very long time, and the sudden outbreaks that began in the late 1990s in disparate locations have been triggered by recent global environmental changes. The synchronous nature of the outbreaks supports this hypothesis, as does evidence from preserved museum specimens and field analyses that show low-level infections without lethal consequences on many amphibian species in the eastern United States.

Examination of preserved museum specimens of North American amphibians reveals that the fungus was in the United States long before it gained the world's attention in the late 1990s. The earliest record dates to 1974 and a leopard frog (*Rana pipiens*) collected in the Rocky Mountains in Colorado. The fungus was also found on a specimen from the Sierra Nevada in California dating to the same year. (It was found on specimens from Québec, Canada, collected in the 1960s.)

Whatever its origins, chytrid can be dispersed on live amphibians sold for food, pets, or research animals. The American bullfrog (*Lithobates catesbeianus*) could be an effective vector, since it is a popular food internationally. Poison dart frogs (*Dendrobates* spp.) from South and Central America could carry the fungus into zoo collections, where they are prized exhibits. Contaminated habitat material such as water and sediment can transfer zoospores to new bodies of water if lodged in tires or on boots or even possibly in the hooves of livestock.

Much research is still needed to unravel the mystery of the distributional history of this fungus.

Habitat. This is an aquatic fungus, and it requires water as well as suitable hosts. It will survive in any body of water from streams and lakes to artificial containers. Chytrid frog fungus prefers cooler temperatures (in laboratory trials, it grows best in water temperatures of 63–73°F [17–23°C] and dies at temperatures above 82°F [28°C]) and permanent flowing water. It is more apt to be found in streams than ponds, because the former are cooler and help to transfer spores over long distances.

Batrachochytrium dendrobatidis lives on the skin of adult frogs and treefrogs, toads, and lungless salamanders. It is most prevalent on the more heavily keritanized skin of the abdomen, especially near the pelvis, on hind limbs, and on feet. It is also found on the oral discs (the structures surrounding the mouths) of tadpoles. Chytrid frog fungus is found in wild populations as well as among captive amphibians held in zoos, aquaria, and aquaculture facilities.

Diet. It apparently feeds on keratin, since it is only found on the keratinized tissues of amphibians. Some research suggests that it may also have non-amphibian hosts or live on dead tissues as well.

Life History. This fungus has two life stages: an attached spherical zoosporangium that is the reproductive stage, and a motile zoospore that disperses to new locations on the same host or to a new host organism. No resting stage has yet been discovered. The zoospore has a single

---

Chytrid Frog Fungus

The global decline in amphibian species is an alarming trend with many possible causes. Among them are habitat destruction, climate change, air pollution, increased UV radiation, chemical contamination of water, and the introduction of nonnative competitors and predators. The chytrid frog fungus may be a major new player. Or it may have been there all along and only now reaches fatal levels of infection when a frog or toad population is already severely stressed by other environmental factors.

---

flagellum with which it moves through water. It attaches to the keratinized outer layers of skin of its host, absorbs its tail, and burrows below the surface. In four days, it will mature into a zoosporangium that has root-like rhizoids that both secrete enzymes to break down keratin and absorb the digested organic products. The zoosporangium forms a single discharge tube that protrudes out of the host's skin and through which each zoosporangium will release as many as 300 zoospores to start the cycle again. Reproduction is asexual or clonal; like many other chytrid fungi, *Batrachochytrium dendrobatidis* seems to lack a sexual stage.

Impacts. Chytrid frog fungus has been implicated as one of several causes of declining amphibian populations around the world. In the United States, regular low-level infections without deleterious effects seem to have occurred since at least the 1960s. Yet particularly in western states, severe chytrid infections have occurred as populations of several frog species have been decreased dramatically. In Arizona, a rapid die-off in 1997 of the lowland leopard frog (*Rana yavapaiensis*) and its subsequent extirpation at many of its past locations may have been due in part to chytridiomycosis. In the Sierra Nevada, die-offs of the mountain yellow-legged frog (*Rana mucosa*) correlated with the presence of *Batrachochytrium dendrobatidis*; but in and around Pinnacles National Monument, California, few dead frogs were encountered during surveys in 2006–2007 even though the fungus was discovered infecting Pacific tree frogs (*Pseudacris regilla*), western toads (*Bufo boreas*), California red-legged frogs (*Rana aurora draytonii*), and foothill yellow-legged frogs (*Rana boylii*). Some populations of the Oregon spotted frog (*Rana pretiosa*), a species in decline in the Pacific Northwest, are persisting even though heavily infected with the fungus. On the other hand, dead and dying boreal toads (*Bufo boreas boreas*) in Rocky Mountain National Park, Colorado, were found to be infected with the chtyrid fungus, a likely cause of their demise. Die-offs linked to chytrid also are reported in Wyoming and Washington. Susceptibility to the disease is highly species specific and perhaps site specific. It may be more virulent at higher elevations with cooler temperatures. Some more tolerant species such as the American bullfrog and some salamanders may act as transmitters of the parasite.

It remains to be discovered how chytrid frog fungus kills its host. Various hypotheses exist, including the possibility that respiration through the frogs' skin may be affected or that the fungus produces a toxin. Recent research showed that the skin of infected frogs was less able to transport sodium and chloride ions and maintain proper balances of sodium and potassium in the blood. Severe electrolyte imbalances lead to heart stoppage.

Much is yet unknown about the impacts of chytrid frog fungus and just how great a risk it is to amphibian populations worldwide.

Management. Little can be done to eradicate the fungus once it invades a body of water. To prevent the spread of this potentially lethal parasite, care should be taken to disinfect all equipment, footwear, and clothing before entering a stream or pond. Do not move amphibians from one source area to another. Never release captive animals into wetlands, ponds, lakes, or streams.

Selected References

Daugherty, Matt, and Kim Hung. "Chytrid Fungus, *Batrachochytrium dendrobatidis*." Center for Invasive Species Research, University of California, Riverside, 2009. http://cisr.ucr.edu/chytrid_fungus.html.

National Biological Information Infrastructure (NBII) and IUCN/SSC Invasive Species Specialist Group (ISSG). "*Batrachochytrium dendrobatidis* (Fungus)." ISSG Global Invasive Species Database, 2006. http://www.issg.org/database/species/ecology.asp?fr=1&si=123.

Ouellet, M., I. Mikaelian, B. D. Pauli, J. Rodriguez, and D. M. Green. "Historical Evidence of Widespread Chytrid Infection in North American Amphibian Populations." *Conservation Biology* 19(5): 1431–40, 2005.

Padgett-Flohr, G. E. "General Amphibian Disease Information." California Center for Amphibian Disease Control, 2002. http://ccadc.us/docs/AmphibianDiseasesPresentation.pdf.

■ | Dutch Elm Disease Fungi

Scientific name: *Ophiostoma novo-ulmi and O. ulmi*
Synonym: *Ceratocystis ulmi*
Class: Sordariomycetes
Order: Ophiostomatales
Family: Ophiostomataceae

Native Range. Uncertain; probably eastern Asia. The disease was first identified and scientifically described in the Netherlands in 1921 and thus became known as Dutch elm disease.

Distribution in the United States. The disease occurs throughout the 48 contiguous states except for the desert regions of the American Southwest.

Description. Dutch elm disease is caused by one of two closely related fungi. *O. ulmi* is becoming less prevalent as it becomes replaced by the more aggressive *O. novo-ulmi*, believed responsible for most elm mortality from the 1950s through the 1970s. The fungi grow inside the xylem of living trees, where white fruiting bodies may be visible if the xylem is exposed in cross-sectioning; they may be positively identified only in a laboratory.

Outward symptoms of this wilting disease of elm trees develop quickly, usually within a month, and are obvious. Leaves on a branch of an apparently healthy tree yellow and wilt, a process known as "flagging." This is caused by the clogging of xylem tubes by the growing fungi and prevention of water transport; flagging usually becomes evident in late spring, when the tree's leaves have reached full size. They will eventually turn brown and drop prematurely. Wilting proceeds from the tips of branches downward through the crown, unless the fungus has entered the tree from its roots. In that case, the signs of infection appear first in the lower crown and quickly envelope the entire crown. Often it takes only one year for the whole tree to be affected; but sometimes it takes two or more years.

Internal symptoms are the result of dead xylem tissue, revealed by a brown streaking of infected sapwood. These streaks run with the grain and are evident when the bark is stripped from a branch. In a cross-sectional cut of the branch, the vascular damage appears as a ring of brown spots.

Two beetles are the vectors for Dutch elm disease, and management often targets these species in order to slow the spread of the fungi. The native elm bark beetle (*Hylurgopinus rufipes*) is a brownish-black coleopteran sparsely covered with stiff yellow hairs and about

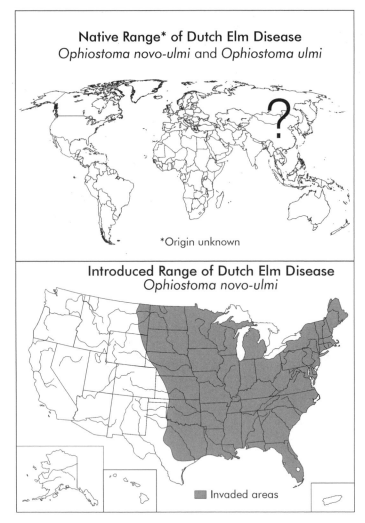

Native Range* of Dutch Elm Disease
Ophiostoma novo-ulmi and *Ophiostoma ulmi*

?

*Origin unknown

Introduced Range of Dutch Elm Disease
Ophiostoma novo-ulmi

■ Invaded areas

Top: Eastern Asia is the likely place of origin for the fungi that cause Dutch elm disease. Bottom: The distribution of *Ophiostoma novo-ulmi*, the chief fungus implicated in Dutch elm disease today. (Adapted from "Pest Distribution Map: Dutch Elm Disease, *Ophiostoma novo-ulmi*." Alien Forest Pest Explorer, USDA Forest Service, Northern Research Station http://www.fs.fed.us/ne/morgantown/4557/AFPE).

0.08–0.12 in. (2–3.5 mm) long. Its wing cases (elytra) are deeply pitted. The larvae are small, white grubs that produce distinctive galleries beneath the bark of elms. Egg-laying galleries of adults run across the grain, while the smaller tunnels made by larvae come off the main gallery parallel to the grain. *H. rufipes* burrows into the bark on branches and trunks of elm trees, so the infections it introduces start in major branches.

The second vector is an introduced insect, the smaller European elm bark beetle (*Scolytus multistriatus*). This beetle is slightly larger than the native elm bark beetle and reddish brown. A prominent spine extends from the underside of its concave abdomen. It constructs galleries in the opposite manner of the native elm bark beetle: the egg-laying gallery runs along the grain of the wood, while the larval galleries are cut perpendicular to it across the grain. Since the smaller European elm bark beetle feeds in the crotches of small twigs, infections transmitted by it occur first in twigs.

Related or Similar Diseases. Two other diseases of elms may be mistaken for Dutch elm disease. Elm yellows (elm phloem necrosis) causes all the leaves of the crown to turn yellow, usually between July and September. Leaves do not wilt or turn brown. No streaking occurs in the sapwood, but the inner bark will be discolored and characteristically develops a wintergreen scent. Bacterial leaf scorch, like Dutch elm disease, infects and clogs the xylem; however, this disease results in a slow decline of the tree over many years. The older leaves on a branch are first to show symptoms of stress. In summer and early fall, the margins of the leaves turn brown, with a yellow area appearing between the green tissue and the scorched outer edge of the leaf. No visible changes to sapwood or inner bark occur.

Introduction History. Dutch elm disease was first identified in the United States in Ohio in 1931. *O. ulmi* had been introduced a few years earlier on diseased elm logs imported from

France to be used as veneers in furniture making. The smaller European elm bark beetle had arrived ahead of it. *O. ulmi* probably had been introduced to Europe from Asia around 1910. It is believed responsible for the first epidemic of Dutch elm disease that swept across eastern North America as well as Europe. The more aggressive *O. novo-ulmi* was likely introduced to the southern Great Lakes region in the 1940s or 1950s; it probably has caused most of the widespread loss of elms since then. The disease spread north and south through the eastern United States and reached the West Coast in 1973. It continues to expand its distribution area and kill off urban and wild elms. However, many seedlings and saplings escape infection long enough to reproduce, so elms are not threatened with extinction.

The fungi are spread short distances by elm bark beetles. These insects fly only a few hundred feet away from the tree in which they hatched to feed on another elm. Dispersal distance may be 2 miles (3 km) or more to breeding sites. The rate of spread seldom exceeds 4–5 miles (6–8 km) a year.

When trees grow close together (within 25–50 ft. [7.5–15 m] of each other) in pure stands, the usual pattern along city streets, spores from the fungi can be transferred from tree to tree via the roots. Where roots from neighboring trees cross each other, they fuse to create a root graft and allow the passage of the Dutch elm disease pathogen. The spores will carried up the new host tree with the sap in the xylem.

Habitat. Urban forests, natural forests, and landscape trees along streets and in yards. *O. novo-ulmi* and *O. ulmi* infect only elms (*Ulmus* spp.). American elm (*U. americana*) is highly susceptible to the fungus, while other native elm species show a range of resistance; but none is immune to Dutch elm disease.

Diet. The fungi that cause Dutch elm disease are both parasites feeding on living tissue of elm trees and saprotrophs living off dead elm tissues. They produce enzymes that digest plant cell walls and perhaps toxins that kill the parenchyma cells of xylem. It is the death of xylem parenchyma that causes the diagnostic discoloration of the sapwood.

Life History. Dutch elm disease fungi are introduced into an elm tree when elm bark beetles contaminated with spores bore into the bark to feed or to create egg-laying tunnels. The spores become dislodged and germinate in the galleries to produce thread-like cells, the hyphae, that penetrate into the xylem, where they grow and form the fruiting bodies of the fungus. The fruiting bodies produce millions of white oval spores. These move through the xylem, where they reproduce asexually by budding and spread the disease through the tree. In dying and recently dead trees, a different type of asexual fruiting structure is produced asexually in the bark and in the galleries just beneath the bark: Round, sticky white spores are attached to the top of dark stalks less than 0.1 in. (1–2 mm) tall. The spores adhere to adult beetles and are carried to new elms, when the beetles exit the tree. Sexually produced spores also occur. These accumulate in sticky droplets that can also attach to elm bark beetles to be carried to new host trees. Newly emerged adult beetles first feed on healthy elms. Later, they move to sick or dead elms to breed. Beetles usually have two broods a year, the second of which overwinters and emerges in early summer. The second brood is responsible for the majority of new infections.

Impacts. Dutch elm disease has destroyed many millions of elm trees in the United States. The stately, vase-shaped American elm once lined the streets of towns and cities in the eastern part of the country, their arching canopies providing welcome summer shade. The elm was favored as a landscape and urban forest tree because it was beautiful, long-lived, and tolerant of poor air quality and compacted soils. The urban landscape is totally altered with the demise of these trees. The American elm was most susceptible to the disease, but other species, including those in natural forests, are also vulnerable to varying degrees.

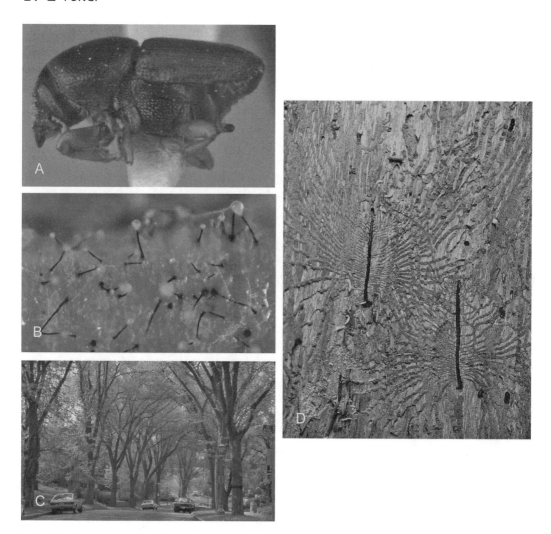

A. Smaller European elm beetle, the primary vector for this pathogen. (Gerald J. Lenhard, Louisiana State Univeristy, Bugwood.org.) B. Fruiting bodies of the Dutch elm disease fungus. (William Jacobi, Colorado State University, Bugwood.org.) C. American elms lined many city streets in days past. The vase-like structure of this native species was one of its desirable characteristics. (Joseph O'Brien, USDA Forest Service, Bugwood.org.) D. Galleries made by the smaller European elm bark beetle. (John A. Williams, USDA Forest Service, Bugwood.org.)

Trees are more resistant under drought conditions, and slower-growing individuals are less susceptible than vigorously growing ones.

Management. Reducing losses of urban trees from Dutch elm disease requires a community-wide campaign to remove and destroy diseased branches and trees. Root grafts also must be destroyed. Application of systemic chemicals can prevent and treat infections in individual specimen trees, but must be repeated every 2–3 years. Insecticides may control beetle populations, but they must be synchronized with the beetles' life cycles and can be expensive, and there are potential hazards in handling these products.

American elms continue to be planted because of their desirable properties. If arborists avoid large plantings of the same species and space elms well apart from other elms, infection rates and tree losses may be reduced significantly.

The best long-term solution is the breeding of resistant strains of the elm. Hybridization between American elms and Asian species resistant to or tolerant of the fungus provided resistance but failed to produce the elegant structure of the American tree. Scientists with the Agricultural Research Service and the U.S. National Arboretum discovered and nurtured some old surviving American elms with the necessary resistance and through careful breeding produced varieties that preserved the tall vase shape of the original species; some are now available commercially. They include such cultivars as American Liberty, Princeton, Independence, Valley Forge, New Harmony, and, most recently, Jefferson elms.

Selected References

D'Arcy, C. J. "Dutch Elm Disease." *The Plant Health Instructor.* American Phytopathological Society, 2000. Revised, 2005. http://www.apsnet.org/edcenter/intropp/lessons/fungi/ascomycetes/Pages/DutchElm.aspx. doi:10.1094/PHI-I-2000-0721-02.

"Fact Sheet: Dutch Elm Disease (*Ophiostoma novo-ulmi*)." Cornell University, Plant Disease Diagnostic Clinic, 2009. http://plantclinic.cornell.edu/FactSheets/dutchelmdisease/DED.htm.

Haugen, Linda. "How to Identify and Manage Dutch Elm Disease." Northeastern Area State and Private Forestry, USDA Forest Service, n.d. http://www.na.fs.fed.us/spfo/pubs/howtos/ht_ded/ht_ded.htm.

National Biological Information Infrastructure (NBII) and IUCN/SSC Invasive Species Specialist Group (ISSG). "*Ophiostoma ulmi sensu lato* (fungus)." ISSG Global Invasive Species Database, 2010. http://www.issg.org/database/species/ecology.asp?fr=1&si=130.

■|Sudden Oak Death

Also known as: Ramorum blight, SOD
Scientific name: *Phytophthora ramorum*
Class: Oomycetes
Order: Pythiales
Family: Pythiaceae

Native Range. Unknown. Two mating types exist, A1 in Europe and A2 in North America. It is presumed that they have a common ancestor, perhaps from Asia. On both continents, the pathogen acts like a new or emerging disease, supporting the idea that it originated somewhere else. (Ramorum blight was first recorded in Europe, on ornamental rhododendrons in Germany in 1993.)

Distribution in the United States. Sudden oak death occurs in 14 counties of coastal California stretching 300 mi. (480 km) from Monterey County north to Humboldt County. It is also in neighboring Curry County in southwestern Oregon.

Description. The fungus is best described according to the symptoms shown on host plants, which manifest themselves in two groups according to the response to the parasite: bark canker hosts and foliar hosts. Tanoaks (*Lithocarpus densiflorus*) fall into both categories. On the first group, which includes coast live oak (*Quercus agrifolia*), California black oak (*Q. kelloggii*), Shreve oak (*Q. parvula* var. *shrevei*), and madrone (*Arbutus menziesii*), the twigs show the first signs of infection, with discolored patches of dead tissue beneath the bark separated from healthy tissue by a black line or reaction zone. This zone is the advancing front of infection. As the fungus grows in the twig, it spreads into larger branches and causes cankers that lead to branch die-off. With continuing expansion, the infection reaches the stem, where cankers may become more than 6 ft. (2 m) long and girdle the cambium, quickly killing the tree. Characteristically, a red or black sap-like fluid oozes from the canker, staining the bark and

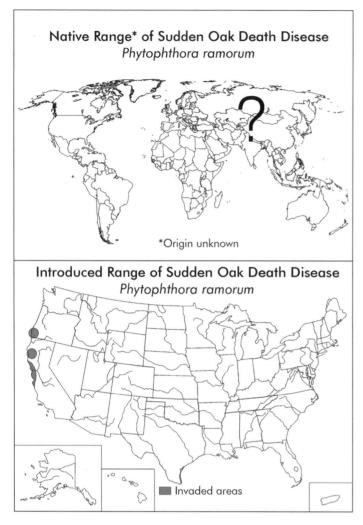

Native Range* of Sudden Oak Death Disease
Phytophthora ramorum

?

*Origin unknown

Introduced Range of Sudden Oak Death Disease
Phytophthora ramorum

■ Invaded areas

Top: The place of origin of the fungus responsible for sudden oak death remains a mystery, although somewhere in Asia is likely. Bottom: Areas known to be affected by *Phytophthora ramorum* in 2009. (Adapted from map by M. Kelly, University of California, Berkeley. http://www.sudden oakdeath.org.)

killing any lichen or mosses growing on the trunk. The canker attracts western oak bark beetles (*Pseudopityphthorus pubipennis*) and ambrosia beetles (*Monarthtum dentiger* and *M. scutellare*) in abnormally large numbers. The abundance of such beetles is one of the distinct signs of sudden oak death. The dead wood of the canker hosts a saprophytic fungus, *Hypoxylon thouarsianum*; its fruiting bodies are flattened khaki-green domes that turn black with age and are another indication of infection by sudden oak disease. On small tanoaks, which are both bark canker hosts and foliar hosts, the first signs of the disease are wilting of the branch tips. When the branch dies, the response of the tree is to resprout with multiple new shoots. Tanoaks often display several cankers at different heights above the ground. As with all bark canker hosts, within a few weeks of girdling of the trunk, the crown suddenly browns. Death of evergreen trees such as live oaks may take one or two years after infection, with the brown leaves remaining on the tree during that period. One does not find single isolated trees infected by the disease; instead, many trees in close proximity will be killed. Another field observation has been that oak and tanoak trees with bark cankers are always growing near infected bay laurels, which are foliar hosts.

Foliar hosts such as coastal redwood (*Sequoia sempervirens*), Douglas fir (*Pseudotsuga menziesii*), bay laurel (*Umbellularia californica*), rhododendrons, and huckleberries are not killed by the fungus; and their infections are often referred to as ramorum blight. The main symptoms of the disease are brown or black spots on green leaves, browning leaves, and sometimes twig dieback.

Despite these several signs, sudden oak death cannot be reliably determined in the field, but requires sophisticated laboratory analyses.

Related or Similar Species. The symptoms of several ailments may resemble those of sudden oak death including canker rots, leaf scorch, freeze damage, and herbicide damage.

In the eastern United States, where sudden oak death is not as yet present outside the occasional infections in nurseries, oak wilt presents without cankers or bleeding bark; it is caused by a different fungus, *Ceratocystis fagacearum*. Oak decline, another disease of eastern oaks, kills many trees, but slowly. It is believed to be brought on by the interactions of multiple stresses such as drought, infections of root fungi such as *Armillaria mellea*, and infestations of inner bark borers such as two-lined chestnut borer (*Agrilus bilineatus*) and red oak borer (*Enaphalodes rufulus*). The red oak borer by itself causes dark, wet stains, but stains on the inner bark show no black zones and are accompanied by beetle burrows filled with fine frass. All other *Phytophthora* species in temperate areas infect roots and root crowns and spread as soilborne or waterborne spores.

Introduction History. The sudden death of large swaths of tanoaks was initially reported in 1995 in Marin and Santa Cruz counties, California. The cause was not identified until 2000, when University of California scientists determined it was as a previously unknown species of the fungus *Phytophthora*. Later that same year, a researcher in the United Kingdom determined that it was the same fungus that had been affecting rhododendrons in Germany and the Netherlands since 1993. On both continents, the origins of the infections are unknown.

In 2001, *P. ramorum* was discovered on ornamental rhododendrons in a Santa Cruz nursery, and Oregon reported a 40-acre outbreak in Curry County. By 2002, the presence of the fungus was confirmed in natural forests in 10 counties in California. Today, it occurs in 14.

In subsequent years, *P. ramorum* was detected in other nurseries in California and Oregon, and many other states established outright bans on the import of plants from California, or temporary quarantines, inspection, and treatment of imported nursery stock. A major scare occurred in 2004 when infected plants from the huge Monrovia nursery showed up in Colorado, Georgia, Louisiana, Maryland, North Carolina, Texas, and Virginia, as well as Canada. *P. ramorum* continues to turn up in nurseries across the country. (It has also spread across Europe.)

Habitat. The disease is most prevalent in the mild, moist habitats exemplified by the fog-swept coasts of California. Temperatures of 64–68°F (18–20°C) and humidity close to 100 percent increase the chances of infection. The fungus that causes sudden oak death is common in the understory of coastal redwood forests on tanoak, in Douglas fir–tanoak forests with understories of huckleberry or other *Vaccinium* species, and in coastal broadleaf forests on coast live oak, California black oak, Shreve oak, madrone, and bay laurel among at least 100 species known to be susceptible to bark canker or—much more frequently—leaf infections. It lives primarily in the phloem tissues of trees, shrubs, and perennial herbs in natural forests, parklands, and residential areas.

Life History. This fungus shows little genetic variability throughout its range in the United States, where it exists as a clone that reproduces only asexually. (Both mating types, A1 and A2, are needed for sexual reproduction.) The fungus forms sporangia, the sacs that produce zoospores, mostly during California's rainy season between December and June. The sporangia break off and may be carried by the wind some 15 ft. (5 m) away from the original infection site. Sporangia apparently form only in infected leaves, and not on the bark of oaks or tanoak. Zoospores are released when the sporangia land on a new host; they are motile and can swim in water or a water film on the leaves and bark. The cankers on susceptible oaks are usually either close to the soil line, suggesting that rain splash may be a means of dispersal; or adjacent to the leaves of infected bay laurels, suggesting airborne dispersal. No wound in the bark appears necessary to allow infection of those plants that are bark canker hosts, nor can they reinfect themselves since the sporangia develop on fungi infecting leaves.

A. Canker with zone lines caused by infection by *Phytophthora ramorum*. B. Canker bleeding. C. Oaks defoliated as a consequence of sudden oak death. (Joseph O'Brien, USDA Forest Service, Bugwood.org.)

Spores can be found in streams all year, so sporulation must occur during the dry season. *P. ramorum* also forms larger, thick-walled resting spores (chlamydospores) that survive through unfavorable environmental conditions. Much remains to be learned about the life history of *P. ramorum*, but in some other *Phytophthora* fungi, the resting spores can remain dormant up to six years.

Impacts. In infested areas, high mortality is experienced by tanoaks; and major losses are incurred by coast live oak, Shreve oak, and California black oak. These trees may be dominants in the canopies of the forests in which they occur and many times make up pure stands. When these trees die, the forest floor is no longer shaded, and soils become less permeable to water, changing hydrologic conditions. Gone too are food (acorns) and shelter for wildlife. The dead trees and litter increase the fuel available for wildfires and may change the fire regime. Possible changes in species composition in affected forests are yet to be determined.

Foliar hosts, although not killed, serve as reservoirs for the disease and are necessary components in the spread of the fungus. This is particularly a problem in ornamental rhododendrons and camellias in California's nursery trade, which sells plants to other states and countries, and also for companies exporting Christmas trees and redwood mulch. The economic toll of import bans, quarantines, and inspections could be significant.

Experiments have shown that northern red oak (*Quercus rubra*), northern pin oak (*Q. palustris*), and mountain laurel (*Kalmia latifolia*) are highly susceptible to sudden oak disease. Many Midwestern and southern forests would be threatened were the fungus to become established in them. Most vulnerable may be forests in the Ozark-Ouachita highlands, live oak stands in Florida, and pin oak sand flats in the Great Lakes area.

Management. No way exists to control the disease in the wild. The main management strategy is to prevent its spread via the nursery trade. Since the resting spores survive in soil and leaf litter, it is also important to clean the tires of vehicles and boots of hikers that have been in affected areas and to prohibit the transport of plant debris, including bark, wood, and mulch, from infected areas.

In nurseries and individual landscape plants, the infected plant is either treated with a special fungicide or cut down and burned.

Selected References

Alexander, J. M., and S. V. Swain, *Pest Notes: Sudden Oak Death* UC ANR Publication 74151, UC Statewide IPM Program, University of California, Davis, Ca, 2010. http://www.ipm.ucdavis.edu/PMG/PESTNOTES/pn74151.html.

Garbelotto, Matteo. "Sudden Oak Death: A Tale of Two Continents." *Outlooks on Pest Management*, 85–89. Research Information Ltd., 2004. http://www.ufei.org/ForesTree/files/collected/Pesticide Outlook.pdf. doi:10.1564/15apl12.

National Biological Information Infrastructure (NBII) and IUCN/SSC Invasive Species Specialist Group (ISSG). "*Phytophthora ramorum* (fungus)." ISSG Global Invasive Species Database, 2008. http://www.issg.org/database/species/ecology.asp?si=563&fr=1&sts=sss.

O'Brien, Joseph G., Manfred E. Mielke, Steve Oak, and Bruce Moltzan. "Sudden Oak Death" *Pest Alert*. U.S. Department of Agriculture, Forest Service, State and Private Forestry, Northeastern Area, 2002. http://na.fs.fed.us/spfo/pubs/pest_al/sodeast/sodeast.htm.

■ | White Pine Blister Rust

Scientific name: *Cronartium ribicola*
Division: Basidiomycota
Class: Teliomycetes
Order: Uredinales
Family: Cronartiaceae

Native Range. Asia. The natural range of this fungus is unknown but presumed to be Asia, since the disease was unknown in Europe prior to the introduction of eastern white pine (*Pinus strobus*) from North America in the 1600s. White pine blister rust was first discovered in Europe in the Baltic provinces of Imperial Russia in 1854. From there, it spread westward across Europe. It is supposed that the rust infected native pines in Asia without significant symptoms but when a naïve nonnative species became widely planted, it found a new host and gained virulence.

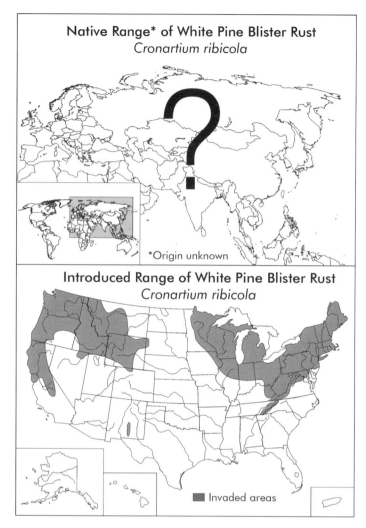

Native Range* of White Pine Blister Rust
Cronartium ribicola

*Origin unknown

Introduced Range of White Pine Blister Rust
Cronartium ribicola

■ Invaded areas

Top: The place of origin of white pine blister rust is unknown but presumed to be in the pine forests of Asia. Bottom: White pine blister rust affects both western and eastern pines in the lower 48 states. (Adapted from map by the Global Invasive Species Team [GIST], the Nature Conservancy. http://www.invasive.org/gist/photos/crori03.gif.)

Distribution in the United States. This parasite occurs throughout the range of white (five-needled) pines in the eastern United States, the Great Lakes area, and the mountains of the northwest. It can be found in at least 33 states and continues to expand its range in the Southwest and in southern California.

Description. The white pine blister rust fungus requires two hosts, white pines and gooseberries or currants (*Ribes* spp.; but see below under Habitat) to complete its life cycle; the most visible signs of its presence appear on the pines. Early symptoms of infection occur in late summer or fall as small yellow spots on live pine needles. The fungus spreads down the needle into the vascular system of the twig and, during the next growing season, produces a slight swelling and yellowing of the bark as a diamond-shaped canker forms on the branch. Within a year or two of infection, yellow blisters (aecia) up to 0.25 in. (3 mm) wide erupt through the bark in early spring. When the blisters rupture, they release vast quantities of yellow-orange spores (aeciospores) and then dry up, leaving a rough patch of bark on the branch. In late spring and early summer, another form of blister (pycnia) appears on the canker. It oozes a sticky yellow-orange fluid that contains pycniospores. Both the blisters and the fluid harden and blacken, and remain on the tree for several weeks. The blisters form again in following years around the margin of the canker, creating a distinctive orange perimeter and expanding the canker until the stem is invaded and girdled and the tree above the canker killed. Rust-infected bark is high in sugar content and attracts rodents; their gnawing at the bark produces copious amounts of pitch (resin) that run down the tree trunk and is another sign of rust infection.

The progression of the disease is visible from a distance: chloritic (yellowed) needles, stunted or dead branches (flagging), and eventually dead crowns or dead trees. On gooseberries and currants, the symptoms of rust infection are less obvious. The lower surfaces of leaves lose their green color and within a few days become covered with minute orange fruiting bodies (uredinia); these tiny bumps produce yellow-orange spores. In late summer, a different type of fruiting body (telium) covers the underside of the leaves with short yellow-brown hair-like structures.

Related or Similar Species. This fungus is the only stem rust of white pines in North America, and it affects only members of the white pine group. Similar symptoms may appear on lodgepole (*Pinus contorta*) and ponderosa (*P. ponderosa*) pines, but these are caused by the western gall rust (*Endocronartium harknessii*). Pine bark adelgids (*Pineus strobi*), an aphid-like insect introduced from Europe, cover themselves in white, woolly, and waxy secretions; heavy infestations could be mistaken for the pitch-dripping cankers on white pines infected with white pine blister rust. So could Armillaria root rot disease, caused by the fungus *Armillaria mellea.* When the lower trunks of conifers host this fungus, large amounts of resin are exuded.

Introduction History. White pine blister rust first entered the eastern United States on eastern white pine seedlings imported from Germany sometime between 1898 and 1908. The U.S. Forest Service had sent seeds to France and Germany, because American nurseries could not meet the demand when replanting cutovers became the vogue, and then encouraged the planting of millions of imported seedlings across the northeastern states from New England to Minnesota. The blister rust invaded the northwestern United States after having been introduced accidentally in 1910 to Vancouver, British Columbia, Canada, on eastern white pine seedlings from France; however, it was not diagnosed until 1921. During this time period, Americans were beginning to understand that they could no longer cut down forests and move ever westward, but had to practice forest management and make efforts to reforest cutover areas. Europeans had faced these problems much earlier and become accomplished nurserymen, producing seedlings of American white pine as well as European species. Europe thus became a valuable source of seedlings for American foresters.

The first outbreak of the disease occurred in 1906 in Geneva, New York, where it was discovered on currants. In 1909, it was found at several eastern locations on eastern white pine seedlings from European nurseries and, by 1915, on native seedlings. By 1922, some areas of New Hampshire saw 50 percent of white pines infected; the rust quickly spread through the forests of the Appalachians and into the Great Lakes region, infecting most major pine regions by 1950. In the west, the blister rust had spread east in western white pine as far as Idaho as early as 1923, and south to Oregon—infecting sugar pine—by 1929. The disease was observed in the northern Sierra Nevada in 1941 and had reached the southern Sierra by 1961. In 1970, it was discovered on southwestern white pine in New Mexico and limber pine in southern Wyoming. It showed up in northern Colorado on limber pine in 1998. The most recently documented range expansion is into central Colorado in 2003, where for the first time it was observed on Rocky Mountain bristlecone pine. It continues to spread through the Great Basin, the southern Rocky Mountains, and the Southwest.

Habitat. White pine blister rust infects pine trees that have five needles per bunch. It attacks trees in natural forests as well as those grown in plantations or as ornamentals. Cool, moist weather in summer and early fall is necessary for the infection of the pine host, so it tends to be restricted to more northerly latitudes or high elevations. In the eastern United States and the Great Lakes area, eastern white pine (*Pinus strobus*) is the sole host. In western states, western white pine (*P. monticola*), sugar pine (*P. lambertiana*), whitebark pine (*P. albicaulis*), limber pine (*P. flexilis*), and southwestern white pine (*P. strobiformis*) are hosts. In 2002, the rust was detected on Rocky Mountain bristlecone pine (*P. aristata*) for the first time.

All currants and gooseberries, both wild species and cultivars, are susceptible to some degree to infection as the alternate hosts required by the rust. Most susceptible is the domesticated European black currant (*Ribes nigrum*). Some red-currant varieties are highly resistant, perhaps immune to the disease. Recently, two western members of the broomrape family (Orobanchaceae), sickletop lousewort (*Pedicularis racemosa*) and giant red Indian paintbrush (*Castilleja miniata*), have also been found to serve as alternate hosts.

Life History. The white pine blister rust has what is known as a macrocyclic, heteroecious life cycle. It is extremely complex and involves five spore types and requires two alternate host species. Germ tubes from basidiospores released from the life stage of the rust

A. Earlier stage of infection with oozing blisters. (John W. Schwandt, USDA Forest Service, Bugwood.org.) B. Fruiting bodies of *Cronartium ribicola*. (USDA Forest Service Archive, USDA Forest Service, Bugwood.org.) C. Well-developed canker on eastern white pine. (Minnesota Department of Natural Resources Archive, Minnesota Department of Natural Resources, Bugwood.org.) D. Uredinia on underside of currant leaf. (Petr Kapitola, State Phytosanitary Administration, Bugwood.org.) E. Dead crowns are among the more conspicuous symptoms of white pine blister rust. (Joseph O'Brien, USDA Forest Service, Bugwood.org.)

that lives on a member of the genus *Ribes* enter the stomata (pores through which a leaf exchanges gases with the atmosphere) of a white pine needle. They grow down the length of the needle and establish mycelia in the vascular tissue of the twig to which the needle is attached, parasitizing living cells. Blister rust cankers expand with time to the branch and eventually the main stem (trunk) of the tree. A year or two after infection, pycnia develop at the margin of the canker and produce pycniospores. Pycniospores result in mycelia that produce aecia (a cuplike fruiting body) the next year. Aecia break through the bark, covered by white membranes that create blisters. When the membrane ruptures, quantities of orange aeciospores are released. Pines cannot directly infect other pines; the aeciospores must find an alternate host for the next phase of the rust's life cycle. The wind can carry them over hundreds of miles, and if conditions are optimal, they will survive for several months until they land on and infect a wild or cultivated currant or gooseberry. Mycelia grow in the *Ribes* leaf, and within a few weeks of the initial infection, the rust produces tiny pustules (uredinia) on the underside of the *Ribes* leaf; from these fruiting bodies urediospores are released, spreading the rust to other currants or gooseberries nearby. These spores cannot infect pines. The rust has entered what is known as the repeating stage, when reinfection of the same alternate host species occurs over and over. In late summer, after two weeks of cool, moist weather, hairlike telial fruiting bodies emerge from the underside of the *Ribes* leaf. These release basidiospores that are wind-dispersed a distance usually less than 1,000 ft. (300 m) and that, if successful, land on a white pine. Their germ tubes enter the needles' stomata to begin the cycle again. The rust overwinters in the pine; it dies in *Ribes* or outside a living host.

The disease progresses from needle to twig, to branch, to the trunk of the tree as the mycelia grow through the vascular system and absorb water and nutrients from the living cells. Cells of the host pine die as they become undernourished or the flow of nutrients is blocked. Once a branch or trunk is girdled by the canker, the tissues beyond the infection point are cut off from water and nutrients, and they die. It may take 15 years for a tree to be killed. *Ribes* shrubs, on the other hand, are not killed by the fungus, but severe infections may lead to defoliation.

Impacts. In the early twentieth century, when the white pine blister rust was introduced, American commercial forestry revolved around the harvest of white pine. When the eastern white pine was logged out of New England and other northeastern states, the industry had shifted westward to the Great Lakes region. By 1900 it was beginning to cut western white pine and sugar pine from the northern Rocky Mountains to the Cascades and Pacific Northwest. Eastern white pine was being replanted in the eastern United States, often with inexpensive seedlings from France and Germany. The rust is especially lethal to young trees of sapling and pole size. Efforts to protect these valuable timberlands began almost immediately and influenced the impacts of the fungus on native forests.

White pine blister rust has caused more damage and had more money spent on its control than any other disease of conifers in the United States. Thousands of stands became unsuited for lumber production or were lost entirely. The initial invasion in the Northeast resulted in high rates of infection and mortality but also selected for existing resistance in eastern white pine. In New Hampshire in the first half of the twentieth century, for example, disease incidence ranged from 20 to 80 percent in sampled stands. By the end of the century, the rate of infection was 7 percent or less. The maturing of surviving white pine probably affects the rate, since young trees are more susceptible to infection. Today, the rust is widespread but not frequent, and not a major concern. Eradication of wild and cultivated *Ribes* as a primary management strategy (see under Management), probably shifted the relative abundance of wild currants in natural forests, although increased forest cover also reduces the abundance of these sun-loving shrubs.

Sugar pine and western white pine are valuable timber trees that have been severely affected by the rust in the western United States. The harvest of western white pines accelerated when white pine blister rust was discovered in the Pacific Northwest. White pines were selectively removed in salvage operations, leaving shade-tolerant conifers such as western hemlock (*Tsuga heterophylla*) and grand fir (*Abies grandis*) to become dominants in the forest canopy. Regrowth of white pines was impeded not only by the rust, but by fire suppression policies, which eliminated the burns pines often require for germination.

High-elevation white pines in the west, especially whitebark pine and limber pine, provide food and cover for a variety of wildlife and plant species. Several produce large seeds that are important elements of the diets of Clark's Nutcrackers (*Nucifraga columbiana*), and other birds, as well as grizzly bear (*Ursus arctos*) and black bear (*U. americanus*). Squirrels could be negatively affected if these forests decline, as would their predators. The demise of white pines that occupy steep, dry slopes may turn these sites into treeless areas. The loss of trees will change slope stability and hydrologic patterns and also alter fire and other disturbance regimes and forest succession. These trees also have scenic value and contribute to the attractiveness of several national parks as well as the tourist economy of the region. Stands of dead whitebark pine occur in Crater Lake, North Cascades, and Glacier national parks. Some scientists believe that all whitebark pines in Glacier National Park may be dead by 2015.

White pine blister rust is an important disease in terms of the history of forest pathology and became one of the most famous tree diseases in the world. Since it struck valuable timber trees, and since it was discovered early on that something could be done to control its spread (unlike the Chestnut blight fungus), it fostered new pest management agencies and techniques including regulations against importing exotic species and establishment of quarantine zones. The devastation wrought by white pine blister rust and chestnut blight in the first decade of the twentieth century led to the passage of the Plant Quarantine Act by the U.S. Congress in 1912. It was the first law aimed at preventing the introduction of foreign pests into the United States.

Management. The earliest strategies to control the spread of white pine blister rust focused on the eradication of *Ribes* as an alternate host. This was somewhat successful in the eastern United States, where cultivated currants and gooseberries were the chief hosts, since only a few wild species in the genus occurred there. Planting, possessing, or importing *Ribes* is still forbidden in some counties that remain in quarantine districts, although such laws are rarely enforced. In the West, where many native species occur in the wild, the strategy was totally impractical and soon abandoned.

Site management for new plantings grew in importance and included the hazard rating of areas according to conditions that support basidiospore production. High-hazard sites usually are low areas where moist, cold air settles at night. Small openings in a forest also have high potential for basidiospore production, because dew often persists on pine seedlings in such areas. A thin overstory can be maintained to prevent dew formation and still allow sufficient sunlight to reach pine seedlings. Pruning the lower branches of saplings up to 10 ft. (3 m) above the ground keeps fungal infections from extending into the trunks of young trees, and it has the added commercial advantage of providing knot-free timber.

In natural forests, the greatest hope lies in the development of resistant trees. Natural selection for such trees seems to have occurred in the Northeast, where the rust entered after mature trees had been removed and many young trees were planted. Breeding programs based on some naturally resistant individuals are progressing throughout the country. Seedlings of resistant western white pine are available to the public; for other species, such as eastern white pine and sugar pine, programs are still in experimental stages.

Selected References

Cox, Sam. "White Pine Blister Rust: The Story of White Pine, American Revolution, Lumberjacks, and Grizzly Bears." 2003. http://landscapeimagery.com/wphistory.html.

Geils, B. W. "Impacts of White Pine Blister Rust." In Proceedings, U.S. Department of Agriculture inter-agency research forum on gypsy moth and other invasive species 2001, ed. S. L. C. Fosbroke and K. W. Gottschalk. January 16–19, 2001, Annapolis, MD. Gen Tech. Report NE-285. Newton Square, PA: U.S. Department of Agriculture, Forest Service, Northeastern Research Station: 61–64, 2001. Available online at http://www.rmrs.nau.edu/rust/Geils2001/text.html.

Laskowski, Michele. "White Pine Blister Rust." High Elevation White Pines. http://www.fs.fed.us/rm/highelevationwhitepines/Threats/blister-rust-threat.htm.

Maloy, O. C. "White Pine Blister Rust." In The Plant Health Instructor. American Phytopathological Society, 2003. Updated, 2008. http://www.apsnet.org/education/LessonsPlantPath/WhitePine/symptom.htm. doi:10.1094/PHI-I-2003-0908-01.

National Biological Information Infrastructure (NBII) and IUCN/SSC Invasive Species Specialist Group (ISSG). "Cronartium ribicola (Fungus)." ISSG Global Invasive Species Database, 2005. http://www.issg.org/database/species/ecology.asp?si=550&fr=1&sts=sss.

Nicholls, Thomas H., and Robert L. Anderson. "How to Identify White Pine Blister Rust and Remove Cankers." North Central Research Station, Forest Service, U.S. Department of Agriculture, 1977. http://www.na.fs.fed.us/spfo/pubs/howtos/ht_wpblister/toc.htm.

"White Pine Blister Rust Factsheet." Plant Disease Diagnostic Clinic, Cornell University, 2005. http://plantclinic.cornell.edu/FactSheets/wpineblister/wpineblister.htm.

Worrall, James J. "White Pine Blister Rust." Forestpathology.com, 2009. http://www.forestpathology.org/dis_wpbr.html.

■ Invertebrates

■ Bryozoan

■|Lacy Crust Bryozoan

Also known as: Sea moss, white lace bryozoan
Scientific name: *Membranipora membranacea*
Family: Membraniporidae

Native Range. Lacy crust bryozoans are native to the temperate, coastal waters of Europe, where they are found in the northeast Atlantic Ocean, Baltic Sea, and Mediterranean Sea. They also occur naturally along the Pacific Coast of North America from Alaska to Baja California.

Distribution in the United States. The lacy crust bryozoan is a nonnative invasive in the coastal bays of New England; it is established in Connecticut, Maine, Massachusetts, New Hampshire, and Rhode Island.

Description. This epiphytic bryozoan forms white, circular colonies that are about 0.04 in. (1 mm) thick and 0.5 in. (1 cm) in diameter. The colonies grow outward, with the oldest member at the center and younger individuals, or zooids, radiating in rows away from it to form lacy or net-like encrustations on seaweeds. Each zooid, about the size of a pin head, is rectangular and box-like and has short knob-like spines at each corner that make the colony feel like sandpaper. The side walls of the zooid are thin and only slightly calcified; the top is transparent, allowing light to reach the seaweed that it encrusts.

Related or Similar Species. Native bryozoans of the Northwest Atlantic do not have rectangular zooids. Sea mat or horn wrack (*Electra pilosa*), the encrusting bryozoan with which lacy crust bryozoans might most easily be confused, has egg-shaped zooids surround by spines (4–12; usually 9); a large central spine protrudes from the zooid, giving *E. pilosa* a spiny appearance. Colonies of sea mat form snowflake- or star-shaped colonies.

Introduction History. This bryozoan first appeared on sugar kelp (*Laminaria saccharina*) in the Gulf of Maine at the Isles of Shoals in 1987. It likely arrived in a ship's ballast water or as a fouling organism on the hull of a ship. It spread north and south from its point of introduction and, by 1990, was the dominant epiphyte on kelp off the shores of Maine and New Hampshire. It has since reached Cape Breton, Nova Scotia.

Habitat. Lacy crust bryozoan colonies encrust kelps and other macroalgae in the shallow subtidal zone of temperate seas. They are most common on *Laminaria* kelps. These bryozoans will also attach to other smooth, hard surfaces such as rocks, glass, and floats. They flourish where there is fast-flowing water or a high tidal exchange.

Diet. These filter-feeders sieve phytoplankton from seawater with a ring of tentacles called a lophophore.

Life History. After the initial zooid becomes attached to a kelp frond, new zooids bud off (asexual reproduction); the colony grows several millimeters a day beginning in late spring. Growth continues throughout the summer, and by fall, large crusts may be apparent as several colonies merge together. Unlike many other bryozoans, in which different types of zooids perform different functions for the colony, only one type of zooid occurs in the lacy crust, and it is involved in feeding, reproduction, and defense. Sexual reproduction usually occurs in spring and summer, when zooids produce eggs and release them into the water. The individual animals may be hermaphroditic in that eggs are fertilized before they are shed. The free-floating eggs quickly develop into tiny triangular larvae (cyphonautes) that become part of the plankton. They settle when they contact an algal frond and grow toward its base, the most stable part of the seaweed. Different colonies compete for space and interact both aggressively and cooperatively, and they can communicate with each other by means of electrical signals.

Sea slugs (nudibranchs) such as *Onchidorus muricata* are the primary predators of lacy crust bryozoans.

Impacts. Lacy crust bryozoan colonies weigh down the kelp fronds that they heavily encrust and make the fronds

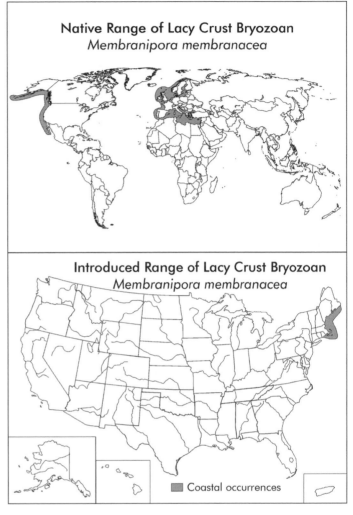

Top: The lacy crust bryozoan is native to the coastal waters of Europe and the Pacific coast of North America. Bottom: The lacy crust bryozoan has invaded coastal waters off New England. (Both maps adapted from "*Membranipora membranacea*," USGS 2009.)

especially susceptible to breakage during storms. This has caused a reduction and, in some cases, total loss of kelp beds in the waters off New England. They apparently also exclude native encrusting animals from kelp and other suitable attachment substrates and interfere with the host kelp's spore production. Kelp beds are important habitat and nursery areas for a variety of marine organisms, including sea urchins, lobsters, and finfish, all of which are therefore threatened by the invasion of this bryozoan. The green or northern sea urchin (*Strongylocentrotus droebachiensis*) in particular has declined in numbers since the introduction of lacy crust. Where kelp beds have been denuded in the Gulf of Maine, a nonnative green algae called oyster thief or dead man's fingers (*Codium fragile tomentosoides*) has invaded and prevented the restoration of kelp beds.

Management. There is no effective way to prevent the spread of this organism.

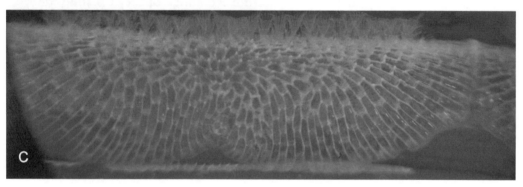

A. Colony of lacy crust bryozoans, showing the box-like zooecia. B. Close-up of zooecia with knobby spines clearly visible at each corner. C. Rings of tentacles or lophophores are extended when the zooecia are filtering food particles from sea water. (Dave Cowles: http://rosario.wallawalla.edu/inverts.)

Selected References

Cowles, Dave, and Jonathan Cowles, "*Membranipora membranacea* (Linnaeus, 1767)," 2007. http://www.wallawalla.edu/academics/departments/biology/rosario/inverts/Bryozoa/Class_Gymnolaemata/Order_Cheilostomata/Membranipora_membranacea.html.

"Lacy Crust Bryozoan." Project UFO (Unidentified Foreign Organisms), n.d. http://www.projectufo.ca/drupal/Lacy_Crust_Bryozoan.

"*Membranipora membranacea* (Linnaeus 1767)." Nonindigenous Aquatic Species Program, U.S. Geological Survey, 2009. http://nas3.er.usgs.gov/taxgroup/Bryozoans/.

"*Membranipora membranacea*." Species Identification Card, Gulf of Maine Research Institute, n.d. http://www.gmri.org/upload/files/VS_Membranipora_membranacea.pdf.

Telnack, Jennifer. "*Membranipora membranacea*: The LacyCrust Bryozoans." Intertidal Marine Invertebrates of South Puget Sound, n.d. http://www.nwmarinelife.com/htmlswimmers/m_membranipora.html.

■ Tunicates

■ | Chain Tunicate

Also known as: Chain sea squirt, orange or red sheath tunicate, violet tunicate
Scientific name: *Botrylloides violaceus*
Family: Styelidae

Native Range. East Asia, from southern Siberia to southern China and Japan. Those in the United States are believed to have come from Japan.

Distribution in the United States. Chain tunicates have invaded both coasts of the continental United States. Along the Pacific Coast, they occur in disjunct locations in Prince William Sound and Sitka, Alaska; Puget Sound and Willapa Bay, Washington; Coos Bay, Oregon; and Humboldt Bay, Bodega Harbor, Tomales Bay, San Francisco Bay, Half Moon Bay, Monterey Bay, Elkhorn Slough, Morro Bay, Santa Barbara, Channel Islands Harbor, Port Hueneme, Marina Del Rey, King Harbor, Santa Catalina Island, Alamitos Bay, Huntington Harbor, Mission Bay, and San Diego Bay, California. On the Atlantic Coast, they are established from Maine south to the entrance to the Chesapeake Bay and possibly to Florida.

Description. The chain tunicate is a sessile, colonial sea squirt consisting of many individual small animals or zooids arranged in elongated clusters or double-rows and chains called systems. Individual zooids are about 0.1 in. (1–2 mm) long; the largest colonies may be about 1 ft. in diameter (0.3 m) and sometimes develop lobes. All the zooids in a given colony are the same color: red, orange, yellow, purple, or tan.

Each zooid is vase-shaped and sits upright on the substrate.

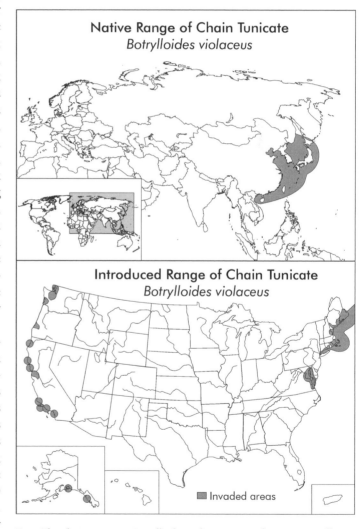

Top: The chain tunicate *Botrylloides violaceus* comes from waters off East Asia. Bottom: Chain tunicates are invasive on both coasts of the continental United States. (Adapted from Fuller, P., "*Botrylloides violaceus*." USGS Nonindigenous Aquatic Species Database, Gainesville, FL. http://nas.er.usgs.gov/queries/FactSheet.asp?speciesID=2418.)

It separately draws water into its body through its own siphon but expels it into a common space from which the waste water leaves the entire colony. The zooids are connected together by a vascular network that ends in small sacs (ampullae) at the edge of the colony. These tiny blobs are pigmented and the same color as the zooids. A clear, firm matrix forms between the systems and surrounds the whole colony. This invertebrate is the only chordate known to be able to regenerate its entire body from pieces of the colony's vascular network.

Related or Similar Species. The golden star tunicate (*Botryllus schlosseri*) is another invasive colonial tunicate found on both coasts of North America. As its common name suggests, its zooids are arranged in star patterns and not in chains. Colonies display two colors rather than just one. On the Pacific Coast, from Baja California to San Francisco Bay, one can find *Botrylloides diegensis*, the individual zooids of which are dichromatic; the ring around the oral siphon is a pale or bright white, yellow, or orange and contrasts with the darker color of the body. A recently introduced colonial tunicate in San Diego and Mission bays, *Botrylloides perspicuum*, is distinguished by a thicker and firmer matrix that forms ridges between systems of zooids. There are also microscopic differences among these tunicates and diagnostic characters in their larvae.

The chain tunicate might be mistaken for various red or orange sponges, but it is more rigid than a sponge.

Introduction History. The earliest record of the chain tunicate in the United States is from the Pacific Coast, where it was reported in 1973 in San Francisco Bay. It likely arrived as a fouling organism on a ship from Japan, but could have been introduced with Japanese oysters (*Crassostrea gigas*). It was first reported in Puget Sound in 1997 and in Prince William Sound in 1999. However, the organism may have arrived earlier. Collection was limited prior to the 1970s, and until 1997, the chain tunicate was not recognized as a form distinct from the very similar *B. diegensis*. It was first reported on the Atlantic Coast in 1981, although again, confusion with *B. diegensis* may have obscured an earlier arrival in the 1970s. In 1981, it was in Great Bay, New Hampshire, in the Gulf of Maine; by the mid-1990s, it was in Penobscot Bay. Early records in the lower Chesapeake Bay date to 2000 and 2001. (In 2004, it had expanded northward and was reported on cultured mussels being grown on Prince Edward Island, Canada, in the Gulf of St. Lawrence.)

Dispersal to new areas along the coasts of North America may occur in contaminated oyster shell and spat associated with the oyster industry, as fouling organisms on the hulls of ships, or by rafting on debris. Since the planktonic larval stage is so brief, it is unlikely that it spreads far if at all in ballast water.

Habitat. Chain tunicates occur in sheltered areas of the shallow subtidal zone. Adults and juveniles attach to submerged surfaces such as rocks, pilings, ropes, boat hulls, macroalgae, eelgrass (*Zostera marina*), and the shells of mussels, oysters, and barnacles. They will also grow over encrusting bryozoans and solitary tunicates. Chain tunicates can live in a range of salinities (24–34 ppt) and temperatures (46–77°F [8–25°C]). It reportedly tolerates polluted water.

Diet. Adults and juveniles are suspension feeders that filter phytoplankton and bacteria from seawater.

Life History. Chain tunicates reproduce both sexually and asexually. The zooid is hermaphroditic and viviparous. It has a brood pouch that extends out from the body wall into the matrix. The egg is ovulated into the brood pouch, where it is fertilized. The embryo develops into a relatively large (0.08–0.12 in. [2–3 mm] long), brightly colored, tadpole-like larva. Gestation is longer than four weeks. Each pouch will hold one egg; each zooid may have one or two pouches.

Upon release from the brood pouch, the larva is free-swimming for a very short time. In just a few hours, metamorphosis takes place, and the juvenile (oozoid) attaches head first to a hard surface. The newly attached oozoid metamorphoses into a zooid and starts to reproduce asexually by budding the next day. A new colony is thus begun.

All the zooids of a given generation grow and die synchronously. As the oldest generation disintegrates, it is absorbed into the colony's vascular system.

Individual vase-shaped zooids of the chain tunicate are arranged in rows and connected by a vascular system that ends in tiny sacs at the edge of the colony. (Dann Blackwood, U.S. Geological Survey.)

Impacts. Chain tunicates may compete for settling space and food with other native and introduced fouling organisms, but generally its ecological impacts are unknown. It has become a nuisance species in mariculture operations producing mussels in Canada, where it overgrows and smothers the bivalve; so it could become a pest in oyster operations in the Chesapeake Bay or wherever other shellfish are cultured. As it becomes more abundant, it will become a problem, fouling fishing and boating gear and requiring time and expense to remove.

Climate change may alter the story. Studies show that chain tunicates settle earlier and grow faster than natives in warming waters. As natives decline in face of winter warming, these tunicates may become dominant.

Management. Similar species have followed the typical pattern of a new invader: rapid population growth followed by a period of decline and adjustment of the fouling community without the loss of native species. It appears that the chain tunicate also will not replace natives, and that management or eradication measures are unnecessary and would likely be unsuccessful anyway.

Selected References

"*Botrylloides violaceus.*" Chesapeake Bay Introduced Species Database NEMESIS: National Exotic Marine and Estuarine Species Information System, Smithsonian Environmental Research Center (SERC), 2005, 2009. http://invasions.si.edu/nemesis/CH-INV.jsp?Species_name=Botrylloides +violaceus.

Cohen, Andrew N. "*Botrylloides violaceus* Oka, 1927." *Guide to the Exotic Species of San Francisco Bay*. San Francisco Estuary Institute, Oakland, CA, 2005. http://www.exoticsguide.org/species_pages/b _violaceus.html.

Pleus, Allen, and Pam Meacham. "*Botrylloides violaceus* (Chain tunicate)," Invasive Species Fact Sheets, Aquatic Nuisance Species, Washington Department of Fish and Wildlife, n.d. http://wdfw.wa.gov/ fish/ans/identify/html/index.php?species=botrylloides_violaceus.

"Tunicate, *Botrylloides violaceus.*" Non-indigenous Aquatic Species of Concern for Alaska, Fact Sheet 15. Prince William Sound Regional Citizens' Advisory Council, 2004. http://www.pwsrcac.org/ docs/d0016000.pdf.

■|Colonial Tunicate

Also known as: Didemnid, colonial ascidian, colonial sea squirt, marine vomit
Scientific name: *Didemnum vexillum*
Synonym (in older literature): *Didemnum* sp., *Didemnum* sp. A.
Family: Didemnidae

Native Range. Uncertain; possibly Japan.

Distribution in the United States. East Coast, from Maine to New Jersey and offshore on Georges Bank; West Coast, California and Washington.

Description. Didemnid colonies take various forms. Where the current is weak, they hang from hard structures in long stringy lobes resembling ropes or beards. This is the usual morphology on pilings, ship hulls, ropes, and the like. Where the current is strong, they grow in dense flat mats with low protruding lobes or tendrils and encrust rock outcrops, pebbles, and cobbles. Colonies are tan, cream-colored, white, or yellow and may attain a diameter of 3 ft. (1 m) or more.

Thousands of individual zooids comprise each colony. Each is about 0.1 in. (2.5 mm) long. They are embedded in a clear, firm matrix, the surface of which is dotted with tiny white calcareous balls, each covered with spines. These may not be visible to the naked eye. All zooids in a given colony are the same color.

Related or Similar Species. On the Pacific Coast, all three related colonial tunicates (*D. carnulentum*, *D. albidum*, and *Tridemnum apacum*) have spiny balls throughout the matrix and not limited to the surface as in *D. vexillum*. Their colonies are never lobed, but always flat and encrusting; they are usually white or gray. On the East Coast, the colonial tunicate could be confused with the chain tunicate, *Botrylloides violaceus* (see Tunicates, Chain Tunicate), but the colonial tunicate is never the red or orange color of chain tunicates.

Native Range* of Colonial Tunicate
Didemnum vexillum
*Approximate: Exact origins only vaguely known.

Introduced Range of Colonial Tunicate
Didemnum vexillum

● Invaded Sites

Top: The colonial tunicate *Didemnum vexillum* may come from the waters off Japan. Bottom: *D. vexillum* has invaded both coasts of the continental United States. (Adapted from "Species *Didemnum vexillum*." USGS 2010.)

Colonial tunicates could also be mistaken for sponges, but they have a smoother texture than a sponge.

Introduction History. The first documented occurrence of the colonial tunicate on the East Coast was in Fort Island Narrows on the Damariscotta River in Maine in 1993. It was likely present but unidentified prior to that time. Reports of a similar organism in Walpole, Maine, date to 1988; and it is suspected that the tunicate was in New England waters by the 1970s. *Didemnum vexillum* was first collected in Shinnecock Bay, Long Island, New York, in late 2004. The first records on Georges Bank date to 2003 and are the first evidence of invasion of offshore habitats.

On the West Coast, *D. vexillum* was collected from San Francisco Bay in 1993. It was recovered from Elkhorn Slough, Monterey Bay, California, in 1998, but originally misidentified as a native species, *D. carnulentum*. Since then, it has been collected in Bodega Harbor, Humboldt Bay, Tomales Bay, Morro Bay, and Port San Luis. On the Washington coast, it was first documented on a sunken wooden boat in Puget Sound in March 2004.

Colonial tunicates were most likely carried to the coasts of North America as fouling organisms on transoceanic vessels. They also may have arrived as colony fragments in ballast water. On the Pacific Coast, it is possible that they were accidentally introduced with oyster or mussel stock or with equipment delivered to aquaculture facilities. Why this species has suddenly become invasive is not known. However, fragments capable of starting new colonies are easily broken off by bottom-fishing dredges, by divers, and by local boat traffic. These pieces may then drift in currents over long distances to new sites.

Habitat. Colonial tunicates are usually associated with man-made structures in the subtidal zone of coastal waters. They have been found on docks, moorings, metal and wooden pilings, ropes and steel chains, discarded automobile tires, and polythene plastic. They also foul shellfish aquaculture gear and ships' hulls. On the seabed, they encrust hard substrates such as rock outcrops, cobbles, pebbles, and boulders and overgrow sessile benthic organisms such as hydroids, solitary tunicates, sponges, barnacles, mussels, and oysters, as well as seaweeds. They occur in shallow coastal waters and also on the continental shelf to depths of about 200 ft. (65 m). These tunicates tolerate water temperatures ranging between 28° and 75°F (−2°–24°C) and apparently require a salinity greater than 26 ppt.

Diet. These filter-feeders siphon seawater into their bodies and consume the phytoplankters and zooplankters, including the larvae of oysters and mussels, held in it.

Life History. The colony broods its larvae within its matrix and then releases them into the water column. They are part of the plankton for only a few hours before they attach head down to a firm surface. Larval settlement occurs in spring and fall. The settled larva quickly metamorphoses into a zooid, which becomes the founder of a new colony. The colony expands through budding (asexual reproduction). Young zooids reach sexual maturity in just a few weeks.

A new colony can also form via fragmentation when a fragile lobe of the colony breaks off and drifts to a new location, where it either settles to the bottom or becomes entangled in marine structures. Such stray pieces may attach to a new substrate within six hours of arrival and start overgrowing it.

Impacts. Colonial tunicates overgrow hard substrates and the native organisms attached to them and thereby threaten to change benthic marine habitats. The high acidity of their tunics may inhibit settling of the larvae of shellfish such as scallops. The overgrowth of gravelly bottoms could reduce habitat in which fish such as Atlantic cod (*Gadus morhua*) and haddock (*Melangrammus aeglefinus*) deposit their eggs. The tunicates can smother bivalves,

Colonies of the colonial tunicate *Didemnum vexillum* take a variety of forms. A. Thousands of individual zooids, each about 0.1 in. long, make up a colony. (Dann Blackwood, U.S. Geological Survey.) B. Stringy lobes occur where currents are weak. Here *Didemnum* covers metal piling and a mussel shell. (Paul Barter, Cawthron Institute, New Zealand.) C. Stiffer lobes protrude when currents are strong. Here the colonial tunicate has encrusted a mussel shell. (Dann Blackwood, U.S. Geological Survey.) D. Mats overgrow the gravelly bottom of Georges Bank. (Dann Blackwood, U.S. Geological Survey.)

so concern exists for oyster (*Crassostrea* spp.), bay scallop (*Argopecten irradians irradians*), sea scallop (*Placopecten magellanicus*), and mussel fisheries, both natural and cultured. Studies on Georges Bank show an increase in two polychaete species in areas infested with the colonial tunicate, suggesting a change in native species composition on the seafloor. Fouling of aquaculture gear, moorings, ships' hulls, and so forth requires expensive maintenance procedures and makes this a major nuisance species.

Management. Wrapping pilings in plastic may suffocate colonial tunicates, but this is only of local significance as a method of control. Vacuuming colonies from the water has been tried, but it did not succeed in eradicating the species. Prevention or slowing the spread of the animal is difficult because of its fragile tunic and ability to rapidly reproduce asexually. Boaters, divers, fishermen, and aquaculturalists could slow the invasion by keeping equipment clean and not discarding bivalve shells in noninfested waters.

Selected References

Cohen, Andrew N. "*Didemnum* sp. A." *Guide to the Exotic Species of San Francisco Bay.* San Francisco Estuary Institute, Oakland, CA, 2005. http://www.exoticsguide.org/species_pages/didemnum .html.

Lengyel, Nicole L., Jeremy S. Collie, and Page C. Valentine. "The Invasive Colonial Ascidian *Didemnum vexillum* on Georges Bank—Ecological Effects and Genetic Identification." *Aquatic Invasions* (4): 143–52, 2009. doi:10.3391/ai.2009.4.1.15 http://www.aquaticinvasions.net/2009/AI_2009_4_1 _Lengyel_etal.pdf.

Morris, James A., Jr., Mary R. Carman, K. Elaine Hoagland, Emma R. M. Green-Beach, and Richard C. Karney. "Impact of the Invasive Colonial Tunicate *Didemnun vexillum* on the Recruitment of the Bay Scallop (*Argopecten irrradians irradians*) and Implications for Recruitment of the Sea Scallop (*Placopecten magellanicus*) on Georges Bank." *Aquatic Invasions* (4): 1: 207–11, 2009. doi:10.3391/ai.2009.4.1.22 http://www.aquaticinvasions.net/2009/AI_2009_4_1_Morris_etal.pdf.

National Biological Information Infrastructure (NBII) and IUCN/SSC Invasive Species Specialty Group (ISSG). "*Didemnum spp.* (Tunicate)." ISSG Global Invasive Species Database, 2007. http:// www.issg.org/database/species/ecology.asp?si=946&fr=1&sts=sss.

"Species *Didemnum vexillum*." Marine Nuisance Species, USGS National Geologic Studies of Benthic Habitats, Northeastern United States, 2010. http://woodshole.er.usgs.gov/project-pages/stell wagen/didemnum/.

■ Cnidarian

■ | Australian Spotted Jellyfish

Also known as: Spotted jellyfish, white-spotted jellyfish
Scientific name: *Phyllorhiza punctata*
Order: Rhizostomeae
Family: Mastigiidae

Native Range. Before the 1950s, the Australian spotted jellyfish was only known from the Indo-Pacific region. Its native range is believed to have extended from the south-central coast of eastern Australia northward into Southeast Asia and the Philippines.

Distribution in the United States. Populations are becoming established along the Gulf Coast from Galveston Bay, Texas, to Orange Beach, Alabama, east of Mobile Bay. Another permanent population occurs on the Atlantic Coast in Indian River Lagoon, Florida. In the summer of 2007, swarms of medusae were sighted off South Carolina, but these may not represent a permanent population. In Hawai'i, populations occur in Pearl and Honolulu harbors and in Kaneohe Bay, O'ahu. Australian spotted jellyfishes are also established in Boqueron Bay, Puerto Rico.

Description. This is a large jellyfish with a somewhat flattened semi-globular gelatinous bell or umbrella. In much of its natural and introduced range, it is bluish white or brown due to symbiotic zooxanthellae; however, in the Gulf of Mexico, it is clear or white due to the absence of these photosynthetic algae. Crystalline opaque white inclusions appear as evenly spaced spots. Eight thick, branching oral arms surround the central mouth area; each ends with a large brown bundle of stinging cells. Transparent ribbons of tissue hang from each oral arm. The average bell diameter of an adult is 14 in. (35 cm), but in the Gulf of Mexico, they are much larger than elsewhere, with an average diameter of nearly 18 in. (45 cm). The largest reported was 24.4 in. (62 cm) in diameter. These large specimens can weigh up to 25 lbs. (11 kg). They are only mildly venomous and not dangerous to humans.

Related or Similar Species. The native moon jelly (*Aurelia aurita*) is of similar size but otherwise very different in appearance.

Introduction History. Australian spotted jellyfish likely entered the Atlantic Ocean and Caribbean Sea on ships passing through the Panama Canal sometime in the early 1950s, if not earlier. Polyps attached to the hulls of ships and/or young medusae held in ballast water could

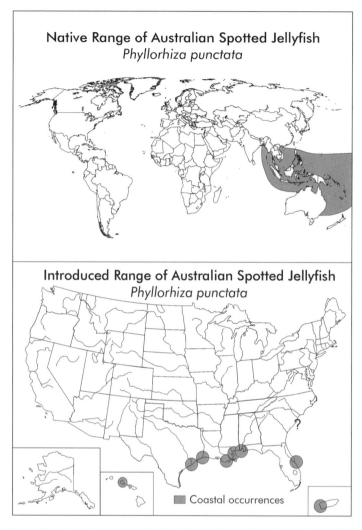

Native Range of Australian Spotted Jellyfish
Phyllorhiza punctata

Introduced Range of Australian Spotted Jellyfish
Phyllorhiza punctata

■ Coastal occurrences

Top: The Australian spotted jellyfish was limited to the Indo-Pacific region prior to the 1950s. Bottom: Australian spotted jellyfish are currently established along the Gulf Coast of the United States and off O'ahu, HI, and western Puerto Rico. (Adapted from Perry 2010.)

have survived the journey from the Pacific. The first colonies appeared off northeast Brazil in 1995. Populations have been established in Puerto Rico and in other lagoons in the Caribbean since at least the early 1970s. The Australian spotted jellyfish was first documented in the Gulf of Mexico by a single specimen collected in 1993. The natural circulation pattern of ocean currents from the Caribbean into the Gulf of Mexico could have carried them northward to the Gulf Coast of the United States, where small populations such as that in Terrebonne Bay, Louisiana, became established in the mid-1990s. They could also have been introduced as attached polyps on ships and towed structures. The proliferation of oil platforms in the Gulf of Mexico may have improved the habitat for these jellyfish and facilitated establishment of local populations by providing attachment sites for the polyp stage and allowing for natural dispersal northward. Overfishing could also have released resources previously tied up by native species.

The spectacular 2000 invasion of thousands of medusae along the Gulf Coast, like the later 2007 one, appeared to have come from the Caribbean though the Yucatán Straits into the Gulf of Mexico, although recent genetic studies do not strongly support this scenario and suggest that the Caribbean jellyfishes may be another species altogether. Gulf Coast jellyfish are more similar to those from Australia or the West Coast, suggesting an invasion via transoceanic shipping. Medusae were observed in Galveston Bay in 2006.

The Australian spotted jellyfish was first identified on the Atlantic Coast in Indian River Lagoon near Melbourne, Florida, in 2001. The 2007 sightings off the coast of South Carolina represent the first known occurrence of the jellyfish north of Florida.

In Hawai'i, the jellyfish entered Pearl Harbor from the Philippines during World War II (1941–1945). It appeared in Kaneohe Bay in 1953–1954. The first reports of Australian spotted jellyfish in coastal waters off the continental United States actually came from the Pacific Coast, from southern California in 1981.

Habitat. Australian spotted jellyfishes usually inhabit warm coastal waters and lagoons and may float into estuaries and bays on flood tides. They appear to prefer water temperatures above 68°F (20°C) and salinities greater than 25 ppt. During the polyp stage of the life cycle, hard substrates are needed for attachment.

Diet. These jellyfish use stinging cells to capture zooplankters, including the eggs and larvae of many fish and shellfish. They feed almost constantly since they can digest the entire content of their gut in two hours. They filter prodigious amounts of water each day, removing most of the small suspended particles. Except in the Gulf of Mexico, they also probably benefit from the primary production of zooxanthellae living in their umbrella tissues.

Life History. Jellyfish undergo alternate generations with asexual and sexual stages in the life cycle. The visible organism is the adult medusa stage. Medusae occur as separate sexes that reproduce via external fertilization and produce free-swimming flat, ciliated larvae (planula). The planulae settle and attach to hard substrates and begin the polyp stage of the life cycle. Polyps develop into a form (strobila) that asexually produces free-swimming immature medusae (ephyrae). The young medusae mature over several weeks. Weak-swimmers, they become part of the plankton and are at the mercy of wind, currents, and tides.

Australian spotted jellyfish in the Gulf of Mexico experienced huge population explosions in 2000 and again in 2007. During these episodes, the medusae appeared in large numbers along the Gulf Coast of the southeastern United States, apparently carried in the Loop Current and eddies of the Gulf Stream. In contrast to these irruptions, small populations are maintained year after year in places such as Terrebonne Bay, Louisiana, and Indian River Lagoon, Florida.

In the Gulf of Mexico, this large jellyfish is white because it lacks the symbiotic algae that make it brown or bluish white elsewhere in its range. (Dwight Smith/Shutterstock.)

Impacts. The 2000 invasion was costly to the Gulf of Mexico shrimp industry. The large gelatinous masses clogged nets and damaged gear and led to the closing of some productive areas to commercial fishing. Concern exists that such large infestations of jellyfish that coincide with the spawning season of many marine organisms could increase predation on fish eggs and the larvae of mollusks and crustaceans to such a degree that forage fish such as bay anchovy (*Anchoa mitchilli*) and oyster and crab populations in Mobile Bay and Mississippi Sound could be affected. With their massive consumption of zooplankton, the jellyfish potentially compete with native shrimp and fish for food and direct energy flow in aquatic ecosystems to themselves, a dead end in the food chain since jellyfish have few predators.

Management. There seem to be no strategies for control or eradication currently available.

Selected References

"Australian Spotted Jellyfish." Field Guide to the Indian River Lagoon. Smithsonian Marine Station at Fort Pierce, n.d. http://www.sms.si.edu/IRLFieldGuide/Phyllo_punctat.htm.

Burkhard, Elizabeth. "A Survey of the Relationship of the Australian Spotted Jellyfish, *Phyllorhiza punctata*, and OCS Platforms." GulfBase.org, n.d. http://www.gulfbase.org/project/view.php?pid=asotrotasjppaop.

Masterson, J. "Species Report: *Phyllorhiza punctata* von Lendenfeld, 1884." Smithsonian Marine Station at Fort Pierce, 2007. http://www.sms.si.edu/irlspec/Phyllorhiza_punctata.htm.

National Biological Information Infrastructure (NBII) and IUCN/SSC Invasive Species Specialist Group (ISSG). "*Phyllorhiza punctata* (Jellyfish)". ISSG Global Invasive Species Database, 2006. http://www.invasivespecies.net/database/species/ecology.asp?si=992&fr=1&sts.

Norris, Scott. "Australian Jellyfish Invade U.S. Waters." National Geographic News, 2007. http://news.nationalgeographic.com/news/2007/08/070827-jellyfish-invasion.html.

Perry, Harriet. "*Phyllorhiza punctata*." USGS Nonindigenous Aquatic Species Database, Gainesville, FL, 2010. Revised June 20, 2006. http://nas.er.usgs.gov/queries/FactSheet.aspx?speciesID=1192.

"*Phyllorhiza punctata* von Lendenfeld, 1884." Hawaii Biological Survey, Bishop Museum, 2002. http://www2.bishopmuseum.org/HBS/invertguide/species/phyllorhiza_punctata.htm.

■ Annelid Worms

■ | European Earthworms

Also known as: Nightcrawlers, red worms, leaf worms, red wrigglers,
grey worms, angle worms
Scientific names: *Lumbricus terrestris, L. rubellus, Aporrectodea caliginosa,*
Dendrobaena octaedra, and others
Family: Lumbricidae

Native Range. The four species described below came from Europe. Of the known alien earthworms in the United States, 25 are European and 14 are of Asian origin. Not all are invasive.

Distribution in the United States. Lumbricus terrestris and *L. rubellus* are found in two disjunct areas. The eastern invasion region extends from the Atlantic coast of North Carolina north through Maine and westward to the edge of the Great Plains. The western part of the distribution area reaches along the Pacific coast of California north through Washington and eastward to the foothills of the Rocky Mountains. *Dendrobaena octaedra* has a continuous distribution across the United States, avoiding the Atlantic and Gulf Coastal plains, Texas, the central Great Plains, most of New Mexico, and the southern parts of California and Arizona. Complete and accurate distribution data for *Aporrectodea caliginosa* are not available, but it is widespread in northern forests. Close relatives occur in grasslands, including the Palouse. Other exotic earthworms, including some from Asia such as the Alabama jumper (*Amynthas agrestis*), occur in southern forests.

Native Range of European Earthworms
Lumbricus terrestris, L. rubellus,
Aporrectodea caliginosa, Dendrobaena octaedra

Many of the alien earthworms in the United States, including the nightcrawler, red worm, angle worm, and octagonal-tail worm came from Europe.

Description. This entry pertains to several different annelid or segmented worms from Europe, only a few of the 45 or more species introduced into the United States. Basic earthworm identification can be based on gross characteristics such as size, color, and the depth at which a species is found or the ecological group to which it belongs (see under Habitat). A more precise and advanced analysis of the species that are described in this entry relies on such anatomical details as the shape, size, and position of the clitellum and the setae, tiny bristles on the body used for locomotion. All of the adult worms are characterized by a smooth band or collar, the clitellum.

The nightcrawler (*Lumbricus terrestris*) is the largest of the common invasive earthworms, attaining lengths of 5–8 in. (12.5–20 cm) or more. This reddish-brown anecic worm makes permanent vertical burrows to depths of 6 ft. (2 m). Its presence may be evidenced on the surface by a pile of casts (fecal material) and leaf stems some 2 in. (5 cm) across and 0.5–0.75 in. (12.5–19 mm) high.

The red or leaf worm (*L. rubellus*) is a medium-sized annelid ranging from 1–4 in. (2.5–10.5 cm) long. Its back is reddish brown and irridescent; the underside is pale. This is an epi-endogeic species that occupies both the leaf litter and the uppermost few inches of the mineral soil. This is the species commonly sold as fish bait.

The grey or angle worm (*Aporrectodea caliginosa*) is 2–6 in. (5–15 cm) long. It is unpigmented, but its internal organs and ingested food gives it a color ranging from gray or brown to greenish. This species is endogeic and makes horizontal burrows in the topsoil layer.

The very small octagonal-tail earthworm (*Dendrobaena octaedra*) reaches lengths of only 0.6–2.5 in. (1.7–6 cm). They are reddish brown, darker on the back and head than the belly. An epigeic species, this earthworm is only found in the leaf litter or duff of the forest floor.

Introduction History. European earthworms arrived in North America with early European settlers beginning in the sixteenth century. They were likely carried in soils used as ballast in ships and on the root balls of plants brought to the New World. Reinforcements continue to arrive in contaminated shipments of ornamental plants and intentional, permitted importations of live bait and composting worms from Canada, nightcrawlers, and red wrigglers (*Eisenia veneta* = *E. hortensis*), respectively.

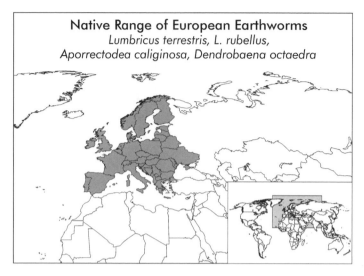

Native Range of European Earthworms
Lumbricus terrestris, L. rubellus,
Aporrectodea caliginosa, Dendrobaena octaedra

Top: Distribution of the octagonal-tail earthworm in the United States. Bottom: Distribution of the nightcrawler (*L. terrestris*) and the red worm (*L. rubellus*) in the United States. (Both maps adapted from Proulx 2003.)

Northern forests developed after the Ice Age without native earthworms and are thus particularly vulnerable to invasion by hardy species from similar climate regions in Europe. On their own, earthworms spread very slowly, by an estimated 35 ft. (10 m) a year. In the 10,000 years that regions glaciated during the Pleistocene have been ice-free, native earthworms have been unable to recolonize most of the area from which they were eliminated by continental ice sheets. However, nonnative species have been deliberately and widely introduced to agricultural areas and garden plots, where they are generally beneficial. Only recently have they invaded undisturbed forests and grasslands. In glaciated regions, numerous lakes attract sport fishermen, and they seem to be responsible for much of the spread of nightcrawlers and leaf worms into northern forests, as they often dump unused live bait on the ground. The construction and logging industries also move worms from place to place when cocoons and young worms become lodged in truck tires. Earthworm concentrations tend to be highest near roads and recreation areas. Homeowners bring composting worms and other garden worms to the forest edge as settlement spreads into new areas. Earthworms only reached central Minnesota, for example, as recently as the 1930s.

Habitat. Many of the European earthworms now found in the United States are closely linked to disturbed and artificial habitats such as lawns, gardens, cultivated fields, and pastures. They have recently been invading hardwood forests. Creatures of the underground, they sort themselves into three main soil habitats. *Anecic* or deep-burrowing earthworms excavate vertical burrows as deep as 6 ft. (2 m), but feed on the surface. *Enogeic* species burrow into the topsoil to depths up to 20 in. (50 cm) and come to the surface to feed. They are responsible for much of the mixing of soil layers. *Epigeic* worms live in the litter and first inch or so of forest soils. They do not excavate burrows.

A succession of earthworm species characterizes invasions of previously undisturbed forests. Epigeic species such as *Dendrobaena* are the pioneers; they are followed by endogenic types such as *L. rubellus* that prepare the soil for latecomers, the anecic nightcrawlers.

Diet. Earthworms eat and decompose leaf litter and also consume decaying organic matter and microorganisms. Nightcrawlers pull leaf debris down into their burrows; stems and leaf fragments accumulate at the burrow entrance along with casts deposited by the burrow's inhabitant and form what are called middens. The leaf worm and grey worm feed on surface litter, on organic material in humus-rich topsoil, and probably on fungi and bacteria in the rhizosphere, that nutrient-enriched area surrounding the roots of plants. The octagonal-tail earthworm feeds mainly on the bacteria and fungi of the litter layer.

Life History. Earthworms are hermaphroditic, each individual possessing both testes and ovaries. They usually mate sexually with another individual, however, lining up "head to toe" so that the clitellum of one worm lies against the segments of the other that contain the male reproductive organs. The clitellum secretes a large amount of slime that encases the two worms in a slime tube. The sperm are released into the slime and carried in special grooves back to the sperm receptacles of the other individual. The worms then separate. A second mucous ring is secreted from the clitellum, and it slides forward over the body. As it passes the openings to the female organs, several ripe eggs are ejected, sticking to the ring. The ring continues to move forward and brings the eggs into contact with the sperm, which fertilize the eggs. The entire ring then passes over the head and away from the worm. Its ends seal to make a cocoon. The cocoon lies in the soil, into which tiny young worms emerge when the eggs hatch.

Impacts. Earthworms are well known as beneficial additions to garden and agricultural soils, where they help decompose organic matter and accelerate nutrient cycles, aerate the soil and increase water infiltration with their burrowing, and mix soil layers to return nutrients to the root zone. They are the focus of a multimillion-dollar vermiculture industry that produces worms for composting and for live bait. It is therefore surprising to many people that they become a major ecological disaster when introduced to northern hardwood forests.

In deciduous hardwood forests, invasive earthworms are "ecosystem engineers" that alter the habitat. They increase decomposition rates of the leaf litter to the degree that they destroy the duff and humus. These organic soil layers not only store nutrients until the spring growing season, but provide insulation and protection from predation for seeds and seedlings. In their absence, nutrients such as nitrates are leached, and seeds do not germinate; bare spots open to erosion develop. Exotic weeds such as garlic mustard (See Volume 2, Forbs, Garlic Mustard) (*Alliaria petiolata*) and Japanese barberry (See Volume 2, Shrubs, Japanese Barberry) (*Berberis thunbergii*) that evolved in the presence of earthworms can invade. The loss of the herb layer of native wildflowers such as trilliums (*Trillium* spp.), bloodroot (*Sanguinaria canadensis*), trout lily (*Erythonium americanum*) and blue cohosh (*Caulophyllum thalictroides*) has been noted in forests in the Great Lakes area as a result

The various introduced earthworms resemble each other but tend to be found at different depths in the soil. (Vinicius Tupinamba/Shutterstock.)

of earthworm invasions. The rare goblin fern or moonwort (*Botrychium mormo*) has been extirpated from some areas

The litter is also habitat to a number of small vertebrates, invertebrates, and microbes. When it disappears, so do they. A more direct link may be made between earthworm invasions and declines in native salamanders that feed on worms. Larger worms such as leaf worms and nightcrawlers will be consumed by adult salamanders, but are too big to be ingested by juveniles. If the exotic earthworms replace native species, salamander recruitment could be significantly reduced.

White-tail deer (*Odocoileus virginianus*) and earthworms may be connected to changes in the species composition of forest communities. Where deer yard up during winter, their feces accumulate and attract and nourish earthworms. An increase in earthworms reduces the litter layer and makes forest herbs more vulnerable to deer browsing. Together, they contribute to the loss of herbaceous plants and tree seedlings.

Their deposition of casts, cementing together of soil particles while burrowing, displacement of other bioturbating organisms, and consumption of bacteria and fungi are some of the means by which earthworms alter the soil environment. With time, earthworm activities lead to soil compaction at depth. Compaction can bind some nutrients, prevent root penetration, reduce aeration and available soil moisture, and increase runoff and removal of phosphorus. The mixing of soil layers may change both the soil structure and its biochemical properties. It may increase weathering rates, but studies have yet to determine if the result will be a greater release of greenhouse gases to the atmosphere or increased storage of carbon in soils. Either negatively or positively, earthworms may be involved in climate change processes.

Management. Prevention of the introduction of new species to the United States and the spread of existing invasive earthworms are paramount. Gardeners should not dispose of compost in natural areas, and fishermen should not dump unused live bait. Inspection of vermiculture products could ensure that the species being sold is truly the one intended and not some other that entered a batch through misidentification or contamination. The horticultural trade in live plants needs better control of accidental migrants in pots and root balls. Vehicle tires could be hosed down or otherwise sanitized before entering areas most vulnerable to invasion.

Selected References

Baskin, Yvonne. *Under-Ground. How Creatures of Mud and Dirt Shape Our World.* Washington, DC: Island Press, 2005.

Blatchford, John. "Earthworms Spread by Human Activity: Worms Improve Soil Fertility but Invasive Species do Damage." Suite 101.com, 2009. http://zoology.suite101.com/article.cfm/earthworms_spread_by_human_activity#ixzz0eJavfI9t.

Dunne, Niall. "Invasive Earthworms—A Threat to North American Forests," *Plants and Gardens News*: 19(1): 2004. http://chicagoconservationcorps.org/blog/wp-content/uploads2/2009/06/L40%20Vermicomposting%20and%20Invasive%20Earth%20Worms.pdf.

"Earthworm Ecological Groups." Natural Resources Research Institute, University of Minnesota, Duluth, 2006. http://www.nrri.umn.edu/worms/identification/ecology_groups.html.

Halsey, Daniel. "Invasive Earthworms: Affects on Native Soils." Soil5125, University of Minnesota, 2009. http://southwoodsforestgardens.blogspot.com/2009/01/paper-on-invasive-european-worms.html.

Proulx, Nick. "Ecological Risk Assessment of Non-Indigenous Earthworm Species." Final draft. Minnesota Department of Natural Resources, 2003. http://www.nrri.umn.edu/WORMS/research/publications/Proulx%202003.pdf.

■ Mollusks

■ | Asian Clam

Also known as: Asiatic clam, gold clam
Scientific name: *Corbicula fluminea*
Family: Corbiculidae

Native Range. East Asia, from southeastern China to Korea and southeastern Russia.

Distribution in the United States. The Asian clam has established populations in all states except Alaska, Maine, Montana, New Hampshire, North Dakota, Rhode Island, South Dakota, and Wyoming. It also occurs in Puerto Rico.

Description. This small freshwater bivalve is distinguished by a rounded to somewhat triangular shell with concentric, evenly spaced ridges. The umbo or raised area above the hinge is high and centrally positioned on the shell. On the inside of the valves, long, straight, serrated ridges, the lateral teeth, emanate from the hinge area, two on each side of the right valve and one on each side of the left valve. Three cardinal teeth project below the umbo on each valve. Two color morphs occur in the United States. In most parts of the country, shells are yellow-brown, but in the southwest, they are black. In the lighter-colored specimens, the inside of the shell is a glossy light purple nacre (mother-of-pearl); in dark morphs, the nacre is deep blue. Adult shell length rarely exceeds 1.5 in. (40 mm); usually it is less than 1 in. (25 mm).

Related or Similar Species. Asian clams could be confused with native fingernail or pea clams (Family Sphaeriidae). Fingernail clams lack both lateral teeth and cardinal teeth and are typically less than an inch (25 mm) long. The Asian clam has a heavier shell than most native species.

Introduction History. Asian clams were first collected in the United States in 1938 in the Columbia River near Knappton, Washington, but it is believed that they had probably occurred in some West Coast drainages since at least 1924. The clams were an important food for Chinese people, who may have deliberately brought them along when they emigrated to North America. It is also possible that they arrived accidentally in contaminated imports of the giant Pacific oyster (*Crassostrea gigas*), a shellfish first imported into Washington from Japan in 1903 and now the basis of a major aquaculture industry.

Asian clams were discovered in the Ohio River in 1957 and quickly spread throughout the Mississippi River drainage system and into Lake Erie and Lake Michigan. On the Atlantic slope, they were first reported from Delaware in 1972, New Jersey in 1982, and Long Island, New York, in 1984. They were not expected to be able to survive New England's cold winters, but were found in the lower Connecticut River in 1990 and in several Massachusetts reservoirs in 2007.

In Hawai'i, the Asian clam first appeared at a farmer's market on O'ahu in 1977. Within a few years, it was established in many streams and reservoirs on O'ahu as well as on the islands of Kaua'i, Maui, and Hawai'i.

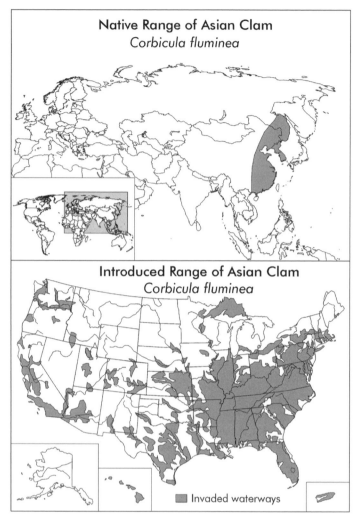

Top: Asian clams came from eastern Asia. Bottom: The Asian clam is invasive in almost all states as well as in Puerto Rico. (Adapted from USGS 2001.)

One of the most recently documented invasions has occurred in Lake Tahoe. From a relatively few individuals spotted there in 2002, the population had exploded by 2009.

These mollusks can disperse naturally within a water body, since juveniles float on currents. Overland transport is accomplished on contaminated boats, trailers, and aquatic sports gear, or, since they are used as fish bait, when unused bait is dumped into a stream, lake, or reservoir.

Habitat. Asian clams inhabit the bottom sediments of streams, rivers, lakes, and ponds of all sizes, and will also invade irrigation canals. The substrate can be silt, sand, or gravel. They require well-oxygenated water, so they prefer flowing water and are intolerant of polluted water. However, they can withstand brackish water with salinities up to 13 ppt for short periods of time; and estuarine populations occur in the Chesapeake Bay and in San Francisco Bay. Water temperatures above 60°F (16°C) are needed for spawning and release of young. Asian clams can tolerate water temperatures as high as 86°F (30°C); they do not survive when temperatures drop below 35°F (2°C). In cold-weather regions such as the Great Lakes, there may be large fluctuations in population size from year to year.

Diet. These filter-feeders are herbivores, consuming mostly phytoplankters sieved from the water column.

Life History. Asian clams are hermaphrodites and capable of self-fertilization. Sperm are released into the water column and are taken in by other individuals. Eggs are fertilized in the inner gills, and there the larvae are incubated. Juveniles are released through the excurrent siphon when about 0.04 in. (1 mm) in size. They are poor swimmers and sink to the bottom, where they begin their adult stage. Reproduction and release of juveniles usually occurs twice a year, in the spring and again in late summer or fall. Juveniles become sexually mature at shell lengths of 0.2–0.4 in. (6–10 mm). Lifespan is one to four years.

A. This small clam, usually less than 1.0 in. across, has a rounded shell with concentric, evenly spaced ridges and a high, centrally positioned umbo. (U.S. Geological Survey.) B. Asian clams with a nickel for scale. (Shawn Liston, Bugwood.org.)

Impacts. Asian clams are notorious biofouling organisms, clogging intake pipes of power plants and other facilities. The weak-swimming juveniles descend to the base of the water column from whence intake pipes usually withdraw water. Live animals, empty shells, and dead clam bodies are sucked in. This has proved to be a particular problem at nuclear power plants, where water is used for fire protection.

Another type of economic consequence occurs where stream beds are dredged for sand and gravel to use as an aggregate in cement, as happens in Ohio and Tennessee. The clams end up in the cement but worm their way to the surface as it starts to set, leaving empty tunnels that weaken the structure.

Large congregations of Asian clams can alter the benthic habitat, which they may come to dominate, and may compete with native fingernail clams for food and space. Recent studies in Lake Tahoe, for example, show an increase in a filamentous green alga (*Zygnema* sp.) in clam beds, probably thriving on the bivalves' waste products.

Management. Prevention of introduction to new bodies of water is the most effective way to slow or stop the invasion of Asian clams. Boats and equipment should be inspected, cleaned of any obvious attached plants and animals, and drained of all water on land before they leave the launch area. Bait buckets should be emptied on land, not in water. Before launching into another body of water, boats and all equipment should be cleaned in hot water and dried for several days.

Chemical (bromine and chlorine) and mechanical (screens and traps) treatments can control infestations in intake pipes.

Selected References

Balcom, N. C. "Aquatic Immigrants of the Northeast, No. 4: Asian Clam, *Corbicula fluminea*," Connecticut Sea Grant College Program, 1994.

Foster, A. M., P. Fuller, A. Benson, S. Constant, and D. Raikow. "*Corbicula fluminea*." USGS Nonindigenous Aquatic Species Database, Gainesville, FL, 2009. http://nas.er.usgs.gov/queries/FactSheet.aspx?speciesID=92.

National Biological Information Infrastructure (NBII) and IUCN/SSC Invasive Species Specialist Group (ISSG). "*Corbicula fluminea* (Mollusc)." ISSG Global Invasive Species Database, 2005. http://www.issg.org/database/species/ecology.asp?fr=1&si=537.

USGS. "Asian clam, *Corbicula fluminea* (Muller, 1774) (Mollusca: Corbiculidae)". Nonindigenous Species Information Bulletin, Florida Caribbean Science Center, 2001. http://fl.biology.usgs.gov/corbicula4.pdf.

■|Asian Green Mussel

Also known as: Green mussel, green-lipped mussel
Scientific name: *Perna viridis*
Synonyms: *Mytilus viridus, Chloromya viridis*
Family: Mytilidae

Native Range. Coastal waters of southern Asia from the Persian Gulf to southern China and the Philippines.

Distribution in the United States. Gulf and Atlantic coasts of Florida northward through coastal Georgia as far as Charleston, South Carolina.

Description. This large marine bivalve has a smooth, elongated, fan-shaped shell. The thin valves clearly show concentric growth rings; the ventral margin is concave on one side. The entire shell of juveniles is bright green, but in adults, it is brown with green margins. The inner shell surfaces are smooth and iridescent blue or blue-green. The beak, at the hinge, has interlocking teeth: two on the left valve and one on the right. Juveniles and adults secrete hairlike byssal threads with which to attach to hard substrates.

Related or Similar Species. A number of native mussels in the same family as the Asian green mussel occur in states infested with this nonnative bivalve. They have similar shapes, but none are green. The scorched mussel (*Brachidontes exustus*), ribbed mussel (*Guekensia demissa*), and hooked mussel (*Ischadium recurvum*), the native species most apt to inhabit the same sites as Asian green mussels, have ribbed rather than smooth shells. The shells of the smaller horsemussels (*Modiolus* spp.) are brown on the outside and white on the inside. The smooth outer shell of the tropical charru mussel (*Mytella charruana*), another nonnative mussel found (rarely) in Florida's waters, has a dark-brown wavy pattern on a lighter background; the inner shell is iridescent purple.

Introduction History. Asian green mussels were first reported in Florida in Tampa Bay in 1999, when they clogged intake pipes at a power plant. This was probably the result of a discharge of ballast water carrying larvae by a ship coming from the Caribbean. Asian green mussels had arrived in Trinidad in 1990 and rapidly spread through the Caribbean. Drifting in currents, larval green mussels have dispersed southward as far as Marco Island, near Naples. A separate invasion event occurred near St. Augustine on the Atlantic side of the Florida peninsula in 2002. They likely arrived on a recreational or fishing boat transported overland from the Gulf coast without proper decontamination of hull, live wells, or gear. Northward-flowing ocean currents dispersed the mussel to Jacksonville, Florida, and along the entire Georgia coast by the end of 2003. In 2006, they were reported at Charleston, South Carolina. Mussels also dispersed southward; their range currently extends to near Titusville, Florida.

Habitat. Asian green mussels inhabit coastal and estuarine waters in the intertidal and subtidal zones to depths near 35 ft. (10 m). They tolerate a wide range of salinities, from hypersaline (80 ppt) to brackish (12 ppt). Optimal salinity is reportedly 27–33 ppt. They also tolerate a wide range of temperatures between 50° and 95°F (10–35°C), but do best in water temperatures of 79–90° F (26–32° C). Juveniles appear to prefer to settle in areas of high water flow about 1 ft. (30 cm) below the low-tide mark. They seem to seek crevices or the undersides of floating objects.

Diet. These mollusks are sessile filter-feeders. An incurrent siphon draws in seawater that is moved by cilia to the branchial cavity. Phytoplankters, zooplankters, and other organic particles are filtered out. Mucus is secreted to bind the food items into a bolus that is directed by cilia to the mouth. Water exits via an excurrent siphon.

Life History. The sexes are separate animals that release eggs and sperm into the water column during spawning episodes that peak in spring and autumn. Fertilization takes place externally. The presence of gametes in the water stimulates other individuals to release eggs and sperm to synchronize spawning within a local population. The dilution of seawater, such as what happens at the onset of a rainy season, also seems to trigger spawning. Within 8 hours of fertilization, the embryos become ciliated, free-swimming larvae and, 8–12 hours later, have a shell and ciliated membrane called a velum. The larvae metamorphose into juveniles and settle 8–20 days later. The juveniles secrete byssal threads and attach to hard surfaces. In Tampa Bay, they reportedly become sexually mature in 1–2 months when shell length is 0.6–1.2 in. (15–30 mm). They live about three years.

Impacts. Asian green mus-

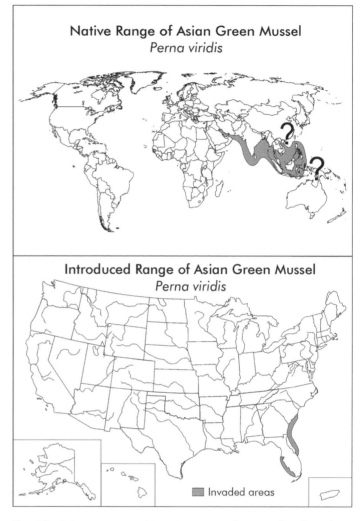

Top: The Asian green mussel is native to coastal waters of southern Asia. Bottom: In the United States, the Asian green mussel is invasive along both the Gulf and Atlantic coasts of Florida and has currently expanded its range as far north as South Carolina. (Adapted from McGuire and Stevely 2009.)

sels are marine biofouling organisms, acting much like the freshwater zebra mussel (see Mollusks, Zebra Mussel). They can clog intake screens and pipes of power and desalinization plants and cover the hulls of boats and submerged parts of buoys, bridges, piling, and seawalls. Not only do they block the flow of water, but they also damage pumps, reduce heat transfer efficiency, and accelerate the rate of corrosion of metal surfaces. Economic impacts of increased plant maintenance can be significant.

Since the Asian green mussel can rapidly build up large populations that form mats on the seafloor, it is suspected that they will complete for space and possibly for planktonic food supplies with native bivalves. In Tampa Bay, green mussels seem to have expanded at the expense of native oysters (*Crassostrea virginica*). Elsewhere, for example in Georgia, where oysters live in habitats where they are exposed at low tide, such replacement does not seem to be happening.

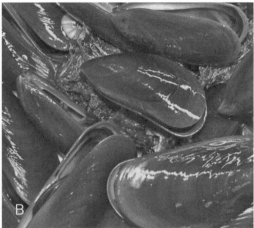

A. Asian green mussel shell shows distinct beak and concentric growth rings. (U.S. Geological Survey.) B. The concave ventral margin and smooth inner shell surfaces are visible on some of these specimens. (U.S. Geological Survey.)

The success of the invasion of Asian green mussels in the coastal waters off Florida suggests they would be equally successful if accidentally introduced to other sites along the Gulf coast or to southern California.

Management. Currently, most management consists of mechanically removing mussels from intake screens and pipes or flushing those systems with chlorinated water. The main focus is to prevent further introductions of the organisms by educating commercial fishermen and recreational boaters and fishermen to inspect and clean the hulls of their vessels and to drain live wells and bilges on land after taking their boats out of the water.

Selected References

Masterson, J. "Species Report: *Perna viridis.*" Smithsonian Marine Station at Fort Pierce, 2007. http://www.sms.si.edu/irlSpec/Perna_viridis.htm.

McGuire, Maia, and John Stevely. "Invasive Species of Florida's Coastal Waters: The Asian Green Mussel (*Perna viridis*)." Publication #SGEF 175, Florida Sea Grant College Program. Institute of Food and Agricultural Sciences (IFAS), University of Florida, 2009. http://edis.ifas.ufl.edu/sg094.

Powell, Cindie. "Asian Invader Musseling in on U.S. Habitats." Texas Sea Grant and NOAA, 2005. http://www.oar.noaa.gov/spotlite/archive/spot_greenmussel.html.

Thornton-DeVictor, Susan, and David Knott. "The Asian Green Mussel: Recent Introduction to the South Atlantic Bight." Species of the Month, Southeastern Regional Taxonomic Center, South Carolina Department of Natural Resources, n.d. http://www.dnr.sc.gov/marine/sertc/The%20Asian%20Green%20Mussel.pdf.

■| Chinese Mystery Snail

Also known as: Asian applesnail, Japanese mystery snail, black snail, trapdoor snail
Scientific name: *Cipangopaludina chinensis malleata*
Synonyms: *Bellamya chinensis, Viviparus chinesis, Viviparous stelmaphora, Paludina malleata et al.*
Family: Viviparidae

Native Range. Southeast Asia. It occurs in Myanmar (Burma), Thailand, Indonesia (Java), Vietnam, China, Korea, Japan, and the Philippines.

Distribution in the United States. Chinese mystery snails are reported in 27 states. In the eastern United States, populations occur in New England and the Great Lakes region, and on the West Coast, they can be found from the San Francisco Bay area north to Seattle. Mid-continent records come from Colorado, Iowa, Nebraska, Oklahoma, Texas, and Utah.

Description. These gastropods are distinguished by their large size, adults reaching lengths of 1.5 in. (60–65 mm) from the tip of the whorl to the lip of the shell. The shell has 6–7 whorls with rounded shoulders and indented sutures; in adults, it is a uniform olive green, greenish brown, brown, or reddish brown without banding on the outside and white to pale blue on the inside. The dark solid operculum, a functioning "trapdoor," is marked with concentric rings. The outer lip is round to oval and black.

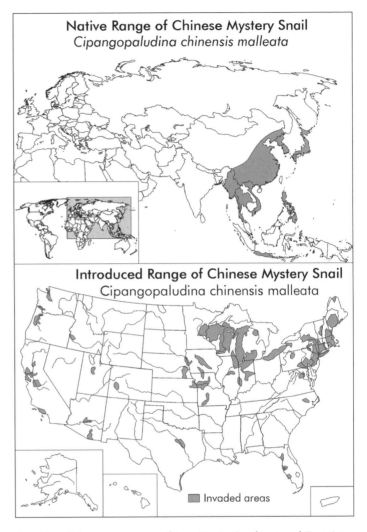

Top: The Chinese mystery snail is native to Southeast and East Asia. Bottom: Chinese mystery snails are reported from 27 states. (Adapted from Kipp and Benson 2007.)

Related or Similar Species. Taxonomic confusion exists as to whether there are two forms of the Chinese mystery snail in the United States, or whether the Japanese mystery snail (*C. japonica*) is a separate species with a more elongate shell and other subtle morphological differences. The banded mystery snail (*Viviparus georgianus*), native to some parts of the United States but an introduced species in the Great Lakes region and eastern states north of the Carolinas, is smaller than the Chinese mystery snail, with a maximum shell length of 1.75 in. (45 mm); its shell is encircled with obvious reddish-brown bands. The brown mystery snail (*Campeloma decisum*), native to the eastern United States, only rarely grows as large as the Chinese mystery snail and is much narrower than either the Chinese or banded mystery snails. Its shell is usually olive green. The brown mystery snail is one of very few native snails with an operculum.

Introduction History. The earliest record of Chinese mystery snails in the United States dates to 1892 and San Francisco, where they were imported for the live-food market.

The shell of the Chinese mystery snail has indented sutures between the rounded whorls. (Pieter Johnson, University of Colorado.)

By 1911, a free-living population was thriving in San Francisco Bay. Snails may have been accidentally introduced into Massachusetts in the early 1900s with goldfish released to control mosquitoes. A population was established in Boston by 1915, perhaps a by-product of the local Asian food market. Snails entered Lake Ontario from the Niagara River between 1931 and 1942. Chinese mystery snails were reported in Florida in 1950 and were established in Texas and Lakes Erie and Michigan and their drainages by 1965. Snail introductions initially seem to have been intentional releases either to develop a local food supply or from the freshwater aquarium trade, in which Chinese mystery snails are used to keep fish tanks clean of algae growths. They may be unintentionally moved from one body of water to another as contaminants in live bait (e.g., minnows and crayfish) or on plants and in water transported on recreational watercraft and boat trailers. It takes only one pregnant female to start a new population.

Habitat. These freshwater snails prefer quiet waters with soft substrates of silt, sand, or mud. They can be found in lakes, ditches, rice paddies, and slow-moving streams at water depths of 1.5 to 15 ft. (0.5 to 5 m). They can tolerate pollution and may thrive in stagnant water, but they cannot survive very low oxygen levels and experience major die-offs under a combination of warm water and algal blooms that reduce dissolved oxygen content.

Diet. Chinese mystery snails feed on organic and inorganic material on the bottom of water bodies and scrape algae from hard surfaces. They also consume zooplankters and phytoplankters.

Life History. These snails are live-bearers. In the eastern United States, embryos develop in the female between May and August, and the young are born in shallow water from June through October. After birthing, females retreat to deeper water for the winter months. Each female may produce as many as 100 tiny snails in a single brood. (It is the sudden appearance of tiny, perfectly developed snails that may be the "mystery" of these snails.) Females live up to five years and tend to have their largest broods in their later years. Males live on average three years.

Impacts. The ecological impacts, if any, of this introduced species remain unknown. It is possible they could compete with native snails for food and space, but little indication exists that this is happening. Chinese mystery snails do carry parasites, including one that can infect humans, but do not seem to be a vector for "swimmer's itch" as some have feared. The species can clog the screens of even large water-intake pipes.

Management. Prevention of new infestations is the best control measure available. Eradication of existing populations is likely impossible. People should refrain from dumping bait and aquarium contents and should sanitize fishing and boating equipment before entering another body of water. Live animals of any sort should never be released into the wild.

Selected References

"Aquatic Invasive Species: Chinese Mystery Snail." Indiana Department of Natural Resources, 2005. http://www.in.gov/dnr/files/CHINESE_MYSTERY_SNAIL.pdf.

Kipp, Rebekah M., and Amy Benson. *"Cipangopaludina chinensis malleata."* USGS Nonindigenous Aquatic Species Database, Gainesville, FL, 2007. http://nas.er.usgs.gov/queries/FactSheet.asp?speciesID=1045.

"Mystery Snail Monitoring Protocol." Aquatic Invasive Species Monitoring Manual—Citizen Lake Monitoring Network, 2009. http://www.uwsp.edu/cnr/uwexlakes/clmn/.

■ | Common Periwinkle

Also known as: Edible periwinkle, wrinkle winkle
Scientific name: *Littorina littorea*
Family: Littornidae

Native Range. Europe, from the coast of northern Spain, throughout the British Isles, and north to Scandinavia and Russia.

Distribution in the United States. Established along the coast of the northeastern United States, from Maine to Virginia and the Chesapeake Bay. On the West Coast, individuals have been collected from California to Washington.

Description. This small marine snail has a conical shell that comes to a point at the apex. Shells of young animals show distinct ridges spiraling up to a prominent point; in adults, the ridges become indistinct and the shell appears smooth. Five or six somewhat swollen whorls are outlined by shallow sutures. The base color of the shells is usually grayish brown or black, getting paler near the top. Dark lines form a spiral pattern on much of the shell, but the central axis or columella is white, as is the inside of the shell. The thick inner lip of the aperture is bent over the columella. Average shell height is 0.6–1.5 in. (16–38 mm), but it may be as great as 2.0 in. (52 mm). Juvenile stages have black barring on their flat, broad tentacles.

Related or Similar Species. The ridged shells of young common periwinkles could be mistaken for those of the rough periwinkle (*Littorina saxatilis*), which is native to the coasts of eastern North America from the Chesapeake Bay northward. Rough periwinkles have 4–5 whorls and grow to only about 0.7 in. (18 mm). The exterior of the shell may bear a checkered pattern; the inside is brown. They tend to inhabit salt marshes, but can be found on rocky coasts. Rough periwinkles have been introduced to San Francisco Bay

Introduction History. Live common periwinkles were first identified in North America in 1840 at Pictou, Nova Scotia, Canada. From this site in the Northumberland Straits of the southern Gulf of St. Lawrence, it moved to the Atlantic Coast, being discovered in Halifax, Nova Scotia, in 1854. By 1861–1862, it had appeared at Eastport, Maine. Soon thereafter, it was found on Cape Cod (1872), and by 1890, it had reached Cape May, New Jersey. It appears that a separate introduction occurred sometime in the early 1950s on the Delmarva Peninsula. Southward expansion from this point of entry was relatively slow; it took about 25 years to reach the entrance to Chesapeake Bay (1978), its current southern range limit.

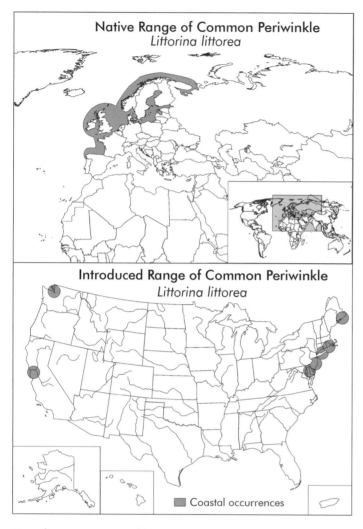

Top: The common periwinkle is native to the Atlantic and Baltic coasts of Europe. Bottom: The common periwinkle is a dominant member of rocky coast communities in the northeastern United States and has been collected in California and Washington State. (Adapted from Benson 2009.)

It is generally agreed that the common periwinkle arrived in Nova Scotia from Great Britain on rocks used as ballast in transoceanic ships, possibly those involved in the early and mid-nineteenth-century timber trade. The initial inoculation was followed by rapid population growth and range expansion along the rocky coasts and salt marshes of New England. However, not all agree with this scenario and raise the possibilities that (a) the periwinkle was actually native to northeast Canada and experienced a range expansion only after European settlement; or (b) that it was carried to Canada by the Vikings in pre-Columbian times. Most evidence to date, including information from prehistoric and historic archaeological sites and modern genetic analyses, support a nineteenth-century introduction.

Habitat. These mobile gastropods inhabit coastal and estuarine environments. The most typical habitat for common periwinkles is on all but the most exposed rocky coasts in the intertidal zone from the spray zone of the upper littoral to the sublittoral zone that is exposed only during the lowest low tides. In sheltered areas, they may found in salt marshes and on mudflats.

Diet. Common periwinkles are herbivores and graze on benthic diatoms and dinoflagellates as well as macroalgae such as sea lettuce (*Ulva lactuca*) and other large green algae and young and ephemeral red and brown algae. While grazing on rocks, they may ingest small invertebrates such as barnacle larvae.

Life History. Common periwinkles have separate sexes and reproduce annually. Females release their fertilized eggs into the ocean in horny capsules about 0.03 in. (1 mm) wide. Each capsule is convex on both sides and usually contains 2–3 eggs, although as many as 9 eggs may be inside. The shedding of egg capsules coincides with spring tides (i.e., when the tidal range is greatest). The larvae exit the capsules and settle within six weeks. Juveniles mature at 2–3 years of age when shell height is about 0.4 in. (10 mm). The lifespan of a common periwinkle is 5–10 years.

A. Common periwinkles are one of the most common mollusks in intertidal communities north of New Jersey. (J. Pederson. Reprinted with permission of the MIT Sea Grant College Program.) B. The conical shell spirals up into a prominent point, while the inner lip of the aperture folds over the columella. (Amy Benson, U.S. Geological Survey.)

Impacts. Common periwinkles quickly became one of the most common mollusks on the Atlantic Coast north of New Jersey and an ecological dominant in intertidal communities. In both exposed rocky coast habitats and estuarine situations densities may reach 20–45/sq. ft. (200–500/m^2). Its consumption of green algae has allowed Irish moss (*Chondrus crispus*), a slow-growing foliose red alga, to proliferate. Constant grazing by periwinkles of young algae prevents the reestablishment of the original algal canopy on rocks and removes a key food item of native marine snails. On rocky coasts, common perwinkles have displaced the eastern mudsnail (*Nassarius obsoletus*) from many habitats and caused a niche shift and population reduction in the rough periwinkle (*Littorina saxatilis*). In estuarine habitats, its bulldozing of soft sediments slows the accumulation of fine materials and stifles the growth of the root mat of salt marsh cordgrass (*Spartina alterniflora*).

Management. Few management practices have been developed to deal with common periwinkles, and little research is conducted on them. They are today well-established members of intertidal communities. Indeed, they may gain value as a bio-indicator of contaminated marine habitats.

Selected References

Benson, A. J. "*Littorina littorea*." USGS Nonindigenous Aquatic Species Database, Gainesville, FL, 2009. http://nas.er.usgs.gov/queries/FactSheet.asp?speciesID=1009.

Chapman, John W., James T. Carlton, M. Renee Bellinger, and April M. H. Blakeslee. "Premature Refutation of a Human-Mediated Marine Species Introduction: The Case History of the Marine Snail *Littorina littorea* in the Northwestern Atlantic." *Biological Invasions* 9: 737–50, 2007. doi:10.1007/s10530-006-9073-x.

National Biological Information Infrastructure (NBII) and IUCN/SSC Invasive Species Specialist Group (ISSG). "*Littorina littorea* (Mollusc)." ISSG Global Invasive Species Database, 2005. http://www.issg.org/database/species/ecology.asp?si=400&fr=1.

■|Giant African Snail

Also known as: Giant African land snail
Scientific name: *Achatina fulica*
Family: Achatinidae

Native Range. East Africa, particularly the coastal areas of Kenya and Tanzania.

Distribution in the United States. Established only in Hawai'i, but its potential to invade the continental United States is a source of major concern. Snails have been collected in California, several southern states, and in Michigan, Ohio, and Wisconsin. A small but rapidly expanding population in Florida was eradicated in 1973.

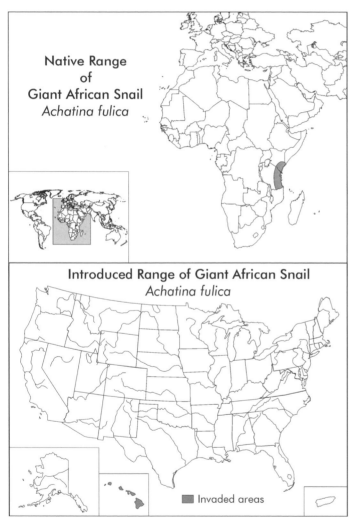

Top: The giant African snail comes from coastal regions of East Africa. Bottom: The giant African snail is an invasive species only in Hawai'i at this time, but it is frequently smuggled into other parts of the United States, risking establishment of new populations elsewhere.

Description. This large terrestrial snail can reach lengths near 8 in. (20 cm). The cone-shaped shell makes up roughly half the total length. An adult giant African snail stands about 3 in. (7–8 cm) high. The shell is usually reddish brown with cream and dark-brown streaks running perpendicular to the whorls, but a light "café au lait" color is also common. The shells of adults narrow toward the apex, which is barely drawn out. They have 7–9 whorls with indented sutures. The semi-elliptical opening takes up less than half the shell length and has a sharp, thin outer lip. The inner lip (columella and parietal callus) is pale blue or whitish. This land snail is much larger than any native species.

Related or Similar Species. Other large African land snails could invade the United States. These include the world's largest land snail, the Ghana tiger snail (*Achatina achatina*); margies or West African land snail (*Archachatina marginata*); and the Nigerian land snail (*Limicolaria aurora*). None of these have been found free-living in the United States to date.

Introduction History. Twice, giant African snails were illegally

A. Darker forms of the giant African snail infest this tree. (David G. Robinson, USDA APHIS PPQ, Bugwood.org.)
B. This large terrestrial snail can reach lengths of 8 in., making it much larger than any native snail species. (Forest and Kim Starr.)

imported into hatcheries on Maui, Hawai'i, in 1936, once by mail and once brought in luggage from Japan. It was bred there and, shortly afterward, on O'ahu. By the time they were discovered by authorities, escapees had established large colonies on both islands. In 1958 the snails appeared on Kaua'i and Hawai'i. By 1963, they had been discovered on Moloka'i and possibly Lana'i.

Between 1948 and 1958, snails were intercepted by quarantine officials on cargo coming into San Pedro and other California ports at least 50 times. However, children returning from visits to Hawai'i were more successful in smuggling the snails into Arizona (Mesa) and Florida (Miami) in 1958 and 1966, respectively. The contraband was released outside when discovered by family members. In Arizona, the snails were quickly eradicated, but the case in Florida became a cautionary tale that still argues against allowing this species anywhere in the United States. The Miami family decided to keep the snails as pets and set them out in the family garden. Within seven years, more than 18,000 giant African snails occupied 42 city blocks in Miami. Another infestation was reported north of the city in Hollywood. It took 10 years and a million dollars for Florida to eradicate them, mostly by hand collection.

Snails still arrive at U.S. ports as part of an illegal pet trade and unintentionally in shipping containers and on plants imported from Hawai'i. Snails appeared in the Midwestern states of Michigan, Ohio, and Wisconsin as part of the pet and shell trades and in response to requests from teachers looking for classroom animals and unaware of the threats posed to agriculture, horticulture, and possibly human health by these attractive, easy-to-care-for gastropods. Local dispersal of eggs, and snails if populations were established, would be facilitated by the movement of garden waste, rubbish, construction materials, vehicles, and other equipment away from contaminated areas. The snails themselves are capable of moving 150 ft. (50 m) a night. It only takes one gravid female to start a new population.

Habitat. Native to humid tropical forests, the giant African snail is highly adaptable and could survive in temperate climates, even those with snow. It can also withstand drier environments because it can seal itself inside the shell and aestivate in loose soil during unfavorable conditions. It is active at temperatures of 48–84°F (9–29°C); at lower temperatures, it becomes dormant. In the tropics, this land snail inhabits disturbed forests and forest edge, shrublands, and wetlands and has invaded agricultural areas, gardens, and urban areas. It is nocturnal; sunlight can be lethal.

Diet. Giant African snails are herbivores and detritivores and reportedly consume some 500 different types of medicinal and ornamental plants as well as food crops, including such temperate-zone species as beans, cucumbers, melons, peanuts, and peas. Young snails with shells up to an inch (3 cm) long are almost completely vegetarian; older individuals increasingly feed on detritus and decaying plant material, although they still feed on live plants.

Life History. The giant African snail is hermaphroditic, each individual possessing both male and female reproductive organs, but must breed with other individuals to produce fertile eggs. Snails lay their first eggs at six months and are able to produce eggs for another 400 days, with as many as three clutches a year. During their first year, they may lay 100 eggs, and in their second year, up to 500. After that time, fecundity decreases. The usual lifespan of a snail is 5–6 years.

The large eggs are about 0.2 in. (4.5–5.5 mm) in diameter and deposited in nests excavated in the soil or under leaves and loose stones. They hatch from a few hours to 17 days after laying, depending upon temperature. They require temperatures above 59°F (15°C). The hatchlings remain below ground for 5–15 days, where they consume their egg shells and organic detritus. They remain close to the nest during this time, but within two months will establish a new home range and begin feeding aboveground on green plants at night.

Impacts. The greatest threat posed by giant African snails is to commercial agriculture because they are capable of devouring tree crops, ornamentals, and vegetables. Florida's tropical plant industry would be particularly vulnerable. The snails could also transmit parasites to humans. They are carriers for rat lungworm (*Angiostrongylus cantonensis*), a nematode that produces eosinophilic meningitis in humans. In natural island habitats, such as in Hawai'i, they may compete with native snails.

Because of its voracious appetite and importation into many tropical islands in the Pacific and Caribbean as well as Southeast Asia, these snails are considered the world's worst pest snail and have been nominated by IUCN as among 100 of the world's worst invaders.

Management. In Hawai'i and Florida, a number of biocontrol measures were tried, the more successful ones involving the importation of carnivorous snails from various parts of the world, including *Euglandina rosea* and *Gonaxis quadrilateralis*. These seemed to depress recruitment in giant African snails but did not eliminate them altogether. Success in Florida came with labor-intensive and thus expensive baiting and hand collection. (In other parts of the world, pesticides are used to control populations.)

Selected References

Davis, C. J., and G. D. Butler Jr. "Introduced Enemies of the Giant African Snail, *Achatina fulica* Bowdich, in Hawaii (Pulmonata: Achatinidae)." *Proceedings, Hawaiian Entomological Society*, 60(3): 377–89, 1964. Available online at http://scholarspace.manoa.hawaii.edu/bitstream/10125/10889/1/18_377-390.pdf.

IUCN/SSC Invasive Species Specialist Group (ISSG). "*Achatina fulica* (Mollusc)." ISSG Global Invasive Species Database, 2006. http://www.issg.org/database/species/ecology.asp?si=64&fr=1&sts=sss.

Robinson, D. G. "Identity: *Achatina fulica* Bowdich, 1882." U.S. Department of Agriculture, Animal and Plant Health Inspection Service, 2002. http://www.aphis.usda.gov/plant_health/plant_pest_info/gas/downloads/achatinafulica.pdf.

Stokes, Heather. "Giant (East) African Snail (*Achatina fulica*)." Introduced Species Summary Project. Columbia University, 2006. http://www.columbia.edu/itc/cerc/danoff-burg/invasion_bio/inv_spp_summ/Achatina_fulica.htm.

■ | Golden Apple Snail

Also known as: Channeled apple snail, applesnail
Scientific name: *Pomacea canaliculata*
Family: Ampullariidae

Native Range. South America, from temperate Argentina into the Amazon basin. It occurs in Argentina, Paraguay, and Uruguay in the La Plata basin and in Brazil and Bolivia in the Amazon basin.

Distribution in the United States. Established on Maui, Kaua'i, O'ahu, and Hawai'i islands in Hawai'i; a couple of isolated sites in southern California; northern Florida near Jacksonville; a three-county area in southeastern Texas at the heart of the state's rice-growing region; and a site near Fort Worth in northern Texas. It has been reported from Mobile, Alabama; Yuma, Arizona; Georgia; Louisiana; North Carolina; and Oklahoma.

Description. This freshwater snail is fairly large; its globular shell measures 2.75–3.6 in. (7.0–9.0 cm) high. Deeply indented sutures or channels separate one whorl from the next. The apical spire progresses in steps to a prominent point on top; the spire has nearly flat shoulders. Shell color in wild forms is olive brown to yellowish brown, often with dark bands; those bred in captivity may have yellow and green shells. The large opening (aperture) may be round or

Top: The golden apple snail is native to the Amazon and La Plata basins of South America. Bottom: Golden apple snails are established on Hawai'i and in isolated locations of southern California and northeastern Florida. They have been found not far from Texas's rice-growing region, where they would pose a major threat to the crop. (Adapted from USGS. "Pomacea canaliculata." USGS Nonindigenous Aquatic Species Database, Gainesville, FL, 2010. http://nas.er.usgs.gov/queries/FactSheet.asp?speciesID=980.)

A. The shell of this immature golden apple snail shows the indented sutures between whorls and the flat-shouldered apex that distinguish this species. B. The large apertures are visible on specimens in this group of adult snails. (Susan Ellis, Bugwood.org.)

oval. The thick, retractable operculum has a concentric structure with the nucleus near the center. Eggs are bright pink.

Related or Similar Species. The taxonomy of apple snails is not completely resolved, and it may be that a complex of closely related species exists. Several common and scientific names have been used in the literature, and misidentification of invaders has added to the general confusion. In Florida, there are four other apple snails, one of which, the Florida apple snail (*P. paludosa*) is native to peninsular Florida. It has a distinctively low, rounded shell spike and is much smaller (shell height 1.6–2.75 in. [4.0–7.0 cm]) than the golden apple snail. The eggs of Florida apple snails are white and relatively large. Genetic studies reveal that the most common introduced apple snail in that state is the island apple snail, *P. insularum*, and not *P. canaliculata* as previously thought. The two are very similar in canaliculata appearance.

Introduction History. Golden apple snails were brought to Southeast Asia in 1980 as a potential food source for local peoples and possibly an export to the gourmet markets of Asia. They were taken to Hawai'i from the Philippines for similar reasons in 1989. Elsewhere in the United States, they entered as part of the pet and aquarium trade, of which they became part in the 1950s. It is likely that infestations in the continental United States are the result of escapes or releases of pet snails. The snail (or a close relative) was in Palm Beach County, Florida, in 1978 and in three other central Florida counties by the late 1990s. A small population existed in a reservoir in Rockingham County, North Carolina in 1993, but did not become established. Golden apple snails were found to be established in Lake Miramar in San Diego County and in a small pond at the Norton Simon Museum in Pasadena, California, in 1997. Live snails and egg masses were retrieved from a canal near the Salton Sea in Riverside County, California, in 2001. Texas Parks and Wildlife discovered the snail in a rice irrigation canal near Houston in 2001, and surveys revealed their presence in canals and bayous in three counties in the southeastern part of that state. Flooding resulting from Tropical Storm Alison in June of that year probably dispersed them to other sites in the area.

Habitat. Native to lakes and swamps, in the United States, the golden apple snail is found in urban ponds, drainage and irrigation ditches, natural streams, bayous, wetlands, and rice and taro fields, which are regularly flooded. It feeds at night and spends the day submerged among aquatic plants. This apple snail withstands periods of drought by burrowing into the bottom and retracting its operculum to seal the shell. They may aestivate as long as five months and can also hibernate for long periods of time during cold weather.

Diet. Golden apple snails are herbivores and consume most types of aquatic vegetation. They show a strong preference for taro and rice. Unfortunately, they will not feed on two invasive aquatic plants, hydrilla (See Volume 2, Aquatic Plants, Hydrilla) and waterhyacinth (See Volume 2, Aquatic Plants, Waterhyacinth).

Life History. Golden apple snails are sexually mature when they are about 1 in. (2.5 cm) in size or between three months and two years of age. Females crawl out of the water at night and deposit clusters of pink eggs on just about any solid surface protruding above the waterline. The eggs are 0.09–0.14 in. (2.2–3.5 mm) in diameter, and each clutch may contain 200–600 eggs or more. Depending upon ambient temperature, the eggs will hatch in 7–15 days. A female will produce a new clutch every few weeks and can breed throughout the year, although reproduction is depressed in cooler months. Golden apple snails will live about four years.

Impacts. The golden apple snail is a major agricultural pest in Hawai'i, where it attacks taro and rice. They feed on the corm and create holes that give access to bacteria and other plant pathogens, either killing the plant or greatly reducing its yield. They also consume young shoots of taro, rice, and other water plants. Elsewhere in the United States, they have not become a problem yet because they are generally in nonagricultural areas or, as near the Salton Sea, in areas where the crops grown are not those likely to be attacked by these gastropods. Only the populations in southeastern Texas, close to rice fields, are cause for concern and have prompted more widespread recognition of the potential harm caused by these invaders. So far, crop damage has not been observed in Texas. (It should be noted that golden apple snails have been very destructive of rice and taro crops in Southeast Asia. For this reason, they have been nominated by the IUCN as one of 100 of the world's worst invasive species.)

In Florida, there is concern that exotic apple snails could outcompete the native apple snail, *P. paladosa*, which is the primary food of the rare and endangered Everglades Kite (*Rostrhamus sociabilis*).

Golden apple snails can transmit rat lungworm (*Angiostrongylus cantonensis*) if they are improperly cooked, causing severe headache, fever, and even death in people who eat them.

Management. In Southeast Asia, where the golden apple snail is such a threat to agriculture, several methods of biological control have been attempted with little success. Among the predators brought in to eat snails are ducks, carp, Nile tilapia, and red ants, which eat snail eggs. Toxic plants grown in rice paddies or left to float on the surface did kill the apple snails, and control of water levels helped. Laborious hand-picking appears to be the most effective means of control, but eradication is nearly impossible. Blocking new introductions is the best way to prevent the spread of this pest.

Selected References

Cowie, R. H. "Ecology of *Pomacea canaliculata*." ISSG Global Invasive Species Database, 2004. http://www.issg.org/database/species/ecology.asp?si=135&fr=1&sts=.

Ghesquiere, Stijn A. I. "*Pomacea (pomacea) canaliculata* (Lamarck, 1819)." The Apple Snail Website, n.d. http://www.applesnail.net/content/species/pomacea_canaliculata.htm.

Howells, Robert G., and James W. Smith. "Status of the Applesnail *Pomacea canaliculata* in the United States." The Seventh International Congress on Medical and Applied Malacology (7th ICMAM) Los Baños, Laguna, SEAMEO Regional Center for Graduate Study and Research in Agriculture (SEARCA), Philippines, 2003. http://www.applesnail.net/pestalert/conferences/icam07/country _report%20_usa.pdf.

Mohan, Nalini. "Apple Snail (*Pomacea canaliculata*)." Introduced Species Summary Project. Columbia University, 2002. http://www.columbia.edu/itc/cerc/danoff-burg/invasion_bio/inv_spp_summ/ Pomacea_canaliculata.html.

Strange, Lionel A., and Thomas R. Fasulo. "Apple Snails of Florida." University of Florida, 2004. http://entnemdept.ufl.edu/creatures/misc/gastro/apple_snails.htm.

■ | Naval Shipworm

Also known as: Atlantic shipworm, great shipworm
Scientific name: *Teredo navalis*
Family: Teredinidae

Native Range. Naval shipworms have been transported around the world in the hulls of wooden sailing ships for centuries. It is difficult to know, therefore, just what their native range is. The Atlantic coast of Europe from Iberia to Scandinavia is the most likely place of origin. Today the species is cosmopolitan, found in tropical and subtropical waters of the Atlantic and Pacific oceans in both the Northern and Southern hemispheres.

Distribution in the United States. Naval shipworms are found in all coastal regions of the United States.

Description. Although called a worm, this animal is mollusk, a somewhat bizarrely formed bivalve adapted to boring into and living in wood. The reddish body is long and wormlike and topped by a very small shell that serves to chip away wood. Most contemporary information on them comes from the Baltic Sea, where they regularly attain lengths of 8–12 in. (20–30 cm) and a width of 0.4–0.8 in. (1–2 cm). In tropical waters, they reportedly become nearly 20 in. (50 cm) long. The shell is ridged and covers but a small part of the body; it is usually 0.5–0.8 in. (12–20 mm) in length.

Frequently, the burrows that shipworms dig in submerged wood are the chief indication that the animals are present. Each burrow is a long cylindrical tube lined with a calcareous coating secreted by the shipworm. It is blocked near the entrance by a pair of calcareous structures or "pallets" that are paddle-shaped. Burrows excavated by adults can be as much as 3 ft. (1 m) long, but larvae with diameters of 0.04 in. (1 mm) also bore into wood. The shipworm extends two siphons from its anterior end through a small opening between the pallets for feeding, respiration, and excretion of wastes.

Related or Similar Species. There are native species of shipworm in several genera, including six species in the genus *Teredo*, found along the Atlantic coast; and others that are introduced or of unknown origin.

Introduction History. The naval shipworm was first reported in the United States in 1839, when it was detected in the sheathing of a foreign wooden ship in Massachusetts Bay. Within 100 years, it was common all along the Atlantic coast north of the bay. Naval shipworms showed up in Long Island Sound in 1869 in the hull of a sailing ship, and within a few decades, they were common in New York Harbor. The first naval shipworms were detected in Chesapeake Bay in 1878, but it remained rare. Later, it was collected from North Carolina south to Florida and along Texas's Gulf coast, as well as in Puerto Rico.

An invasion on the Pacific coast in San Francisco Bay occurred in 1913 with drastic results (see under Impacts). Today, the naval shipworm is less abundant as a result of many fewer wooden-hulled ships and the widespread use of chemically treated timbers in waterfront construction.

Habitat. Shipworms live in timbers submerged in seawater, including the hulls of wooden ships and pilings, piers, and other wooden waterfront structures. They tolerate a wide range of salinities (5–45 ppt), so they also thrive in estuaries. Larvae probably die in salinities less than 5 ppt. Shipworms reproduce at water temperatures ranging from 52–86°F (11–30°C), but are known to survive in water temperatures as low as 33°F (0.7°C). They are found in tropical seas as well as cool temperate waters. They survive up to six weeks in oxygen-poor situations by sealing themselves into their burrow and metabolizing stored glycogen.

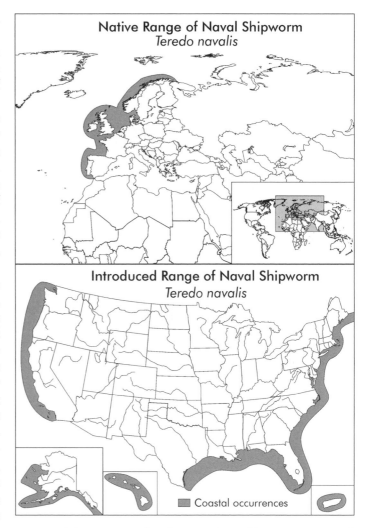

Top: Naval shipworms are found around the world as a consequence of global trade and explorations, especially in the days of wooden-hulled ships. Their origins are obscured but are likely to have been in the Atlantic waters off Europe. Bottom: Shipworms are found in all coastal waters of the United States.

Diet. Shipworms feed almost exclusively on wood with the aid of symbiotic bacteria that digest cellulose and fix nitrogen. The bacteria reside in special cells in the gills. The specially sculptured shell rasps away wood that is then transported to the mouth by cilia. Shipworms may also be filter-feeders of plankton, drawing in water through the incurrent siphon.

Life History. Adult shipworms occur as separate males and females. Spawning apparently depends on temperature. It reportedly occurs from April to September in New Jersey and from May to October at Woods Hole, Massachusetts. Males release sperm into the water column. Other individuals take them in through their incurrent siphons, and fertilization

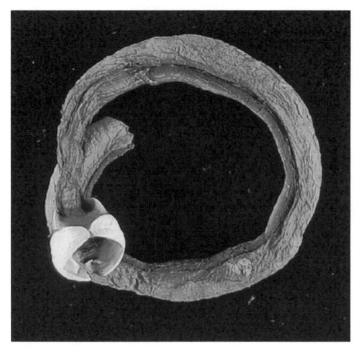

This peculiar bivalve has a long wormlike body and a tiny shell used to bore into wood. (U.S. Geological Survey.)

occurs internally in the chamber housing the gills. Larvae are brooded in the gills until the velum (a ciliated organ that provides locomotion) and a straight-hinged shell form. Then the larvae are released into the water column, where they begin a planktonic stage lasting up to four weeks. During this phase, they form siphons, gills, and an obvious foot. The tiny (0.04 in. or 1 mm) final-stage larvae detect wood chemically and swim toward it, attaching themselves with byssal threads. It is not known how the soft-shelled larvae bore into wood; maternal enzymes may aid in softening the surface. The settled larvae rapidly metamorphose into juvenile shipworms. They become sexually mature adults six weeks after settlement.

Impacts. In the past, this animal was so abundant in harbors around the world that they determined the lifespan of a wooden ship. It was reported in Wales that submerged ship timbers were destroyed in eight years. The presence of shipworms necessitated covering ships' hulls in tar or cladding them in copper. The outbreak in San Francisco Bay from 1919 to 1921 destroyed untold numbers of wharves, piers, and other waterfront structures. On average, a major structure was lost every two weeks with enormous economic consequences. In the Baltic Sea area, there is concern today that thousands of ancient shipwrecks lying on the seafloor and of great archeological value could be destroyed if naval shipworms invaded the sites.

If and how this species affects native or other introduced shipworms has not been studied, but competition for food is possible.

Management. The naval shipworm is less of a problem today than in the past as a result of many fewer wooden-hulled ships and the widespread use of chemically treated timbers in waterfront construction. However, in the absence of these defenses, it will still destroy wooden structures below water.

Selected References

Masterson, J. "*Teredo navalis.*" Smithsonian Marine Station, 2007. http://www.sms.si.edu/irlSpec/Teredo_navalis.htm.

NIMPIS. "*Teredo navalis* Species Summary." National Introduced Marine Pest Information System, edited by C. L. Hewitt, R. B. Martin, C. Sliwa, F. R. McEnnulty, N. E. Murphy, T. Jones, and S. Cooper, 2002. Available at http://www.frammandearter.se/0/2english/pdf/Teredo_navalis.pdf.

"Shipworm (*Teredo navalis*)." Wreck Protect, n.d. http://wreckprotect.eu/fileadmin/site_upload/wreck_protect/pdf/shipwormspdfnew.pdf.

■ | New Zealand Mud Snail

Also known as: Jenkin's spire shell
Scientific name: *Potamopyrgus antipodarum*
Synonyms: *Hydrobia jenkinsi, Potamopyrgus jenkinsi*
Family: Hydrobiidae

Native Range. Freshwater streams and lakes and brackish habitats in New Zealand and adjacent islands.

Distribution in the United States. Populations are found in all western states except New Mexico and in four of the Great Lakes: Ontario, Michigan Erie, Michigan and Superior.

Description. This very small aquatic snail typically measures 0.25 in (4–5 mm) in the western United States, although it may be twice that large in its native range. The cone-shaped shell has right-handed coils and contains 5–6 whorls separated by deep indentation. A retractable plate (operculum) covers the oval opening to the shell and protects the animal when it retreats into its shell. Shell color ranges from gray to light or dark brown.

Different morphs appear among introduced populations. Many from the Great Lakes and western states have a slight keel in the middle of each whorl, whereas others from the Great Lakes have spines on the whorls. One morph found in the West is wider than usual and so pale and transparent that internal structures are visible.

Related or Similar Species. The New Zealand mud snail could be confused with a number of native snails in the western United States and may be confidently identified only by experts. The shell is longer and narrower than most native snails of the same family and has more whorls. No native western snails have keels. Most native species have fewer whorls than the New Zealand land snail. (The Aquatic

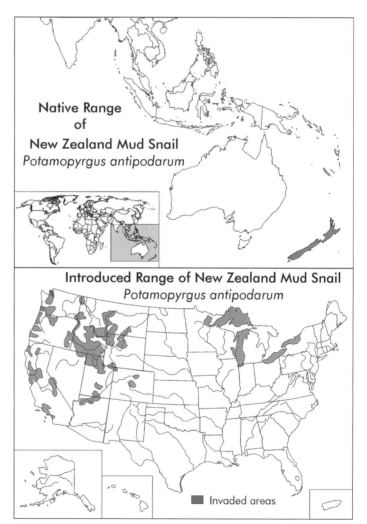

Top: The New Zealand mud snail is native to fresh and brackish water habitats in New Zealand. Bottom: New Zealand mud snails have invaded almost all western states and four of the Great Lakes. (Adapted from Benson and Kipp 2009.)

A. The retractable plate or operculum and deep indentations separating whorls are clearly visible on this common form of the New Zealand mud snail. B. Spines occur on the whorls of many individuals from the Great Lakes region. C. The living animal (D. L. Gustafson, Montana State University.)

Nuisance Species Research Program of the U.S. Army Corps of Engineers presents comparative photos of native snails at http://el.erdc.usace.army.mil/ansrp/potamopyrgus_antipodarum.pdf.)

Introduction History. The exact means of introduction to U.S. waters is unknown, but the snails likely arrived with shipments of fish eggs and live game fish from New Zealand to some western states. Introduction to the Great Lakes, on the other hand, probably occurred with the discharge of ballast water by ships coming from Europe. First records of the snail came from the Snake River, Idaho, in 1987. Populations were reported near the mouth of the Columbia River in Oregon in 1997, and they were established in the lower Columbia in Washington by 2002. First reports from California come from the Owens

River; it is now widespread in that state. In 2002, the snails were in the Colorado River in northern Arizona; in 2004, they were discovered in a small creek near Boulder, Colorado.

An established population existed in Lake Ontario in 1991 and in Lake Erie by 2005. They were also found in Duluth harbor on Lake Superior in 2005.

The tiny New Zealand mud snail may be inadvertently transferred from one body of water to another on boats, boots, waders, clothing, and other recreational gear. They may possibly be spread by animals, because they can pass unharmed through the guts of fish and birds. Even cattle wading in streams and ponds could transport these mollusks, since they are able to withstand desiccation for a short period time by withdrawing into their shells behind a closed operculum.

Within a river or lake system, the snails may float downstream or become attached to floating mats of algae. They have also been shown to be capable of moving upstream under their own volition.

Since most New Zealand mud snails reproduce asexually, it takes only one snail to begin a new population. They spread rapidly and now have access to the eastern United States via the Great Lakes.

Habitat. The New Zealand mud snail is found in a variety of freshwater and brackish-water habitats at depths up to 150 ft. (45 m). It inhabits streams, lakes, estuaries, lagoons, canals, ditches, and water tanks on just about any type of substrate, from mud and sand to concrete and within many types of vegetation. It prefers the littoral zone of lakes and slow-moving streams, but because it can burrow into sediments. It tolerates fast-moving water. These snails thrive in salinities of 0–15 ppt and for short periods of time will tolerate salinities near that of seawater. They also tolerate a wide range of temperatures from 32° to 110°F (0–34°C). They are preadapted to live in degraded water systems with high nutrient content and/or heavy siltation.

Diet. This small gastropod is a scraper/grazer that feeds nocturnally on plant and animal detritus and on living algae and other organisms attached to the substrate (the periphyton). Diatoms are prominent in its diet.

Life History. Most New Zealand mud snails in the United States are members of all female populations, meaning that individuals in a population are clones and genetically identical. Males are not needed for reproduction; indeed, the female is born with 20–120 developing embryos already in her reproductive system. Sexual reproduction does occur, but sexual males and females are very rare. Females generally release live young in summer and autumn. Average lifespan is about one year. Sexual males and females are mature when about 0.1 in. (3 mm) in size.

New Zealand mud snails have a high reproductive potential and it is not unusual for population densities to exceed 5,290/ft.^{2} (100,000/m^2). In the Madison River near Yellowstone National Park in Montana, densities three times greater are reported.

Impacts. The New Zealand mud snail has had negative impacts in other parts of the world and is expected to have them in the United States as well. Concerns center on its potential to decrease populations of native herbivorous invertebrates—including native snails—through competition for food and space and to reduce the periphyton cover that serves as nourishment and substrate for some aquatic organisms. Evidence from Montana suggests a decrease in larval mayflies, stoneflies, caddisflies, and black flies when the snails become abundant. These insects are important food for native trout and other fish, and their loss could negatively impact the important recreational fishing industry in many western states. In the Snake River, Idaho, five native mollusks (Utah valvata snail [*Valvata utahensis*], Idaho springsnail [*Pyrgulopsis idahoensis*], Snake River physa [*Haitia natricina*], Banbury

Springs lanx [*Lanx* sp.], and Bliss Rapids snail [*Taylorconcha serpenticola*]) are listed as threatened or endangered, possibly as a consequence of competition from New Zealand mud snails.

A decrease in the periphyton would alter the physical environment and affect ecosystem processes beginning with primary production. Very dense populations of the mud snails could clog intake filters at facilities withdrawing lake or river water.

Management. Little can be done to eradicate established populations of the New Zealand mud snail. Management is focused on the prevention of its spread to additional bodies of water. Boats and fishing and other gear should be scrubbed before leaving an infested body of water and thoroughly dried for at least 24 hours or treated with heat (>85°F [30°C]) before being used elsewhere.

Selected References

Benson, A. J., and R. M. Kipp. "*Potamopyrgus antipodarum*." Nonindigenous Aquatic Species Fact Sheet, U.S. Geological Survey, 2009. http://nas.er.usgs.gov/queries/FactSheet.asp?speciesID=1008.

Crosier, Dani, and Dan Malloy. "New Zealand Mudsnail (*Potamopyrgus antipodarum*)." United States Federal Aquatic Nuisance Species Task Force, 2005. http://www.anstaskforce.gov/spoc/nzms.php.

Crosier, Danielle, and Daniel P. Molloy. "New Zealand Mudsnail–*Potamopyrgus antipodarum*." Aquatic Nuisance Species Research Program, U.S. Army Corps of Engineers, n.d. http://el.erdc.usace.army.mil/ansrp/potamopyrgus_antipodarum.pdf.

Richards, David C., Billie L. Kerans, and Daniel L. Gustafson. "New Zealand Mudsnail in the Western USA." Montana State University-Bozeman, 2004. http://www.esg.montana.edu/aim/mollusca/nzms/.

■ | Quagga Mussel

Scientific name: *Dreissena rostriformis bugensis*
Family: Dreissenidae

Native Range. Dnieper and Bug rivers and Dnieper-Bug Estuary, Ukraine.

Distribution in the United States. Established in all of the Great Lakes. Also found in inland bodies of water in Arizona, California, Colorado, Illinois, Iowa, Kentucky, Michigan, Minnesota, Missouri, Nevada, New York, Ohio, and Pennsylvania.

Description. The quagga mussel is a small freshwater bivalve with asymmetrical valves and a convex ventral side that prevents it from maintaining a stable upright position. The byssal groove on the ventral side is small and positioned near the hinge, and the angle between the ventral and dorsal surfaces is rounded. There are usually dark concentric bands on the shell that fade near the hinge, but shell patterns are quite variable and the bands may be black, cream, or white. In Lake Erie is a type that is completely white. Two distinct forms are found in the Great Lakes. One type, the so-called "epilimnetic" or shallow water form, has a high flat shell; the other, the "profunda" or deep, cold-water form, has a somewhat elongated, globular shape. The epilimnetic type attaches to hard surfaces with hairlike byssal threads and forms dense colonies or druses; they may attach to the shells of other quagga mussels and create mats 4–12 in. (10–30 cm) thick that encrust or clog pipes and other manmade features. The profunda type, while it can attach to objects with byssal threads, may also partially bury itself in soft sediments and extend a long incurrent siphon above its shell to draw in suspended organic matter. Quagga mussel shells grow to lengths of about 1.5 in. (4 cm).

Related or Similar Species. Quagga mussels are quite similar to the closely related zebra mussel (*Dreissena polymorpha*; see Mollusks, Zebra Mussel). In fact, the two were not immediately identified as separate species. The quagga mussel shell has a convex ventral side in contrast to the flat or convex ventral side of zebra mussels. The quagga mussel shell is rounded, whereas that of the zebra mussel is more triangular. The zebra mussel also is distinct by virtue of bilateral symmetry along a midventral line and a long byssal groove in the middle of the ventral side.

The Asian clam (*Corbicula fluminea*), another small invasive bivalve (see Mollusks, Asian Clam), lacks byssal threads and has a thicker, ridged shell. It is light-yellowish brown to dark brown and usually is not striped.

Introduction History. The first known occurrence of quagga mussels in the Great Lakes dates to September 1989, although at the time the collected specimen was believed to be a form of zebra mussel. The specimen was obtained near Port Colburne, Lake Erie. In 1991, a mussel taken from the Erie Canal was positively identified as a new species and dubbed the quagga mussel after a less-striped relative of the Plains zebra that is now extinct. The mussel was most likely introduced in discharged ballast water from transoceanic freighters. The mussel spread through

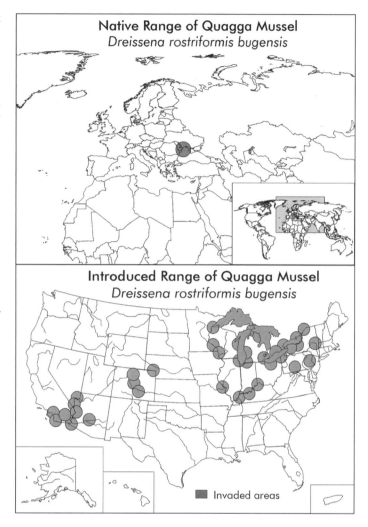

Top: Quagga mussels are native to the Dnieper and Bug rivers and estuary in Ukraine. Bottom: Quagga mussels are established in all of the Great Lakes and in many inland bodies of water across the lower 48 states. (Adapted from Benson, Richerson, and Maynard 2010.)

the lower Great Lakes and reached Duluth Harbor, Lake Superior, by 2005. Quagga mussels were discovered in the Mississippi River between St. Louis, Missouri, and Alton, Illinois, in 1995. Larvae can float downstream in rivers or on currents in lakes. They can leapfrog from lake to lake attached to boats and boat trailers or in anything that holds water. Adults are able to withstand exposure to air for 3–5 days.

In January 2007, quagga mussels showed up in the major impoundments on the Colorado River, first in Lake Mead, near Boulder City, Nevada, and then in Lakes Mohave and Havasu along the lower Colorado River. The same year, they entered the Colorado River Aqueduct, which diverts water from the Colorado River to southern California. Since then, they have been found in 15 reservoirs in southern California and also in six reservoirs in the state of Colorado. They are also in the Nevada State Fish Hatchery on Lake Mead and at Willow Beach National Fish Hatchery, just downstream from Hoover Dam on the

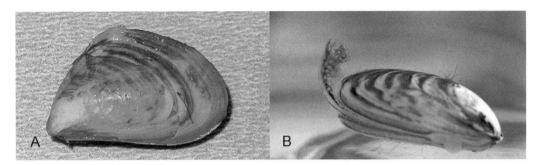

A. Quagga mussel shells are asymmetrical; pattern and color vary. (U.S. Geological Survey.) B. The living animal. (M. Quigley, GLERL.)

Colorado, which raises trout and threatened native fishes to stock Lakes Mead and Mohave and to restock the Colorado River.

Habitat. Quagga mussels are found in the upper reaches of estuaries, in lakes and reservoirs and connecting waterways. They do best in freshwater with salinities less than 1 ppt, but can still reproduce in brackish water with salinities near 3 ppt. (They die in salinities greater than 6 ppt). Quaggas apparently have a broader range of temperature tolerance than zebra mussels and hence inhabit a wider range of water depths. As a consequence they occupy both shallow warm waters and deeper cold-water habitats. In North America they are colonizing waters where the temperature remains between 39° and 48°F (4–9°C) all year. Adults attach to natural hard surfaces of rock, shell, and wood or to large aquatic plants as well as to manmade structures of concrete, metal, nylon, fiberglass, and wood. Some morphs also bury themselves in soft lake bottoms.

Diet. Filter-feeders, quagga mussels ingest food particles by pulling water into the shell cavity with cilia. The water passes into an incurrent siphon where phytoplankters, zooplankters, and other suspended organic matter are extracted. Indigestible particles are cemented together with mucus and ejected from the incurrent siphon as pseudofeces; water is discharged from an excurrent siphon.

Life History. Individuals are either male or female and reproduce externally when eggs and sperm are released into the water column. Microscopic embryos develop into larvae in a few days. The larvae soon sport tiny bivalve shells and drift as part of the plankton for 3–4 weeks. When suitable substrates are found, juveniles settle and attach themselves with byssal threads. Adults are also sessile, but able to detach and move to new sites.

Spawning usually peaks in spring and autumn in the Great Lakes, but elsewhere may occur throughout the year. In Lake Havasu on the Colorado River, quaggas reportedly spawn six times a year. The average lifespan is five years.

Impacts. Major alteration of aquatic ecosystems can result from heavy infestations of quagga mussels. Their removal of much of the phytoplankton and other suspended organic matter from the water column reduces the food supply of zooplankters and thereby affects the rest of the food chain. Their production of large amounts of pseudofeces transfers much of the energy in the system from the upper parts of the water column to benthic habitats. Bottom-feeding fish increase at the expense of plankton-feeders. In addition, when large accumulations of pseudofeces decompose, oxygen is withdrawn from the water. In the central basin of Lake Erie, dead zones of oxygen-depleted water have appeared for decades in late summer at depths greater than 40 ft. (12 m). Even with the general amelioration of severe nutrient pollution problems that affected the lake in the 1970s, these dead zones persist. One hypothesis is that changes in nutrient cycles brought on by the introduction of quagga and zebra mussels may be a cause.

The reduction of planktonic green algae increases water clarity and allows greater growth of macrophytes in shallow waters, changing the species composition of the plant community and the physical structure of the aquatic environment as well as available food sources. Emergent and floating plants become nursery areas for a variety of aquatic organisms.

Indications are that quagga mussels outcompete their close relative the zebra mussel and are replacing them in shallow waters. They also foul native mollusks and other hard-shelled invertebrates, but their presence in American waters is too brief for any impacts to be realized.

Mussels bioaccumulate pollutants in the water. These pollutants collect in pseudofeces and from there can pass up the food chains to game fish and other wildlife. The bivalves themselves are consumed by some crayfish, fish, and diving ducks, offering another pathway for concentrated toxins to move into food chains.

Quagga mussels biofoul organisms and clog the pipes and screens of water intake structures. This can reduce pumping capacity and damage equipment and cost industries and communities millions of dollars to mitigate. Fouling of water-intake screens and sluice gates has rapidly become a major problem along the Colorado River, which supplies drinking and irrigation water via long aqueducts to the major cities and agricultural regions of southern California, southern Nevada, and southern Arizona.

Management. A variety of control methods are available, but so far none has proved both effective and environmentally sound. Biological controls that would target quagga (and zebra) mussels and interrupt their reproductive cycle or interfere with the settling of the larvae are being researched.

Selected References

Benson, A. J., M. M. Richerson, and E. Maynard. "*Dreissena rostriformis bugensis.*" USGS Nonindigenous Aquatic Species Database, Gainesville, FL, 2008. http://nas.er.usgs.gov/queries/FactSheet.asp?speciesID=95.

National Biological Information Infrastructure (NBII) and IUCN/SSC Invasive Species Specialist Group (ISSG). "*Dreissena bugensis* (Mollusc)." ISSG Global Invasive Species Database, 2006. http://www.issg.org/database/species/ecology.asp?si=918&fr=1&sts=sss.

Richerson, Myriah. "*Dreissena* species FAQs, a closer look." USGS, 2009. http://fl.biology.usgs.gov/Nonindigenous_Species/Zebra_mussel_FAQs/Dreissena_FAQs/dreissena_faqs.html.

■ | Veined Rapa Whelk

Also known as: Asian rapa whelk
Scientific name: *Rapana venosa*
Synonym: *Rapana thomasiana*
Family: Muricidae

Native Range. Estuarine and marine waters of the western Pacific Ocean. It is known from Vladivostok, Russia, south to Taiwan in the Sea of Japan, the Yellow Sea, the Bohai Sea, and the East China Sea. Genetic studies of whelks from the Chesapeake Bay suggest they arrived in American waters via the Black Sea, to which they were introduced in the 1940s.

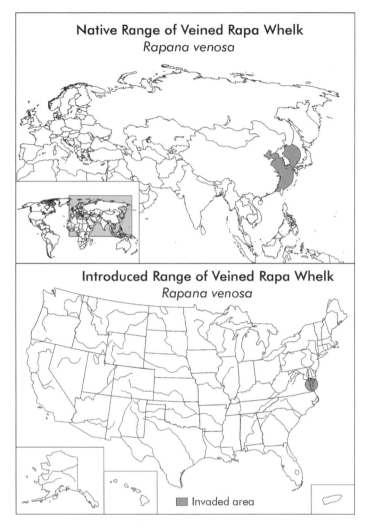

Top: The veined rapa whelk is native to estuarine and marine waters from southeastern Russia to Taiwan. Bottom: Veined rapa whelks have been found in the Chesapeake Bay. (Adapted from International Council for the Exploration of the Sea 2004.)

Distribution in the United States. Established in Chesapeake Bay from the mouth of the Rappahannock River south to the bay's entrance. Based on water temperatures similar to that of their native distribution area, and the existence of a high volume of coastwise trade out of Hampton Roads, their potential range extends from Cape Cod, Massachusetts, to Charleston, South Carolina.

Description. This large marine snail is most readily identified by the bright orange color of the inside of its shell, visible at the large, oval aperture that opens on the right side of the shell. The heavy shell is globular in shape and has a short spire, broad flat columella, and deep umbilicus (the hollow on the ventral side). Smooth spiral ribs terminate in blunt knobs adorning the shoulder, and the edge of the outer lip is finely toothed. The outer shell ranges from gray to brown and bears dark brown lines that create an interrupted pattern over the entire shell. Dark veining may occur on the inside of the shell. In their native range, they reach a length of 7 in. (180 mm). Adults slightly over 6.5 in. (170 mm) have been collected in the Chesapeake Bay.

The egg cases of the veined rapa whelk are also distinctive, usually described as resembling a yellow shag rug. They are about 1.2 in. (30 mm) high and individually attached to the seafloor to form a mat.

Related or Similar Species. Two large native whelks inhabit the potential range of the veined rapa whelk along the East Coast of the United States. Neither the knobbed whelk (*Busycon carica*) nor the channeled whelk (*Busycon canliculatum*) has a broad flat columella or the dark veining on the shell. Both native whelks are more elongated, have thinner shells, and display less ribbing on the outside of their shells than does the veined rapa whelk. Egg cases of the knobbed whelk are strung together in a long chain, a "mermaid's necklace."

Introduction History. The first scientific evidence of veined rapa whelks in Chesapeake Bay came in 1998, when researchers from the Virginia Institute of Marine Science collected a specimen in Hampton Roads, Virginia. The large whelk was probably 10 years old, strongly suggesting the presence of whelks in the bay since 1988, a fact confirmed anecdotally by local watermen. The species likely arrived in ballast waters in empty coal vessels returning from the eastern Mediterranean, since Norfolk, Virginia, is a major coal-exporting port. Genetic studies support this contention. The long larval period when the whelk floats in the water column facilitates the long distance dispersal of this gastropod by currents in the bay and by container ships and other vessels that take up ballast water before heading to ports all along the East Coast.

Adult veined rapa whelks found in the Chesapeake Bay have reached just over 6.5 in. long. Larger specimens occur in their native range. (George Chernilevsky.)

Habitat. Subtidal zones of coasts and estuaries. Adults prefer sandy bottoms into which they burrow, but juveniles tend to dwell on hard substrates, including oyster reefs, and only migrate to areas with sandy bottoms when shell length is about 2.75 in. (70 mm). In its native range, it tolerates a wide annual range of temperatures, from 39–80°F (4–27°C); in cooler climates, it may migrate to deeper and warmer water during winter. Adults and larvae can withstand low salinities and polluted and/or oxygen-poor water.

Diet. Larvae consume phytoplankters, but juveniles and adults are carnivores. Prey includes other shellfish that live in soft sediments (the infauna), such as clams and mussels, and oysters. Most vulnerable in the Chesapeake Bay are native hard clams (*Mercenaria mercenaria*), but mussels (*Mytilus edulis*), soft-shell clams (*Mya arenaria*), and oysters (*Crassostrea virginica*) are also taken. The whelk engulfs its prey whole; and, when the bivalve opens, it sucks out the soft body of the victim, leaving behind a clean shell with few telltale scars.

Life History. In its native range, mating occurs during winter and spring, and eggs are laid from April to late July. Each egg case is attached by its base to a hard substrate. From 200 to 1,000 eggs are contained in each case; each mat may consist of 50- to 500-egg cases. A female can lay several mats a year from a single mating. Initially, the egg cases are white, but as the larvae become visible swimming inside, they turn yellow. After an incubation period of 14–17 days (depending upon temperature), the egg cases blacken, open, and the larvae swim free out of the top of the case to become part of the plankton. The larval stage lasts for a variable length of time, extending as long as 80 days. Upon metamorphosis, they settle onto attached invertebrates such as bryozoans and barnacles and become tiny, hard-shelled, cryptically colored whelks. The young whelks grow quickly so that within three weeks, their shell's length approaches 0.02 in. (0.5 mm); within a year, they are more than 2.4 in. (60 mm) long. Veined rapa whelks begin to reproduce at two years of age, when they have a shell length of about 3 in. (74 mm). The largest whelks collected from the Chesapeake Bay are estimated to be about 10 years old.

Impacts. The veined rapa whelk has a history of negative impacts in European waters, especially in the Black Sea, where its predation was implicated in the rapid decline of native edible bivalves and near extinction of oysters (*Ostraea edulis*) on the Guadata oyster bank. Concern is therefore high that similar impacts could ensue at commercially important fisheries in the Chesapeake Bay and elsewhere along the East Coast of the United States. In the Chesapeake Bay, it appears that hard clams are most threatened, since oysters are confined by disease to very low-salinity sites.

These nonnative whelks might compete with the native oyster drill (*Urosalpinx cinerea*), a species struggling to recover from a massive die-off caused by an influx of freshwater related to Hurricane Agnes in 1972. A shift in relative abundance of hermit crab species is also possible. The shells of the veined rapa whelk are well suited to house the striped hermit crab (*Clibanarius vittatus*), but less so to the currently dominant flat-clawed hermit crab (*Pagurus pollicaris*). The former is predator of oysters, and its increase would add to pressures already affecting the bay's remaining oyster populations.

Management. It appears unlikely that established populations can be eradicated, so the emphasis of research is on identifying sites vulnerable to invasion and trying to prevent the organism's spread. Several states on both coasts of North America have issued alerts for their citizens to be on the lookout for veined rapa whelks and report any sightings to authorities. For a time, researchers on Chesapeake Bay offered a bounty on whelks caught by fishermen, in part to monitor the status and spread of the whelk population and in part to discourage fishermen from throwing the by-catch back into the water.

Selected References

Chesapeake Bay Introduced Species Database. "*Rapana venosa*." NEMESIS (National Exotic Marine and Estuarine Species Information System). Smithsonian Environmental Research Center, 2009. http://invasions.si.edu/nemesis/CH-TAX.jsp?Species_name=Rapana%20venosa.

International Council for the Exploration of the Sea. "Alien Species Alert: *Rapana venosa* (Veined Whelk)," edited by Roger Mann, Anna Occhipinti, and Juliana M. Harding. ICES Cooperative Research Report No. 264, p. 14, 2004. http://www.ices.dk/pubs/crr/crr264/crr264.pdf.

Richerson, Myriah. "*Rapana venosa*." USGS Nonindigenous Aquatic Species Database, Gainesville, FL, 2006. http://nas.er.usgs.gov/queries/FactSheet.asp?speciesID=1018.

Southeast Ecological Center. "Veined Rapa Whelk, Asian Rapa Whelk, *Rapana venosa* (Valenciennes, 1846) Mollusca: Gastropoda, Muricidae." Nonindigenous Species Information Bulletin, U.S. Geological Survey, 2009. http://fl.biology.usgs.gov/Nonindigenous_Species/Rapa_whelk/rapa_whelk.html.

■|Zebra Mussel

Scientific name: *Dreissena polymorpha*
Family: Dreissenidae

Native Range. Eastern Europe and western Asia in the drainage systems of the Black and Caspian seas and the Sea of Azov. U.S. populations originate from mussels transported from the southern limit of their range.

Distribution in the United States. Zebra mussels are now found in waterways in or bordering at least 30 states in the continental United States. They are established in all of the Great Lakes and in most navigable drainages in the Mississippi River system, including the Missouri River in Nebraska and South Dakota. They are also established in the Hudson

River and found in numerous inland lakes in the Midwest. The most recent introductions have occurred in San Justo Reservoir in central California, two reservoirs in Colorado, Lake Texoma on the Texas-Oklahoma border, and a lake in western Massachusetts.

Description. Zebra mussels are small freshwater bivalves with triangular shells. The bottom of the shell where the hinge lies is flat or concave, and the shell margin is sharply angled. When placed upright on a flat surface, they are stable. Many bear dark stripes on the shell, but this pattern can be quite variable and even absent. Some will be plain and cream-colored; others black. Filaments or byssal threads extend from the bottom of the shell and bind them to each other or to hard surfaces. Usually zebra mussels occur in dense clusters or colonies known as druses. Shells of adults generally range in length from 0.2–1.8 in. (6–45 mm), with a maximum length of about 1.9 in. (0.5 cm).

Related or Similar Species. Two other dreissenids occur in North America, the native false dark mussel (*Mytilopsis leauco-*

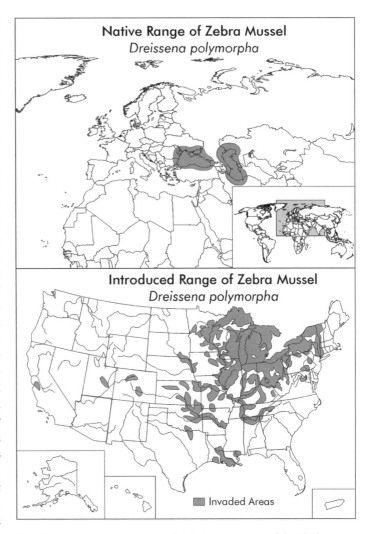

Top: Zebra mussels are native to the drainage systems of the Black, Azov, and Caspian seas. Bottom: Zebra mussels are a scourge in at least 30 states and are established in all the Great Lakes and much of the Mississippi River system. (Adapted from Benson and Raikow 2009.)

phaeata) and the introduced quagga mussel (*Dreissena rostriformis bugensis*; see Mollusks, Quagga Mussel). Each has a convex bottom edge and a rounded shell margin, which prevents it from balancing in an upright position.

Introduction History. The introduction of zebra mussels probably stems from a single exchange of ballast water by a commercial ship entering the Great Lakes from the northern Black Sea. The first reports came from Lake St. Clair, which connects Lake Erie and Lake Huron, in 1988. It is likely they had been in the lake for 2–3 years before having been detected. By 1990, the passive drifting of larval mussels and the ability of juveniles and adults to attach to barges and other vessels had allowed them to disperse into all the Great Lakes. In 1991, they were in the Illinois River, gateway to the Mississippi River, and within a year they were established in the Arkansas, Cumberland, Ohio, and Tennessee rivers. They were also in the Hudson River and the Finger Lakes of New York by 1991. Overland

A. Zebra mussels overgrow the shells of native mollusks. (Eric Engbretson.) B. A living zebra mussel is in a stable position when placed upright on its hinge. (M. Quigley, GLERL.) C. The dense clusters in which zebra mussels usually congregate are called druses. (Eric Engbretson.) D. Some of the variation found in shell patterns of *Dreissena polymorpha*. (U.S. Geological Survey.)

dispersal has undoubtedly been facilitated by recreational boaters trailering their watercraft from infested bodies of water to pristine waters without decontaminating hulls, trailers, live-wells, and engines. Zebra mussels were first recorded in Virginia in a quarry pond in 2002. Other states with first records in the twenty-first century include Kansas (Perry Lake and the Kansas River) in 2007; California (San Justo Reservoir, San Benito County) and Colorado (Pueblo Reservoir—an impoundment on the upper Arkansas River near Pueblo and four reservoirs west of Denver) in 2008; and Texas and Oklahoma (Lake Texoma), and Massachusetts (Laurel Lake, Berkshire County) in 2009.

Habitat. Zebra mussels can be found in freshwater lakes, ponds, and rivers. Usually they occur in shallow, algae-rich water at depths of 6–30 ft. (2–9 m). They require hard sub-strates for settling and attachment. They thus may occur on rocks; in and on metal, concrete, or other manmade structures; on the shells of other mollusks and the exoskeletons of crustaceans; and even on the cases of caddisfly larvae. In soft sediments, they may attach to pieces of shell or wood and other plant material or to stones and still be able to establish a druse. Adults do not survive freezing temperatures and die when water temperatures are above 75°F (24°C) for extended periods of time. They also do poorly in oxygen-poor waters. Optimal calcium-ion concentrations for zebra mussels are 45–55 mg/l, and optimal pH for adult growth lies between 7.4 and 8.0.

Diet. Larval, juvenile, and adult zebra mussels are filter-feeders, capturing primarily algae and zooplankters in the water column. They also feed on bacteria and detritus.

Life History. Female mussels release eggs into the water column, where they are fertilized by sperm released by males. Most spawning occurs between June and September, when water temperatures are at an optimal 54–68°F (12–20°C). Three distinct periods characterize the life cycle of the zebra mussel. The microscopic larvae are part of the plankton and undergo four stages when they are essentially clam-like in shape. In the final larval stage, a larva settles on a substrate and crawls to a suitable attachment site. There it undergoes metamorphosis and begins its juvenile period with a small (1–3 mm), triangular shell. Gametes become fully developed juveniles in 8–180 days; the colder the water, the slower the development time. The adult period starts when the mussel becomes sexually mature, usually at one year of age and a shell length of 0.3–0.35 in. (8–9 mm). A zebra mussel lives on average 2–3 years after attachment.

Impacts. Dense populations of zebra mussels have the potential to alter freshwater ecosystems and change the physical environment by eliminating or reducing the numbers of native organisms. The great filtering capacity of these small mussels reduces the phytoplankton and increases water clarity. This allows large, rooted aquatic plants to increase and changes not only the food supply, but the physical structure of the system. In some instances, the sessile juveniles and adults even outcompete zebra mussel larvae, which rely upon phytoplankters for nourishment. A diminished supply of phytoplankters such as diatoms leads to a reduction in the zooplankters upon which many fishes depend. Larval stages of fish such as the bluegill (*Lepomis macrochirus*) in inland lakes may be strongly affected. At the same time, the prodigious amount of feces and pseudofeces produced in large druses increases the food supply for macroinvertebrate bottom feeders and thus for fish that feed on these animals. Bacterial productivity also increases to break down the mussels' wastes, which may increase food available to the mussels themselves. Zebra mussels are eaten by native mollusk-eating fish such as freshwater drum (*Aplodinotus grunniens*), lake sturgeon, yellow perch, and catfish, which have been shown to increase in heavily infested waters. Thus, a change in the composition of the fish community occurs as benthic-feeding fish replace planktivorous species.

Zebra Mussels

The invasion of zebra mussels in the 1980s sparked new interest in the problems associated with exotic species. The new science of invasion biology developed within the discipline of ecology, the Nonindigenous Aquatic Nuisance Prevention and Control Act was passed, and public perception of the threats posed by introduced species to both aquatic and terrestrial ecosystems grew. Invasions became a common topic in the media and increasingly a subject of scientific research. For some, the zebra mussel is the "poster child" of biological invasions.

The attachment of zebra mussels to the shells of native clams (unionids) makes it more difficult for the clams to move through sediments in search of food and optimal concentrations of oxygen. In contrast, encrusted bivalves in wave-impacted areas are easily dislodged and washed away from prime sites. Furthermore, mussels can obstruct the openings of the clams' shells and prevent food intake and the release of gametes or prevent defensive closing of the shell. During the very early invasion of Lake Erie, reports came of thousands of zebra mussels on a single unionid. Native clam populations in the western basin of Lake Erie and in Lake St. Clair have plummeted.

The zebra mussel is the most serious biofouling organism to have been introduced into American waters. It clogs water intake and distribution pipes in industrial and power plants that withdraw water from lakes and rivers, increases the corrosion rates of iron and steel pipes and rivets, and obstructs pumps, valves, weep holes, screens, and the like, damaging equipment and entire facilities. After storms on the Great Lakes huge windrows of mussel shells accumulate on beaches; the decaying mollusks release methane and produce a noxious stench. The razor-sharp shells are hazards to beachgoers.

Management. Control of zebra mussels is problematic. Chemical molluscicides; manual removal; ultraviolet, hot water, and CO_2 treatments; filters; screens; and electrical currents all have been tried. Preventing their spread into new waterways is the target of most management schemes. This can be best achieved through the decontamination of diving and fishing gear and of boats and boat trailers, and stopping the dumping of unused bait in lakes and rivers.

Selected References

Benson, A. J., and D. Raikow. "*Dreissena polymorpha.*" USGS Nonindigenous Aquatic Species Database, Gainesville, FL, 2009. http://nas.er.usgs.gov/queries/FactSheet.asp?speciesID=5.

Nichols, S. Gerrine. "Zebra Mussel Identification." Zebra Mussel Research Program. Zebra Mussel Information System, Environmental Laboratory, Engineer Research and Development Center, United States Army Corps of Engineers, 2002. http://el.erdc.usace.army.mil/zebra/zmis/zmishelp4/zebra_mussel_identification.htm.

■ Crustaceans

■| Chinese Mitten Crab

Scientific name: *Eriocheir sinensis*
Family: Varunidae (formerly Grapsidae)

Native Range. Subtropical East Asia, from Fukien Province, China, north to the Korean Peninsula.

Distribution in the United States. The Chinese mitten crab is established only in California, where it occurs in the Sacramento-San Joaquin Delta and tributary streams. Specimens have been collected in Delaware, Louisiana, Maryland, New York, Ohio, and Washington. Until 2007, all those collected outside of California were males; but recently females with eggs have appeared in the Mid-Atlantic states, so breeding populations may now exist there.

Description. The Chinese mitten crab is most easily identified by the character that gives it its name: light brown, hairy claws of equal size that have white tips and make it look like the animal is wearing mittens. The brown hairs (setae) are evident in juveniles when the carapace is about an inch (25 mm) wide. The carapace is round in outline and convex, but bumpy on the dorsal surface. There is a distinct notch between the eyes. On the edges of the front part of the carapace are four spines, the fourth of which is quite small. Females have a wide abdominal flap shaped like a beehive; males have a narrow, bell-shaped abdominal flap. These differences are apparent in animals larger than 0.4 in. (10 mm). The legs are more than twice as long as the carapace is wide. The crab is brownish orange to greenish brown in color. Carapaces of adults are 1.2–4.0 in. (30–100 mm) wide. In California, adults reach about 3 in. (80 mm).

Juvenile Chinese mitten crabs will dig burrows into vertical, clay river banks within the intertidal zone. The burrows have oval openings and may be 1–3 in. (2.5–7.6 cm) wide and as much as 8 in. (20 cm) deep. They slope downward and may have two entrances.

Related or Similar Species. The Harris mud crab (*Rhithropanopeus harrisii*), similar in size and general appearance and native to the Atlantic coast of North America, could be mistaken for a young Chinese mitten crab. However, it lacks the hairy claws and the notch between the eyes and is most commonly found in estuaries. The Harris mud crab has ridges on the back of the carapace, which is brown or black. Maximum carapace size is 0.75 in. (19 mm). The entrances to its burrows are circular and about 1.5 in. (38 mm) wide.

Occasionally, the very similar Japanese mitten crab (*Eriocheir japonicus*) is reported in the United States, but this species is not widely established outside its native range. (Some scientists think it is the same species as the Chinese mitten crab.)

The Chinese mitten crab is the only crab found in freshwater in the United States. Its carapace shape differs from all other true crabs that occur here.

Native Range of Chinese Mitten Crab
Eriocheir sinensis

Introduced Range of Chinese Mitten Crab
Eriocheir sinensis

■ Invaded areas

Top: The Chinese mitten crab is native to subtropical regions of east Asia. Bottom: The Chinese mitten crab is invasive only in California, although individuals have been collected elsewhere. Females with eggs have been reported from Mid-Atlantic states, suggesting populations may be becoming established there. (Adapted from Benson and Fuller 2007.)

Chinese mitten crabs get their name from their hairy claws. (© N. Sloth/ Biopix.)

Introduction History. A single Chinese mitten crab was taken by a shrimp trawler in South San Francisco Bay in 1992. They were captured in San Pablo Bay in 1994, Suisun Marsh in 1996, and the Delta in the fall of 1996. Several dozen were trapped upstream at the U.S. Bureau of Reclamation's Tracy Fish Collection Facility at the Delta Mendota canal in 1996. Migrating crabs in the fall of 1997 numbered in the tens of thousands; and the following year, the population was estimated to be 775,000 and threatened to shut down fish salvage operations at the canal pump station. By 2000, Chinese mitten crabs were reported 30 mi. (50 km) upstream from the Bay-Delta, with no halt to their expansion expected.

It is likely that this popular Asian delicacy was deliberately introduced as live crabs for human consumption or to develop a local food resource, although the release of untreated ballast water could also disperse these animals. Specimens collected in other parts of the United States suggest that the illegal import and release of live crabs continues, although so far, populations have not become established outside of California. The crab was found in Hawai'i in the 1950s; and Chinese mitten crabs have been reported from the mouth of the Columbia River, Washington, from Yaquina Bay in central Oregon, and from the Great Lakes region since the 1960s. More recently they have shown up in Chesapeake Bay (2005–2007), Delaware Bay (2007), the Hudson River in New York, (2007–2009), and the Shrewsbury River in New Jersey (2008–2009) on the East Coast of the United States. Only in New York have both male and female adult and juvenile crabs been collected. Live crabs have not been detected in Chesapeake and Delaware bays since 2007.

Habitat. The Chinese mitten crab is catadromous and spends most of its life cycle in freshwater. Juvenile and adult crabs are also able to walk on dry land. Adults live in the bottoms and banks of freshwater streams, but they reproduce in the brackish water of estuaries. Late larval stages float in the upper part of the water column and are transported by currents toward the mouths of estuaries. Settling and metamorphosis into juveniles occurs on the seafloor along coasts and in embayments. A temperate-zone species, it is adapted to changing water temperatures and salinities and thus tolerates pollution. Optimal temperatures are 75–82°F (24–28°C) for juveniles, 59–64°F (15–18°C) for larvae. Larvae in early stages of development tolerate a wide range of salinities.

Diet. Usually described as opportunistic omnivores, Chinese mitten crabs consume algae, macrophytes, detritus originating on land, and invertebrates. They scavenge dead fish and are notorious for stealing fishermen's bait.

Life History. In San Francisco Bay, Chinese mitten crabs mate in the late fall and winter when water salinity is greater than 20 ppt. Some 24 hours later, a female spawns eggs, which are affixed underneath her abdominal flap. A female will carry 250,000 to one million eggs through the winter until they hatch in the spring or summer. Hatchlings become part of the plankton for 1–2 months. They settle in salt or brackish water in late spring or early summer, and the juveniles migrate to freshwater, perhaps aided by tidal currents. When living in tidal freshwater rivers, the juveniles congregate in dense colonies and burrow into the bank for protection against predators and desiccation at low tide. In nontidal areas, they apparently do not burrow. Older juveniles occur farther upstream than younger ones. While migrating upstream, crabs will leave the water and walk across banks and levees to bypass dams and other obstacles. When 1–4 years old, the males and females migrate downstream in late summer or fall to brackish water, where they become sexually mature. The males arrive first; mating begins as soon as the females arrive. Adults die soon after mating.

Impacts. In the San Francisco estuary, Chinese mitten crabs may compete for food and shelter with such commercially important species as the red swamp crayfish (*Procambarus clarkia*) and the signal crayfish (*Pacifasticus leniusculus*). However, it is primarily a nuisance species for shrimp trawlers and both commercial and recreational fishermen, stealing bait, tearing nets, getting entangled in gear, and eating the catch. During downstream migrations, the large numbers of crabs involved will clog water intakes, reducing water flow and potentially causing power plant systems to overheat. The burrowing of juveniles can accelerate river bank and levee erosion and collapse. Chinese mitten crabs are intermediate hosts of a mammalian lung fluke (*Paragonimus* spp.) that could affect humans. However, neither the parasite nor its primary hosts, Asian freshwater snails, have been found in the United States.

In China, the mitten crab is an agricultural pest in rice paddies, eating young shoots and damaging levees. There is concern that should the crab become established in the rice-growing areas of the Gulf coast, it could have similar negative impacts.

Management. In Germany, where these crabs have become a major problem, some success in controlling them has been achieved with traps on the upstream side of dams capturing juveniles during their upstream migration. Similarly, the use of traveling screens and trash racks at water intakes can capture large numbers of migrating crabs. Periodic back-flushing can remove crabs from water intakes.

Selected References

Benson, A. J., and P. L. Fuller. "*Eriocheir sinensis.*" USGS Nonindigenous Aquatic Species Database, Gainesville, FL, 2010. Revised July 20, 2007. http://nas.er.usgs.gov/queries/FactSheet.aspx?speciesID=182.

"Chinese Mitten Crab." ProjectUFO, 2009. http://www.projectufo.ca/drupal/Chinese_Mitten_Crab.

Chinese Mitten Crab Survey Program. "Chinese Mitten Crab Update, U.S. Atlantic Coast Bays and Rivers." Smithsonian Environmental Research Center, 2009. http://www.serc.si.edu/labs/marine_invasions/news/CHINESE_MITTEN_CRAB_UPDATE_APR21_09.pdf.

Chinese Mitten Crab Working Group. "National Management Plan for the Genus *Eriocheir* (Mitten Crabs)." Aquatic Nuisance Species Task Force, 2003. http://www.anstaskforce.gov/Species%20plans/national%20mgmt%20plan%20for%20mitten%20crab.pdf.

"Life History and Background Information on the Chinese Mitten Crab." California Department of Fish and Game, 1998. http://www.dfg.ca.gov/delta/mittencrab/life_hist.asp.

■|Green Crab

Also known as: European green crab, European shore crab
Scientific name: *Carcinus maenas*
Family: Portunidae

Native Range. The green crab is native to the western Baltic Sea and the Atlantic coasts of northwest Europe and North Africa from Iceland and Norway to Great Britain and south along Spain and Portugal to Morocco and northernmost Mauritania.

Distribution in the United States. On the Pacific coast, the green crab is established in most bays and estuaries from Monterey Bay, California, north to Grays Harbor, Washington. Populations, however, have remained small. Along the Atlantic seaboard, green crabs are established along the coast of New England. They are found south along the Atlantic coasts of Maryland and Virginia, but have not been collected in Chesapeake Bay. A single green crab was collected in Hawai'i in 1973.

Description. The green crab has a somewhat hexagonal carapace. Five "teeth" or blunt spines edge the carapace behind each eye. Three bumps or rounded teeth lie between the eyes. The carapace has a granular texture and is usually a mottled dark green or brown with white or yellowish spots. The carapace is broader than it is long; its width is 2.4–3.9 in. (6–10 cm). The ventral surface varies in color from green to yellow, orange, or red, depending on molt status. The second and third pairs of walking legs are the longest and are almost twice as long as carapace length. The fourth pair is the shortest; they are relatively flat and bear hairs (setae).

Related or Similar Species. On the West Coast, some native crabs are green and could be mistaken for green crabs. One

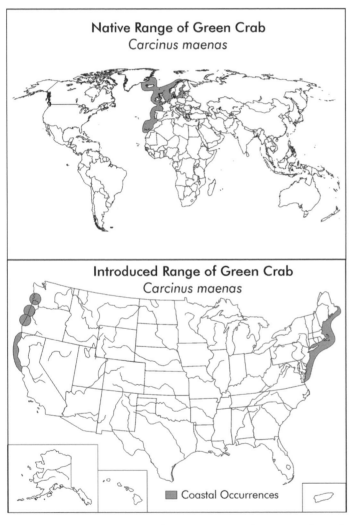

Native Range of Green Crab
Carcinus maenas

Introduced Range of Green Crab
Carcinus maenas

■ Coastal Occurrences

Top: Green crabs are native to the Atlantic coast of Europe and North Africa. (Adapted from map by National Introduced Marine Pest System [NIMPIS], Australia.) Bottom: The green crab is established in bays and estuaries from Monterey Bay, California, northward along the west coast and from Maine to the mouth of the Chesapeake Bay along the east coast. (Adapted from Perry 2010.)

such native is the helmet crab (*Telmessus cheiragonus*), which is distinguished by six spines behind each eye, a pair of long antennae and a body covered in stiff hairs. The green crab is the only shore crab on the Pacific coast with five spines at the edge of the shell behind each eye.

Introduction History. Green crabs arrived in the United States in the early 1800s at or near Cape Cod, Massachusetts. The first known report dates to 1817. They began to extend their range northward in the early twentieth century, entering Maine's waters in the 1950s, and reaching Nova Scotia, Canada, in the 1960s. They were first collected in New Jersey in 1929 and later in Maryland and in Chincoteague Bay, Virginia. East Coast introductions could have come in ballast water or by crabs living amongst fouling organisms on the hulls of transoceanic vessels or in the burrows of shipworms living in wooden sailing ships.

The green crab occupies a wide variety of habitats on sheltered coasts. (P. Erickson. Reprinted with permission of the MIT Sea Grant College Program.)

The green crab was first reported on the Pacific coast in 1989 in Estero Americano, Sonoma and Marin counties, California. About the same time, it appeared in San Francisco Bay. Genetic analyses suggest they came from the east coast of North America and did not represent a direct introduction from Europe or North Africa. It is probable that they arrived in algae used to pack New England baitworms (*Nereis virens* and *Glycera dibranchiata*). Green crabs were first collected in Bodega Harbor in 1993 and Humboldt Bay, 1995. They appeared in Oregon, first in Coos Bay (1997), then in Tillamook, Netarts, and Yaquina bays in 1998. Also in 1998, they were first collected in Grays Harbor and Willapa Bay, Washington. It is believed that strong El Niño–derived currents transported green crab larvae northward in 1997–1998. In many of the invaded areas, populations have remained small or have died out. The highest densities occur in Bodega Harbor, but these are slight compared to East Coast infestations.

Habitat. Green crabs can be found on protected and semi-protected coasts under both marine and estuarine conditions. They will occupy both rocky and soft sediment shores, seagrass meadows, and tidal marshes. In the United States, they are important members of rocky intertidal communities on the East Coast, but on the West Coast, they do not occupy exposed rocky shores. Juveniles and adults occur in intertidal and subtidal zones to depths of 180 ft. (55 m). Adult green crabs tolerate wide ranges of salinity (4–52 ppt) and temperature (32–86°F or 0–30°C), but larvae require salinities of 26–29 ppt and temperatures of 52–77°F (11–25°C) for successful development.

Diet. Green crabs are primarily predators that consume bivalves, including mussels, clams, oysters, and scallops; snails; other crabs; barnacles; and isopods. However, they also eat algae. They may dig nearly 6 in. (15 cm) into sandy or muddy bottoms in search of prey.

Life History. Mating occurs after the females molt. The time of the molt varies with geographic location, but most commonly is between June and October. A male carries a smaller female underneath his body until she molts. At that time, she turns over and releases a mass of eggs that she will keep on her abdomen. The male then releases sperm and fertilizes them. The eggs are held by the female through the winter until spring or early summer. When the larvae hatch, they become part of the plankton and drift in the water column for several weeks or months. It appears that tidal currents transport the larvae away from shore and into the open sea. During the planktonic stage, larvae molt and pass through several stages of the life cycle. During the final larval stage, tides bring them back to shore, and they metamorphose and settle to the bottom as juvenile crabs. The juveniles grow to a carapace width of about 0.25 in. (6 mm) before their first winter and 0.5–1.0 in. (13–25 mm) before their second winter. They are sexually mature at 2–3 years of age and may live for 5 years.

Impacts. The green crab is believed largely responsible for the decline of New England's soft shell clam (*Mya arenaria*) fishery in the 1950s. It is also implicated in declines of the northern quahog (*Mercenaria mercenaria*) and a scallop (*Argopecten irradians*).

The cultivated Manila clam (*Venerupis philippinarum*) harvest in Tomales and Humboldt bays, California, declined by 40 percent after the establishment of green crabs. In Bodega Harbor, decreases in the density of small clams (*Nutricola* spp.) and shore crabs (*Menigrapsus oregonensis*) are attributed to the introduction of the green crab. It is feared that the commercially important Dungeness crab (*Cancer magister*) could be similarly affected if the green crab were to expand its range and numbers along the west coasts of the United States and Canada, since it could prey on juveniles.

Management. Fencing, trapping, and poisoning were generally ineffective where tried on the East Coast. Washington is attempting to stem the influx of green crabs legislatively by declaring it a deleterious exotic species and prohibiting the possession and transportation of the animal in the state. Aquaculture in Washington is under restrictions that limit the transfer of shells, living animals, and equipment from infested Willaba Bay and Grays Harbor to waters the crab has not yet invaded. Imports of shellfish seed from out-of-state waters where green crabs occur are also restricted.

Selected References

"*Carcinus maenas* (European Green Crab)." Invasive Species Fact Sheet. Washington Department of Fish and Wildlife, 2005. http://wdfw.wa.gov/fish/ans/identify/html/index.php?species=carcinus _maenas.

Cohen, Andrew N. "*Carcinus maenas* (Linnaeus, 1758)." *Guide to the Exotic Species of San Francisco Bay.* San Francisco Estuary Institute, Oakland, CA, 2005. http://www.exoticsguide.org/species_pages/ c_maenas.html.

Perry, Harriet. "*Carcinus maenas*." USGS Nonindigenous Aquatic Species Database, Gainesville, FL, 2010. Revised April 25, 2008. http://nas.er.usgs.gov/queries/FactSheet.aspx?SpeciesID=190.

■| Rusty Crayfish

Also known as: Crawdad, crawfish
Scientific name: *Orconectes rusticus*
Family: Cambaridae

Native Range. Rusty crayfish are native to the Ohio, Tennessee, and Cumberland drainage systems in the states of Kentucky, Indiana, Illinois, Ohio, and Tennessee.

Distribution in the United States. Outside their native range, rusty crayfish are native transplants in the Great Lakes region, Iowa, Maryland, Missouri, New Jersey, New York, Pennsylvania, and all of the New England states except Rhode Island.

Description. The rusty crayfish has relatively large claws that are smooth and have black tips. When closed, the pincers leave an oval gap between them. Color varies from gray-green to reddish brown. The most distinctive feature is a pair of dark-red or rusty spots on either side of the carapace just in front of the abdomen ("tail"); however, these spots are not always present. Maximum length of the body exclusive of the claws is about 4 in. (10 cm). Males are larger than females.

Related or Similar Species. Some 65 species of the genus *Orconectes* are native to the United States. The rusty crayfish has larger, smoother claws than most others. Among those most apt to be found where rusty crayfish have been introduced, the northern clearwater crawfish (*O. propinquus*) is most similar. However, it lacks the two spots and instead has a dark patch on its tail. Another close relative, the virile crayfish (*O. virilis*), an invasive in part of its range, can be distinguished by the white bumps on the claws and a very narrow gap when the claws are closed. The calico or papershell crayfish (*O. immunis*) shows a distinct notch in the closed claws. Neither the virile nor the calico crayfish has black bands at the tips of their claws.

Introduction History. The expansion of the rusty crayfish out of its native range is not well documented, but it is assumed that nonresident fishermen, familiar with them in their homes states within the Ohio River basin, brought them to new areas as bait and released them intentionally or accidentally. Rusty crayfish are also popular in school laboratories and sold by biological supply companies. Releases from classrooms are another way rusty crayfish may have entered nonnative waters. Rusty crayfish are commercially harvested for human food, and this provides an economic incentive to introduce them to new locations. Since female crayfish store sperm, a single animal could start a new population.

Their introduction to Wisconsin lakes and streams occurred sometime in the 1960s. The first report from southern Minnesota came in 1967. In both of these states, the range has increased dramatically in ensuing years. The most recent introduction was in Maryland, where they were found in the Monocacy River and Susquehanna drainage in 2007.

Habitat. These freshwater crustaceans require permanent bodies of water and adequate cover such as rocks, tree limbs, or logs. They inhabit lakes and streams, where they can be found in both still pools and fast-flowing stretches. A variety of bottom types from silts and sands to gravel or rock are used. They do not dig deep burrows as do some of their relatives and so cannot escape when intermittent streams dry up.

Diet. Rusty crayfish have unusually high metabolic rates and are therefore voracious and opportunistic feeders. They consume aquatic plants; bottom-dwelling invertebrates such as aquatic worms, snails, bivalves, insect larvae, and crustaceans; detritus; fish eggs; and small fish. Juveniles concentrate on invertebrates such as mayfly, stonefly, and midge larvae, and freshwater shrimp.

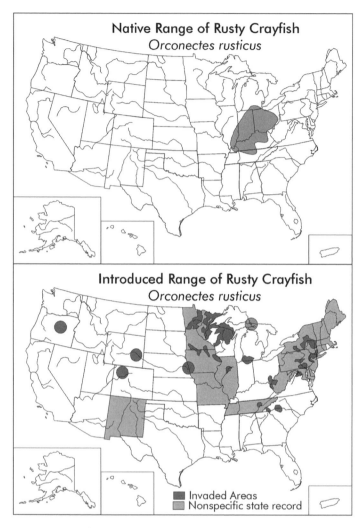

Native Range of Rusty Crayfish
Orconectes rusticus

Introduced Range of Rusty Crayfish
Orconectes rusticus

Invaded Areas
Nonspecific state record

Top: Rusty crayfish are native to the Ohio, Tennessee, and Cumberland river systems. Bottom: Rusty crayfish are invasive native transplants in many drainages of the Great Lakes region, the central United States, and the Northeast. (Both maps adapted from USGS 2008.)

Life History. Rusty crayfish mate in late summer or early fall, and sometimes in early spring. The male transfers his sperm to the female, which stores them in a seminal receptacle on her belly until the water begins to warm in late April or May. External fertilization takes place when eggs and sperm are simultaneously released by the female. As this happens, she secretes a mucus-like substance called glair that forms white patches on the underside of the tail fan. The fertilized eggs attach to the swimmerets on the underside of the female's abdomen by means of the glair. Females produce 80–575 eggs.

It takes 3–6 weeks for the eggs to hatch. The young continue to hold on to the mother's swimmerets for several weeks as they go through 3–4 molts. After the young leave the female, they undergo another 8–10 molts before they become adults, usually the year after they hatch. Sexual maturity occurs at a total body length of 1.4 in. (3.5 cm). Growth slows once the crayfish is mature. Females molt only after releasing their young, usually in June or early July. Males molt twice a year, once in the spring into a sexually inactive form, and again in summer into a reproductive form. The twice-annual molt allows males to become larger than females. Rusty crayfish live 3–4 years.

Impacts. Rusty crayfish are aggressive and displace native crayfish, such as the virile crayfish and the northern clearwater crayfish, from shelter and compete with them for food. When the smaller natives are forced from their hiding places, they become vulnerable to increased predation by fish. Declines in these two native crayfish have occurred in lakes in Wisconsin (and parts of Ontario, Canada). Another close relative, Sanborn's crayfish (*Orconectes sanbornii*), has been displaced in Ohio in waters where it was not native. Hybridization with the northern clearwater crayfish contributes to the decline of that species.

The high metabolic rate of the rusty crayfish means it consumes a lot of food for its size. The destruction of submerged vegetation that provides shelter for invertebrates, young game fish and panfish, and forage species for other fish can be particularly damaging to the aquatic ecosystem as it eliminates important and sometimes scarce habitat. They may directly harm native fish by consuming their eggs and reducing the amount of invertebrate prey items upon which juvenile fish depend. Declines have been documented in larval midges, mayflies, and stoneflies and in fish such as bluegill (*Lepomis macrochirus*), pumpkinseed (*Lepomis gibbosus*), smallmouth bass (*Lepomis gibbosus*), largemouth bass (*Micropterus salmoides*), lake trout (*Salvelinus namaycush*), walleye (*Sander vitreus*), and northern pike (*Esox lucius*) when rusty crayfish invade.

The rusty crayfish is so named because of the rusty patches of color on either side of the carapace just in front of the tail. (Jeff Gunderson, Minnesota Sea Grant.)

Large numbers of rusty crayfish become a nuisance to swimmers, who are in danger of stepping on them and being pinched by their large claws.

Management. Few environmentally sound means of control are available. Commercial harvests may reduce numbers and keep them in check. Some researchers believe that population explosions of rusty crayfish are in part due to the overfishing of predatory fishes, and they recommend restoring healthy populations of sunfish and bass as a way of reducing the impacts of rusty crayfish. The best management practice is to prevent further introductions by educating people about the ways in which this little animal threatens their local waters.

Selected References

Gunderson, Jeff. "Rusty Crayfish: A Nasty Invader." Minnesota Sea Grant, 2008. http://www.seagrant.umn.edu/ais/rustycrayfish_invader.

Pappas, J. "*Orconectes rusticus*." Animal Diversity Web, 2002. http://animaldiversity.ummz.umich.edu/site/accounts/information/Orconectes_rusticus.html.

U.S. Geological Survey. "*Orconectes rusticus*." USGS Nonindigenous Aquatic Species Database, Gainesville, FL, 2008. http://nas.er.usgs.gov/queries/factsheet.aspx?SpeciesID=214.

■| Spiny Water Flea

Scientific name: *Bythotrephes longimanus*
Order: Cladocera
Family: Cercopagidae

Native Range. Northern Eurasia, from Great Britain and Scandinavia across Russia to the Bering Sea. Since the spiny water flea has been widely dispersed by humans in Europe and Asia to lakes in which it was not native, its true natural range is difficult to reconstruct.

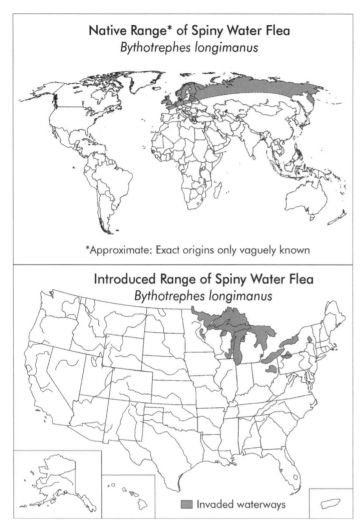

Native Range* of Spiny Water Flea
Bythotrephes longimanus

*Approximate: Exact origins only vaguely known

Introduced Range of Spiny Water Flea
Bythotrephes longimanus

■ Invaded waterways

Top: Spiny water fleas occur across northern Eurasia. The exact native range is uncertain because people have introduced the species to many lakes in both Europe and Asia. Bottom: The spiny water flea is invasive in all of the Great Lakes as well as inland lakes in adjacent states. (Adapted from Liebig and Benson 2007.)

Distribution in the United States. This invader is established in all the Great Lakes and has been found in inland lakes in Ohio, Michigan (Long Lake), Minnesota (Greenwood and Flour lakes), New York (Great Sacandaga Lake) and Wisconsin (Stormy Lake and Gile Flowage).

Description. The spiny water flea is not an insect, but a crustacean. Many of its features are difficult to discern without magnification. Because they assemble into gelatinous clumps, they often resemble wet cotton batting full of tiny black spots. Each individual in the mass has a well-developed abdomen and a long, thin tail spine. The head is clearly separated from the trunk and bears a single large black compound eye. Two swimming antennae used to propel the animal through water are attached just behind the head. The spine is straight and accounts for more than 70 percent of total body length. The spine bears 1–4 pairs of barbs depending upon the animal's age. Juveniles have a single pair; additional pairs are added above the spine when the exoskeleton covering the trunk of the body is shed to allow for growth. Two morphs representing different reproductive modes (see under Life History) are present. Parthenogenically reproduced individuals, all females, have longer tail spines with an obvious "kink" in the middle of the spine and, when fully developed, have three pairs of barbs resulting from two molts. Sexually reproducing females will acquire four pairs of barbs during three molts. (The parthenogenically reproduced females were once believed to be a separate species and identified as *B. cederstroemi*.) Females are also identifiable because they carry their eggs in a brood pouch that balloons from the back of the body. Males will not gain a new pair of barbs during their final molt and end up with only two pairs.

Spiny water fleas have four pairs of legs. The first pair is longer than the others and used for snaring prey. The others are used to grasp the prey while it is being eaten. Average adult size is 0.4 in. (1 cm).

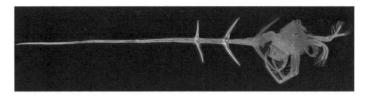

The spiny water flea is a crustacean barely visible with the naked eye. (Microscopy by Howard Webb.)

Related or Similar Species. This tiny crustacean has a unique body shape with a long tail spine that clearly distinguishes it from freshwater zooplankters native to the Great Lakes. Another exotic cladoceran, the fishhook water flea (*Cercopagis pengoi*), also collects into masses, but individuals can be separated from spiny water fleas by virtue of their angled tails that end in a distinct loop or "fishhook" and their tiny eyes.

Introduction History. The first live specimens of spiny water flea were retrieved from Lake Huron in 1984. The next year, they were found in Lakes Erie and Ontario. By 1986, they had made their way to Lake Michigan, and by 1987, they had invaded Lake Superior. By early 2002, they had been reported from 66 lakes in the northern United States and southern Canada. In 2007, spiny water fleas were discovered in Stormy Lake, Wisconsin, and in 2008, they were found in Great Sacandaga Lake, in New York's Adirondack Mountains.

Genetic analysis has traced the source of the invasion to the port of St. Petersburg, Russia. Spiny water fleas are native to nearby Lake Lagoda. During spring snowmelt, they are carried down the Neva River to the harbor, where the water becomes fresh or only slightly brackish. Freighters, which had brought wheat from the United States, return empty and therefore take up ballast water and any zooplankton contained in it at St. Petersburg. They then discharge that water when they next take on grain in American ports in the Great Lakes.

Local dispersal from the Great Lakes to inland lakes probably occurs by means of contaminated fishing gear and boats.

Habitat. Spiny water fleas do best in deep lakes in temperate climates. They will inhabit both large and small lakes, and may live in shallow lakes and rivers as well as in brackish waters. They are limited to water temperatures of 39–86°F (4–30°C) and salinities of 0.04–8.0 percent. Optimal conditions are temperatures of 50–75°F (10–24°C) and salinities of 0.04–0.4 percent. Females are conspicuous to predatory fish during daylight hours because of their large eyes and brood pouches, and thus undergo daily vertical migrations. During daylight hours, they move lower into the water column and are usually at depths of 30–65 ft. (10–20 m). They swim up closer to the surface where food is more abundant and at night, when most will be found within the top 30 ft. (10 m) of the water column.

Diet. Spiny water fleas are tiny predators that feed on smaller, herbivorous zooplankters. *Daphnia* spp. and other water fleas are preferred prey, but they also eat copepods and rotifers. Studies show that a single spiny water flea can consume 20 prey items a day.

Life History. Spiny water fleas have a complex life cycle involving both sexual and asexual reproduction. Most of the time, females reproduce by cloning (parthenogenesis) and produce 1–10 eggs that develop into new females without fertilization by males. A new generation may be produced every two weeks in the summer, when temperatures are warm and food abundant. Under these conditions, males are rare in the population. However, sex of the young is determined not by genetics, but by the environment. When conditions deteriorate in the fall, males begin to be produced. They mate with females that then produce "resting eggs." At first, the resting eggs are carried in the brood pouch, but eventually they are released into the water and settle to the bottom of the lake, where they go into a near-dormant phase (diapause) in order to survive the cold of winter. As water temperature rises to 39°F (4°C) and above in the spring, the eggs hatch into females that will reproduce parthogenetically and make possible rapid population growth during favorable conditions. The brief sexually reproducing part of the life cycle promotes both genetic diversity and dispersal capability.

Impacts. The initial impact of the introduction of spiny water fleas was as a nuisance to fishermen. Hundreds of these cladocerans can clump onto fishing lines and downrigger cables, especially at connections and swivels, and clog the first guide of an angler's rod. Sometimes it becomes impossible to reel in the catch, and the line must be cut. With time, ecological impacts became evident. The community structure of the zooplankton community can be altered by heavy predation on small herbivorous crustaceans such as *Daphnia*, copepods, and rotifers. In Lake Michigan, both *D. retrocurva* and *D. pulicaria* populations collapsed after the introduction of water fleas, and populations of another tiny cladoceran, *Holopedium gibberum*, declined dramatically. In consuming these animals, spiny water fleas also compete for food with the plankton-feeding larvae of several fish species as well as with other cladocerans, such as the giant water flea (*Leptodora kindtii*), which declined in some lakes when spiny water fleas were introduced. Such results are not always duplicated in smaller lakes.

Impacts on fish populations are not clear. Juvenile fish of species that feed heavily on *Daphnia* may suffer, affecting annual recruitment rates of species such as yellow perch (*Perca flavescens*). The barbed tail of the spiny water flea cannot be accommodated in the small mouth gape and throats of young fish, so they cannot shift their diets to include the invader species. On the other hand, adult yellow perch, as well as white perch (*Morone americana*), bass (*Micropterus* spp.), and alewife (*Alosa pseudoharengus*) selectively prey on spiny water fleas and could be helped by a new, seasonally abundant forage species. These fish in turn are eaten by larger game fish such as Chinook salmon (*Oncorhynchus tshawytscha*) and lake whitefish (*Coregonus clupeaformis*), so economically important fisheries could be affected in a positive manner.

Management. Control measures are aimed at preventing the spread of the spiny water flea to other inland lakes. Educational campaigns encourage fishermen to clean gear properly and empty bilges, livewells, and bait buckets responsibly before moving from one lake to another.

Selected References

Caceres, Carla E., and John T. Lehman. "Spiny Tailed *Bythotrephes*: Its Life History and Effects on the Great Lakes." Minnesota Sea Grant College Program. University of Minnesota, 2004; last modified January 26, 2010. http://www.seagrant.umn.edu/exotics/spiny.html.

Liebig, Jim, and Amy Benson. "*Bythotrephes longimanus*." USGS Nonindigenous Aquatic Species Database, Gainesville, FL, 2010. Revised January 25, 2007. http://nas.er.usgs.gov/queries/FactSheet.aspx?SpeciesID=162.

National Biological Information Infrastructure (NBII) and IUCN/SSC Invasive Species Specialist Group (ISSG). "*Bythotrephes longimanus* (Crustacean)." ISSG Global Invasive Species Database, 2005. http://www.issg.org/database/species/ecology.asp?si=151&fr=1&sts=sss.

Sikes, Benjamin. "Spiny Water Flea." Invader of the Month. Institute for Biological Invasions, University of Tennessee, Knoxville, 2002. http://invasions.bio.utk.edu/invaders/fleas.html.

■ Arachnids

■ | Honeybee Tracheal Mite

Scientific name: *Acarapis woodi*
Order: Trombidiformes
Family: Tarsonemidae

Native Range. Uncertain. Honeybee tracheal mites were first described from collections made in dying bee colonies on the Isle of Wight in the English Channel in 1921. They were responsible for wide-ranging honey bee mortality in Europe in the early part of the twentieth century.

Distribution in the United States. Throughout the continental United States, except Alaska, in both managed and feral honey bee colonies.

Description. Invisible to the naked eye, tracheal mites have white oval bodies with a shiny, smooth cuticle. The body and legs have several long fine hairs. The mouth parts are beak-like and elongated. Females are 143–147 microns long and 77–81 microns wide; males are 125–136 microns long and 60–70 microns wide.

Few outward symptoms are evident in infected bees until the infestation is severe. Then bees may develop disjointed, "K" wings and become unable to fly. Honey production may decline. Sudden death of the hive during winter is a sign that the honeybee tracheal mite may be present. Verification requires dissection of the dead bees and examination of the tracheae, which will be brown instead of clear or white as in healthy bees.

Introduction History. Honeybee tracheal mites were unknown in North America prior to 1980, when they were detected in Mexico, some 200 miles (320 km) south of the U.S. border. They were first discovered in the United States at a commercial beekeeping enterprise in Weslaco, Hidalgo County, Texas, in early July 1984. The following month, they were found in New Iberia, Louisiana, and by October 1984, they were being reported in Florida, Nebraska, New York, North Dakota, and South Dakota. By August 1985 they infested bee colonies in 17 states. The rapid spread of the mite was facilitated by migratory beekeepers, who truck bee colonies from southern states northward to pollinate various crops. The commercial bee business also contributed to the spread by selling queens and package bees.

Mites move quickly through a colony via bee-to-bee contact in the hive. Workers and drones drifting from hive to hive disperse the mite through entire apiaries or from one apiary to another. Bees also encounter mites when they rob honey from other hives; and colonies weakened by heavy mite infestations are particularly vulnerable to robbing. Since mites cannot live more than a day outside the bee host, they readily leave dead bees and hitchhike to new colonies inside robber bees. Normal swarming of bees can also spread the tracheal mite to new areas.

Habitat. Honeybee tracheal mites are internal parasites that live within the breathing tubes (trachea) of adult bees. They prefer the larger tubes at the base of the bees' wings that provide oxygen to the flight muscles. Occasionally, they occur in the air sacs. They infest only European honey bees (*Apis mellifera*), Africanized honey bees (*Apis mellifera scutellata*; see Insects, Africanized Honey Bee) and Asian honey bees (*Apis cerana*).

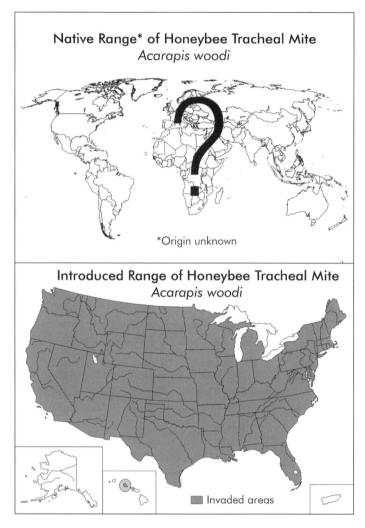

Top: The origins of the honeybee tracheal mite are unknown. It was first reported affecting domestic honey bees in England. Bottom: The honeybee tracheal mite is found throughout the lower 48 states and on O'ahu, Hawai'i.

Diet. Tracheal mites puncture the walls of the breathing tubes and feed on the blood (hemolymph) of workers, drones, and queen bees.

Life History. The life cycle consists of four stages: egg, larva, a resting stage nymph, and adult. The female mite enters a bee shortly after it emerges from its cell by moving through the first thoracic spiracle into the breathing tubes. She remains in the host for the rest of her life or until the bee dies. Three or four days after arrival, she lays 5–7 eggs, which hatch in another three or four days. Adult male mites are mature 11–12 days after the eggs are laid; females are mature in 14–15 days. Gravid females crawl to the tip of a bee hair and "jump" to a new host, entering through the breathing pores (spiracles) to lay their eggs. They tend to attack young bees less than 24 hours old. Mite populations tend to be cyclical; heaviest infestations occur in winter, when the bees are confined and crowded in the hive, and then decline in summer.

Impacts. Initially, the introduction of the tracheal mite caused widespread losses of bee colonies and even entire apiaries throughout the United States. Problems were especially severe in the more temperate parts of the country, where bees cluster in confinement within the hive during the winter months. Highest bee mortality usually occurs in late winter. Infected bees often show no signs of a problem, though pollen collection and honey production may decline as a bee weakens.

The mites kill by clogging the breathing tubes with their bodies and waste products. Normally elastic tracheae become stiff and brittle, and flight muscles atrophy. Furthermore, hemolymph contains agents that act as antifreeze, so as the mites consume the bee's body fluid, it reduces its ability to withstand low temperatures.

The effects of mite infestation, known as acarine disease, may remain in a colony for years with little damage. Worker bees and queens become less susceptible to infestation as they age. It appears that most honey bees in the United States have developed some resistance

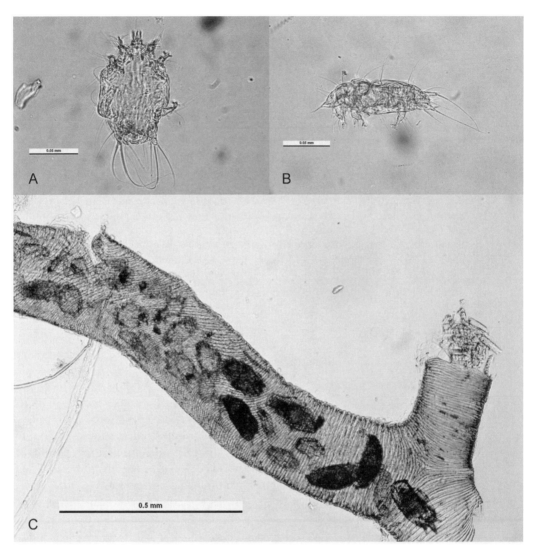

A. View of honeybee tracheal mite from above. B. Lateral view of honeybee tracheal mite. C. Mites clogging a bee's trachea. (Pest and Diseases Image Library, Bugwood.org.)

to or tolerance of mite infestations and that the problem of acarine disease is not as severe as it was in the 1980s and 1990s. Its impacts may have been overshadowed by the introduction of the varroa mite (*Varroa destructor*) in 1987 (see Arachnids, Varroa Mite).

Management. After the initial discovery of mites in Texas in 1921, the federal Honey Bee of Act of 1922 was passed. It prohibited the importation of any honey bee into the United States. The law has been relaxed somewhat since then to allow the import of bees from Canada, where mite infestations are unknown, and a few other mite-free countries.

Chemical control may involve fumigation with menthol crystals. Grease patties, made of vegetable shortening and sugar and available commercially, keep mites from infesting young bees because the oily bees apparently do not "smell" right to mites. Evaporating formic acid is also very effective in controlling tracheal mites and has the added bonus of reducing infestations of the varroa mite.

Several stocks of honey bees, including the Buckfast bee, have been bred to resist tracheal mites. The bees still become infested, but only at levels too low to cause significant damage. Use of these bees eliminates the need to treat the colony with chemicals.

Selected References

Ambrose, John T., and Michael Stanghellini. "Tracheal Mites." Note 2.02. Beekeeping. Insect Pest Management, Department of Entomology, North Carolina State University, 2001. http://www.cals.ncsu.edu/entomology/apiculture/PDF%20files/2.02.pdf.

Collison, Clarence H. "Honey Bee Tracheal Mite." Publication 1753, Extension Service of Mississippi State University, 2009. http://msucares.com/pubs/publications/p1753.htm.

Hunt, Greg. "Parasitic Mites of Honey Bees." *Beekeeping*. E-201-W. Department of Entomology, Purdue University Cooperative Extension Service, 2006. http://extension.entm.purdue.edu/publications/E-201.pdf.

Sammataro, Diana. "An Easy Dissection Technique for Finding the Tracheal Mite *Acarapis woodi* (Rennie) (Acari: Tarsonemidae), in Honey Bees, with Video Link." *International Journal of Acarology*, 32(4), 2006. Available online at http://entnemdept.ufl.edu/HoneyBee/files/speaker_notes/exp_wksp_Tracheal_Mite_Dissection_Notes.pdf.

■ | Varroa Mite

Scientific name: *Varroa destructor*
Order: Parasitiformes
Family: Varroidae

Native Range. Mainland Asia. It is known to be endemic to Japan, Korea, and Thailand, where its host is the Asian honey bee *Apis cerana*. At some point it shifted hosts to the European honey bee; and, in 1963 in Singapore, it was first identified on European honey bees as a new species. Previously it had been classified as *Varroa jacobsoni*, a parasitic mite of the Asian honey bee known from much of mainland and insular Asia.

Distribution in the United States. In all 50 states, in both wild and managed honey bee colonies.

Description. This small arachnid is an external parasite of honey bees. Tick-like, it has a flattened oval body and eight legs. Adult females are reddish brown and measure about 0.06 in. (1.5 to 1.99 mm) in width, approximately the size of the head of a pin. Their bodies are curved to allow them to squeeze into the abdominal folds of a bee and thereby be protected from the bee's normal cleaning habits. Males are smaller and spherical in shape; they are yellowish and have tan legs. Both sexes are visible to the unaided eye.

Symptoms of severe varroa infestations include an accumulation of dead bees at the entrance to a hive, the uncapping and destruction of brood cells by worker bees, and deformed legs and wings on adult bees.

Related or Similar Species. The bee louse (*Braula coeca*), a tiny wingless fly, is similar in size and color, but has six legs instead of eight. It is rare in most hives today since it succumbs to treatments used to control the varroa mite.

Introduction History. A single varroa mite was discovered in Maryland in 1979, its source unknown. It showed up again in 1987, this time in Florida and Wisconsin. It has since spread rapidly throughout the United States. The last state invaded was Hawai'i, where the first infestation occurred in O'ahu in 2007; the following year, the varroa mite was found in honey bee colonies on the Big Island.

Mites are transported from hive to hive by drifting workers and drones. Worker bees also pick up mites when they rob honey from smaller colonies. People can move mites from place to place when they transport infested colonies of bees to fields and orchards for pollination and through the importation of infested package bees. Swarming bees may carry mites to other apiaries or to feral populations. It is also possible that mites are spread short distances on bumblebees and other nectar-feeding insect hosts.

Habitat. Varroa mites are usually found on the thoraxes and abdomens of larvae, pupae, and adults of all races of the European honey bee (*Apis mellifera*) and the Africanized honey bee (*A. mellifera scutellata*; see Insects, Africanized honey bee). All life stages inhabit the brood cells of honey bees; adult females will emerge with the young bee and live on drone and worker bees. The varroa mite also occurs on the American bumblebee (*Bombus pennsylvanicus*), flower fly (*Palpada vinetorum*), and rainbow scarab beetle (*Panaeus vindex*), but cannot reproduce on them. It is also a parasite of Asian honey bees (*A. cerana* and *A. koschevnilovi*).

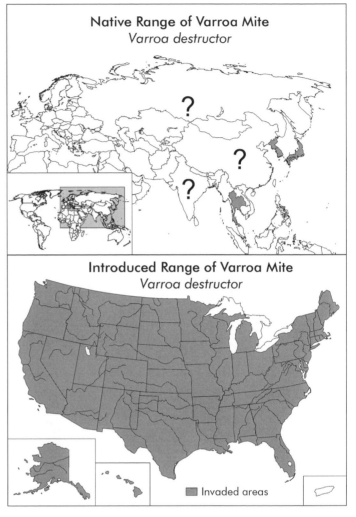

Top: The varroa mite is known to be native to Japan, Korea, and Thailand and perhaps other parts of Asia. It originally infested the Asian honey bee but later adopted the European honey bee as its host. Bottom: Varroa mites are found in all 50 states, where they infest feral and managed bee colonies.

Diet. Varroa mites suck the hemolymph ("bee blood") of developing honey bee larvae and mature adults. They pierce the soft skin of larvae and the tougher integument of adults to obtain this fluid. Like other arachnids, they inject enzymes that predigest a bee's tissues so they can consume them also.

Life History. The life cycle of varroa mites is synchronized with that of its host, the honey bee. Females enter the brood cells of honey bees just before worker bees cap the cell, when the bee larva is five days old and about to pupate. About three days later, she lays her first egg, which usually is unfertilized and becomes a male. Later, she will lay a fertilized egg every 30 hours; these become females. All immature mites feed on the bee larva and must mature and mate before the bee emerges from its cell. It usually takes five to eight days for the females to mature and a few days less for the male. The male mite dies in the cell, but

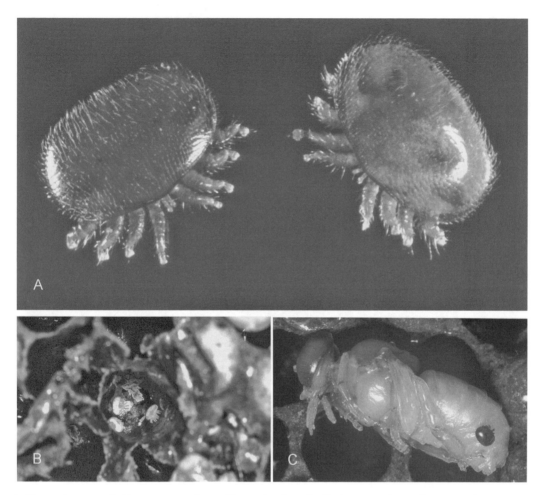

A. *Varroa destructor.* B. Varroa mites at base of honey bee brood cell. C. Mite on honey bee pupa. (Scott Bauer, USDA Agricultural Research Service, Bugwood.org.)

mated females will leave the cell on the host bee to seek new brood cells, preferably those of drone larvae, where the process begins again. The average lifespan is about 50 days during the breeding season. In winter, when brood-rearing by bees declines, mites live solely on adult bees in the hive. Adult mites can survive only a few days without bees.

Impacts. Heavy infestations of varroa mites cause young bees just emerging from their brood cells to have malformed wings, legs, and bodies. On adult worker bees, they may affect the flight behavior, orientation, and success in returning to the hive laden with pollen and nectar. As worker bees die off, fewer bees are available to tend to the brood and collect nectar, and the colony weakens. Weak colonies are susceptible to having their honey stores robbed by stronger hives. Eventually, the colony dies. In temperate climates, the death of a colony may take 3–5 years; but in the milder climate of Florida, infested bee colonies have died within seven months.

Mites transmit several RNA viruses that kill bees by compromising their immune systems. These include deformed wing virus (DWV), acute bee paralysis virus (APV), and slow paralysis virus (SPV). They result in a condition known as parasitic mite syndrome, which can destroy a colony in a few months.

Without treatment, colonies—indeed, entire apiaries—can be wiped out by varroa mites. In the United States, most wild (feral) bee populations, where treatment is not an option, have been decimated by this invasive arachnid. Pollination of numerous field and orchard crops depend upon honey bees. Their loss would have devastating economic impacts.

Management. Chemical means are available to detect varroa mites and control them in a bee colony. A common treatment involves hanging strips impregnated with fluvalinate (Apistan) in the brood nest area for about a month. As dead mites fall off bees, they can be collected on sticky paper or a fine mesh placed screen on the bottom of a hive. This product allows detection of low-level infestations. Such strips can also be used as a control measure. A problem with this method is the development of fluvalinate-resistant mites. Another chemical treatment using coumaphos (CheckMite+) is also very effective, but has the disadvantage of employing a dangerous organophosphate. Two formulations of thymol (oil of thyme—Api-Life VAR and Api-Guard) and one using fenpyroximate (Hivastan) are also available to beekeepers. All should be applied with careful adherence to instructions.

Research suggests biological controls may be implemented in the future. One candidate is a strain of the fungus *Metarhizium anisopliae* that is lethal to varroa mites. Another possibility is genetic engineering to develop a bee resistant to mite infestation. Asian bees (*Apis cerana*), the original host for varroa mites, are aggressive groomers and cleaners and remove mites from other bees by grooming and chewing the mite. They decap infested brood cells and clean them out, discarding dead and dying larvae at the entrance to the colony. Bees that develop faster in the brood cell can outpace the developing mites and emerge with fewer of them. Bee breeders can select for queens that produce workers that are good at grooming, that clean out brood cells, and that have quicker development times than average European honey bees.

It is essential to stem the spread of infested colonies through constantly surveying bees for the presence of the mite. Captured swarms, package bees, and other new colonies should be quarantined away from the rest of the apiary and examined for mites before being allowed contact with existing colonies.

Selected References

Bessin, Ric. "Varroa Mites Infesting Honey Bee Colonies." ENTFACT-608. College of Agriculture, University of Kentucky, 2001. http://www.ca.uky.edu/entomology/entfacts/ef608.asp.

Hunt, Greg. "Parasitic Mites of Honey Bees." *Beekeeping.* E-201-W. Department of Entomology, Purdue University Cooperative Extension Service, 2006. http://extension.entm.purdue.edu/publications/E-201.pdf.

National Biological Information Infrastructure (NBII) and IUCN/SSC Invasive Species Specialist Group (ISSG). "*Varroa destructor* (Arachnid)." ISSG Global Invasive Species Database, 2006. http://www.issg.org/database/species/ecology.asp?si=478&fr=1&sts=sss.

Reid, Brendan. "Varroa Mite (*Varroa destructor*)." Introduced Species Summary Project, Columbia University, 2004. http://www.columbia.edu/itc/cerc/danoff-burg/invasion_bio/inv_spp_summ/varroa_destructor.html.

Sanford, M. T., H. A. Denmark, H. L. Comroy, and L. Cutts. "Featured Creatures: Varroa mite (*Varroa destructor*)." University of Florida Institute of Food and Agriculture, 2007. http://www.entnemdept.ufl.edu/creatures/misc/bees/varroa_mite.htm.

■ Insects

■|Africanized Honey Bee

Also known as: AHB, Killer bee
Scientific name: *Apis mellifera scutellata*
Order: Hymenoptera
Family: Apidae

Native Range. The African parent of the hybrid Africanized honey bee derives from eastern and southern Africa, where *Apis mellifera scutellata* occurs from Kenya south to the Indian and Atlantic coasts of South Africa. Bees from Tanzania were taken to Brazil, where hybrids or *Africanized* honey bees were developed in Rio Claro in the state of Sao Paulo. From there, they spread throughout tropical South America and Central America and into the United States (see under Introduction History).

Distribution in the United States. As of July 2009, Africanized honey bees were established from Texas and Oklahoma west through New Mexico and Arizona to southern Nevada and southern California. Isolated swarms have been reported in southern Utah, Arkansas, Louisiana, and Florida.

Description. The Africanized honey bee is a hybrid species and nearly indistinguishable from the common domesticated European honey bee, one of its parents. The hybrid form requires inspection and analysis in a laboratory to positively identify it. Behavioral traits do set it apart, however. It is usually recognized by its extremely aggressive nature, the guarding of a large area around its nest, frequent swarming, and nest location. Worker bees, all sterile females, are about 0.75 in. (19 mm) long, imperceptibly smaller than European honey bees. Drones, all males, have narrower bodies and are somewhat longer. Queens are larger still and more robust, with enlarged abdomens.

The bees' bodies are covered with a yellow-brown fuzz and marked with black stripes. They have four clear wings and six legs, all attached to the middle body segment (thorax). The abdomen is larger than the thorax and ends in a stinger. The venom is not as toxic as that of European bees, but since large numbers of bees mass to defend their nest, the multiple stings that result can be dangerous, indeed lethal, especially to sensitive persons.

Related or Similar Species. The European honey bee (*Apis mellifera*) is more a docile domesticated animal selected for characteristics that suit it for easy handling by humans. It produces and hoards more honey than Africanized honey bees and swarms less often. European honey bees nest in large cavities in trees, in hollow walls, and in beehives, where they develop from egg to adult a bit more slowly than their hybrid descendents. European honey bees have adapted to the cold winters of the mid-latitudes, in part by storing large amounts of honey in their hives.

Introduction History. African honey bees were brought to test sites in Rio Claro, Brazil, in 1956 by a Brazilian geneticist, Dr. Warwick Kerr. European honey bees had not fared well in subtropical and tropical parts of South America. The hope was that crossing them with bees from tropical Africa would improve both their honey production and survival rates by selecting for the climatic adaptations of the African parents and the docility and honey-producing attributes of the European ones. In 1957, 26 African queens swarmed with European worker bees and escaped the experimental apiary. These bees hybridized with commercial and feral populations of European honey bees, but in the wild retained most of the traits of the African line, particularly the aggressive defense of their nests. They rapidly spread through all of subtropical and tropical South America except Chile, and north through Central America. They were first identified in Mexico in 1985. The rate of spread averaged 125–180 mi. (200–300 km) a year. This dispersal was largely unaided by humans. In 1990, Africanized honey bees reached Hidalgo, Texas, on the Mexican border. By 1993, they were in Arizona, and by 1994, they were discovered near Blythe in southern California. In 1996 they were found to the north in Lawndale, California. Since that time, the dispersal rate has, for unknown reasons, slowed.

Africanized honey bees dispersed across South and Central America by swarming and by hybridizing with existing feral and domestic colonies of European honey bees. Anecdotal evidence suggests European queens breed preferentially with Africanized drones and that Africanized queens will enter the hives of European honey bees and kill and replace the European queen.

It is believed that cold winter temperatures will prevent their establishment very far north of the present U.S. distribution limits, and that high rainfall will restrict their spread east of Texas. Traits adaptive to tropical environments will be

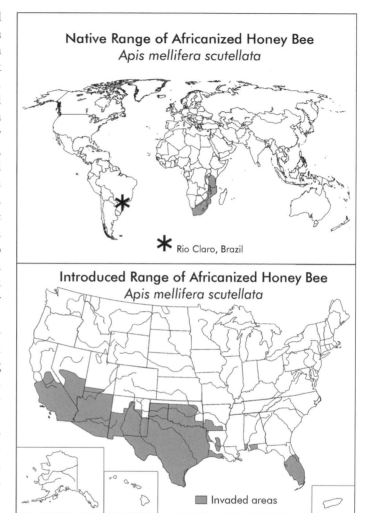

Top: The African parent of the Africanized honey bee has its origins in southern and eastern Africa. The hybrids between African bees and the European honey bee were developed in Rio Claro, Brazil. Bottom: Africanized honey bees are established in southern states from California to Florida. Short-lived colonies may appear as far north as Virginia. (Adapted from "Spread of Africanized Honey Bees by Year, by County." USDA Agricultural Research Station, 2009. http://www.ars.usda.gov/Research/docs.htm?docid=11059&page=6.)

selected against in colonies living in temperate climates. Disjunct populations in Florida may have resulted from swarms arriving on ships; Africanized honey bees have been trapped regularly at Florida's deep-sea ports. Occasionally, temporary infestations occur as far north as Virginia, perhaps the result of bees being transported accidentally on trains.

Habitat. The hybrid Africanized bees are less selective in habitat than the European honey bee and nest in smaller numbers. Their nests may hang from exposed tree limbs or under the eaves of buildings, or may be constructed in old tires and empty containers, cement blocks, rotted logs, animal burrows, rock piles, and so forth—all places avoided by European honey bees. They are apparently restricted to tropical and subtropical climates, particularly those with distinct wet and dry seasons, and are seldom found poleward of 34° latitude. Along the Gulf coast, their failure to become well established east of Texas has been attributed to the higher annual precipitation in the Gulf coast states (greater than 50 in. or 1,270 mm a year) and possibly to greater rates of parasitism by varroa mites (*Varroa destructor*; see Arachnids, Varroa Mite) in those more humid areas.

Diet. Africanized honey bees consume nectar and pollen. Nectar is converted to honey and becomes the bees' main source of carbohydrates. Pollen constitutes the main source of protein. Most flowering plants serve as food sources.

Life History. A queen mates with a drone to produce a fertilized egg, which develops into a female worker bee. Unfertilized eggs become male drones. If larvae are fed royal jelly—a mixture of glandular secretions—they develop into queens. The life cycle involves egg, larva, pupa, and adult. Eggs are attached to the bottom of cells in the comb and hatch in approximately 60 hours. The larvae are then fed secretions from the heads of young nurse bees. If the egg is to become a worker, it is sealed into the cell eight days after hatching. Within 24 hours, the larva spins a cocoon, and the next day enters the prepupal stage. A day later, the white, immobile pupa has formed. Adult worker bees emerge 18.5 days after the egg was laid; queens about 16 days after laying.

Worker bees live about 30 days, drones 5–10 weeks, and queens 1–3 years.

As a colony grows, it will split into two or more colonies. A queen and a group of workers leave the nest together in a "swarm." A new queen hatches in the original nest, mates, and begins egg-laying. Africanized bees swarm frequently (as often as every six weeks) and at any time of year. Their survival strategy, evolved in an African cultural environment in which people were and still are bee hunters and honey thieves rather than beekeepers, is to invest energy into large numbers of offspring and produce many reproductive swarms. They fiercely defend their nests, but they also abandon them readily and start new colonies elsewhere.

Impacts. The most direct impacts of Africanized honey bees are on the beekeeping industry and agriculture. These hybrid bees compete with domestic European honey bees and, by mating with European queens, take over hives. The Africanized bees not only are more

A. The Africanized honey bee is a hybrid species and can only be distinguished from the European honey bee by laboratory analysis. (Jeffrey W. Lotz, Florida Department of Agriculture and Consumer Services, Bugwood.org.) B. Africanized honey bee (left) and European honey bee (right). Normally color differences do not allow distinction between the two. (Scott Bauer, USDA Agricultural Research Service, Bugwood.org.)

difficult to handle and much more apt to abandon their hives, but they also produce less honey than European honey bees. Honey bees are extremely important pollinators of many orchard and field crops, and beekeeping is a commercial enterprise of its own. One estimate has honey bees adding $10 billion or more each year to the value of some 90 crops in the United States. The problem in handling the aggressive hybrid bees has led some to give up beekeeping altogether and others to adopt more labor-intensive and therefore costly management practices (see under Management). Africanized honey bees are best suited to subtropical areas of the United States, the very states where the packaged bee and queen industry is located.

Africanized honey bees are dangerous to people who venture too close to their nests or otherwise disturb them because of the multiple stings victims receive from the hundreds of bees that attack en masse at the slightest provocation. Their preferred nesting locations bring them into close proximity to humans, increasing the likelihood of bee-human interaction. The bees can sense intruders more than 50 ft. (15 m) away and be irritated by the vibrations of power equipment more than 100 ft. (30 m) distant. They may chase people whom they feel are threats for 0.25 mi. (0.4 km). Some people have died from as few as 100–300 stings, but usually it takes 500–1,000 stings for the attack to be fatal. Livestock and pets are at risk as well as humans.

Effects on natural ecosystems and native pollinators are unknown. Competition from Africanized honey bees could displace other nectar- and pollen-feeding invertebrates from their more important food resources.

Management. As the invasion spread through South and Central America, beekeeping and honey production suffered significant declines at first. Then beekeepers learned new skills to better manage their Africanized hives and reduce the frequency of bee attacks. Among the benefits to agriculture of these new practices was the production of pollinators many farmers feel are superior to the European honey bee. A simple change was to keep each bee colony on its own, separate hive stand so that one hive could be worked without disturbing neighboring colonies. More smoke is used well ahead of working with the bees to calm them and possibly mask alarm pheromones. In addition, a change from traditional black bee veils to white ones as part of the beekeeper's protective gear minimizes attacks.

In the United States, the strategy has been to try to diminish the infusion of African genes into bee populations by either "drone-flooding" or frequent requeening with known European queens. Drone-flooding involves adding European drones to a colony to lessen the chances of the queen mating with an Africanized one.

To avoid painful or even deadly encounters with Africanized bees, people in areas where these bees now occur should remain alert to their potential presence when outdoors and stay away from known nests or swarms. Anyone knowing the location of hives or swarms should obtain professional help to remove them. If an attack occurs, immediately run away in a straight line, covering your face if possible, and seek shelter in a car or building. Do not swat at them or try to hide in a pool of water. Scrape stingers off the skin; do not squeeze them, as that will cause them to release more venom. Wash and apply ice to stings and see a doctor if breathing becomes labored. People can bee-proof their homes and yards by sealing openings, regularly inspecting eaves, and removing other potential nest sites.

Selected References

"Africanized Honey Bee." Oklahoma Invasive Species Site. Oklahoma State University, 2007. http:// oklahomainvasivespecies.okstate.edu/africanized_honey_bee.html.

"Africanized Honeybee Pest Profile." California Department of Food and Agriculture, 2010. http:// www.cdfa.ca.gov/phpps/pdep/target_pest_disease_profiles/ahb_profile.html.

Ellis, Jamie, and Amanda Ellis. "African Honey Bee." Featured Creatures, Publication Number EENY-429, University of Florida Institute of Food and Agricultural Sciences, 2008 http://entomology.ifas.ufl.edu/creatures/misc/bees/ahb.htm.

National Biological Information Infrastructure (NBII) and IUCN/SSC Invasive Species Specialty Group (ISSG). "*Apis mellifera scutellata* (Insect)." ISSG Global Invasive Species Database, 2005. http://www.issg.org/database/species/ecology.asp?si=325&fr=1&sts.

Sanford, Malcolm T., and H. Glenn Hall. "African Honey Bee: What You Need to Know." Fact Sheet ENY-114, Entomology and Nemotology Department, Florida Cooperative Extension Service, Institute of Food and Agricultural Sciences, University of Florida, 2005. http://edis.ifas.ufl.edu/mg113.

Villa, J. D., T. E. Rinderer, and J. A. Stelzer. "Answers to the Puzzling Distribution of Africanized Bees in the United States or 'Why Are Those Bees Not Moving East to Texas?'" *American Bee Journal* 142(7): 480–83, 2002. http://www.ars.usda.gov/research/publications/Publications.htm?seq_no_115=133427.

■ | Argentine Ant

Scientific name: *Linepithema humile*
Synonym: *Iridomyrmex humilus*
Family: Formicidae

Native Range. South America: The Argentine ant is native to northern Argentina, southern Brazil, Uruguay, and Paraguay. Genetic studies reveal that those ants in the continental United States derive from the southern Paraná River region of Argentina.

Distribution in the United States. Argentine ants are widely established in California. They are also found in southern Arizona. In the Gulf coast states, they are common from east Texas to Florida. They have also been reported in Georgia, South Carolina, and North Carolina. In Hawai'i, they are spreading through Haleakala National Park on Maui. Populations also exist on the Big Island.

Description. The Argentine ant is medium to dark brown and about 0.1 in. (2–3 mm) long. The workers all have the same morphology. Their bodies are smooth and shiny without hairs on the back of the head or thorax. The petiole or waist segment consists of a single upright scale. When viewed head-on, the eyes do not protrude beyond the outline of the face. These ants lack stingers or acidopores. They move quickly along strong foraging trails, often in large numbers. When crushed, they give off a musty odor.

Related or Similar Species. Most other ants give off an acidic odor when crushed.

Introduction History. The Argentine ant entered the United States at New Orleans in the early 1890s. The first record dates to 1891. It probably arrived in shipments of coffee or sugar from Argentina and may have spread across the southern United States on trains. It was first documented in California in 1907. Reintroductions to the Southeast may have occurred in the early 1900s in cargo arriving at Gulf ports from California. Argentine ants came to Hawai'i during World War II, presumably with shipments of goods on troop ships from California.

Local dispersal may be facilitated by the ants' predilection to move their nests frequently. Therefore, they can rapidly take advantage of new opportunities for new nest sites in potted plants or in refuse, items that humans will further disperse. They also "raft" when their habitat is flooded and thereby move to new areas. New colonies can be established with as few as 10 workers and a queen.

The invasive success of Argentine ants has been attributed to the genetic similarity of all individuals, a consequence of either just one or very few colonizing events. The argument has been that the ants fail to recognize individuals born of different queens as "others," so all can live together in massive supercolonies formed by budding. In other words, they lack population-control measures that would have resulted from competition among colonies. Others maintain that the reason supercolonies in areas to which the Argentine ant has been introduced are so much larger than in their native range is due to a reduced parasite load and reduced competition from other ant species in the introduced range. Those two factors allow supercolonies to survive longer and attain great sizes, an aid when invading new territory.

Habitat. Argentine ants prefer moist environments with moderate temperatures. They can be found in a variety of natural, disturbed, and man-made habitats, from agricultural areas (especially citrus orchards) to natural forests and grasslands, and in urban areas both indoors and outdoors. In drier regions, such as southern California, they prefer to be near perennial streams or irrigated fields and human dwellings. They are the most common "house ant" in California, particularly in urban San Diego and Los Angeles. Usually restricted to lower elevations, they are invading alpine areas in Hawai'i. Argentine ants nest in mulch or soil, in rotting wood, and in garbage. Nests are often found in soil under objects or near tree roots and in potted plants or under walkways.

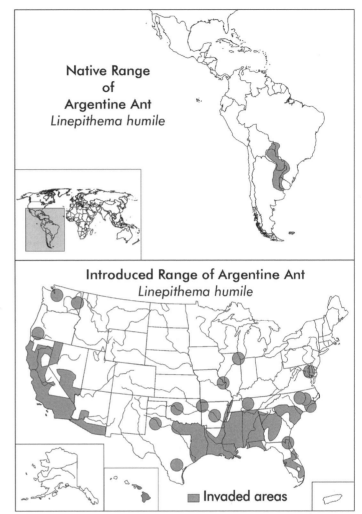

Top: Argentine ants are native to the Paraguay-Paraná River basin of northern Argentina, southern Brazil, Paraguay, and Uruguay. Bottom: Argentine ants are established in California and the Gulf coast states; disjunct populations occur north of these areas. In Hawai'i, they occur on Mau'i and the island of Hawai'i. (Adapted from "Sources of Introduced Argentine Ants Tied to Reduced Genetic Variation." University of California, San Diego, news release, n.d. http://ucsdnews.ucsd.edu ?rewsrel/science/mcants.htm.)

Diet. This ant prefers sweet substances. It tends aphids and consumes the honeydew that these and other insects such as mealybugs and scales secrete. It also feeds on nectar and the body fluids of dead animals. Occasionally, it will capture small, slow-moving insects or take their eggs. In homes it seeks sugar, proteins, and water.

Life History. Queens and drones mate in the nest, unlike many ants that perform nuptial flights. All worker ants are sterile. Nonetheless, worker ants can rear eggs and direct the development of early instar larvae into reproductive females without queens being present.

0.5 mm

The petiole or waist of the Argentine ant has a single upright scale, clearly visible in this photograph. (Eli Sarnat.)

A colony expands and disperses by budding new subcolonies. The result can be a huge supercolony with many queens. The production of males apparently is determined by the quantity of food available to larvae. Workers may kill the queen after a year and accept a newly mated queen in her place.

Impacts. Argentine ants are aggressive and have displaced native ant species in various parts of the world. In California, local extinctions of natives such as army ants (*Neivamyrmex* spp.) and harvester ants (*Messor* spp. and *Pogonomyrmex* spp.) have occurred.

Ants and other native insects are often important seed dispersers and plant pollinators. If the natives are severely reduced by invading Argentine ants, native plants may also decline. This is a major concern in Haleakala National Park, Hawai'i, where Argentine ants are encroaching upon the habitat of the endemic giant silversword (*Argyroxiphium sandwicense*), which is pollinated by yellow-faced bees (*Hylaeus* spp.). It is feared ant predation on bee larvae will decimate the native bee.

Competition between Argentine and native ants for food and habitat in California has affected the coastal horned lizard (*Phrynosoma coronatum*), which has a diet specializing on native harvester and carpenter ants. The loss of native ants has contributed to the decline of this once-common lizard. Decreases in the numbers of the native desert shrew (*Notiosorex crawfordi*) are also attributed, in part, to the presence of the Argentine ant.

Argentine ants can be serious pests in commercial apiaries. Ants invade honey bee (*Apis mellifera*) hives, causing the bees to abandon it. The ants then remove the honey and bee larvae to their own nest. They can destroy 1–2 hives a day and soon overwhelm an entire bee yard.

This ant has been nominated as by IUCN as among "100 of the 'World's Worst' invaders."

Management. Once a colony is established, it is difficult if not impossible to eradicate it. Chemical controls such as DDT, chlordane, and dieldrin that once helped control ant populations are no longer legal to use in the United States. Bait traps can help, since the poison works slowly and ants bring it back to their nests, where it will kill all workers and queens. Prevention, through the inspecting of cargo and refuse, sealing cracks and other points of entry to buildings, and maintaining landscapes that offer no nest sites, coupled with swift action to eliminate any new infestations, is the most effective management strategy to limit the further spread of the Argentine ant.

Selected References

Krushelnycky, Paul, Andrew Suares, and IUCN/SSC Invasive Species Specialty Group (ISSG). "*Linepithema humile* (Insect)." ISSG Database, 2006. http://www.issg.org/database/species/ecology.asp?fr=1&si=127.

Sarnat, E. M. "*Linepithema humile*." PIAkey: Identification Guide to Ants of the Pacific Islands, Edition 2.0, Lucid v. 3.4. USDA/APHIS/PPQ Center for Plant Health Science and Technology and University of California–Davis, 2008. http://keys.lucidcentral.org/keys/v3/PIAkey/Fact_Sheets/Linepithema_humile.html.

Spring, Joe. "Argentine Ant (*Linepithema humile*) (Mayr, 1868)." Introduced Species Summary Project, Columbia University, 2004. http://www.columbia.edu/itc/cerc/danoff-burg/invasion_bio/inv_spp_summm/Linepithema_humile.html.

Westervelt, David, and Eric T. Jameson. "Argentine Ant, *Liniepithema humile* Mayr (Hymenoptera: Formicidae)." Florida Department of Agriculture and Consumer Services, 2009. http://www.doacs.state.fl.us/pi/pest_alerts/liniepithema_humile.html.

■ | Asian Longhorned Beetle

Also known as: ALB
Scientific name: *Anoplophora glabripennis*
Order: Coleoptera
Family: Cerambycidae

Native Range. East Asia: It is widely distributed in China from Sichuan and Gansu provinces north and east to the vicinity of Beijing. It is also found on the islands of Hainan and Taiwan, throughout Korea, and in Japan.

Distribution in the United States. The major infestations of Asian longhorned beetles are in urban trees in Brooklyn, Queens, Manhattan, Long Island, Staten Island, and Prall's Island, New York; and in suburban Chicago. They have also been found in Hoboken, Jersey City, and Middlesex/Union counties, New Jersey; and in and around Worcester, Massachusetts.

Description. The adults are large, glossy black beetles with irregular white spots on the wing covers (elytra). The scutellum, a small triangular plate between the attachment points of the forewings, is black. The antennae are longer than the body and marked with distinct black and white bands. Their bodies are 0.75–1.0 in. (20–35 mm) long and 0.25–0.5 in. (6–12 mm) wide. Adults are seen in the open from late spring into fall.

The presence of Asian longhorned beetles is often recognized by signs on infested trees. Shallow oval borings in the bark of trees ooze sap. Sawdust-like debris collects at the base of trees or in the crotches of large branches. Circular holes the width of an adult are the exit holes of newly emerging beetles.

Eggs are off-white and oblong, with slightly concave ends. They are 0.2–0.3 in. (5–7 mm) long. Larvae are pale yellow elongate grubs that reach lengths of nearly 2.0 in. (50 mm) before pupating. The pupae are off-white and about 1.2–1.3 in. (30–33 mm) long and 0.4 in. (11 mm) wide. All of these life-cycle stages are hidden deep inside living trees.

Related or Similar Species. The whitespotted sawyer (*Monochamus scutellatus*), a native beetle found in most forested areas of the United States, is about the same size as the Asian longhorned beetle and also has long antennae. The body, however, is a dull or bronzed black, and the antennae are only faintly banded. The scutellum is white. Larvae are very similar to those of the Asian longhorned beetle. Whitespotted sawyers infest conifers, whereas Asian longhorned beetles prefer hardwoods.

Another large native long-horned beetle is the boldly black-and-white cottonwood borer (*Plectrodera scalator*) of the eastern United States and the Great Plains. Its antennae are solid black and equal in length to the body. Larvae dwell in and near the roots, and pupation takes place below ground level in roots; therefore, no exit holes appear in tree trunks or branches.

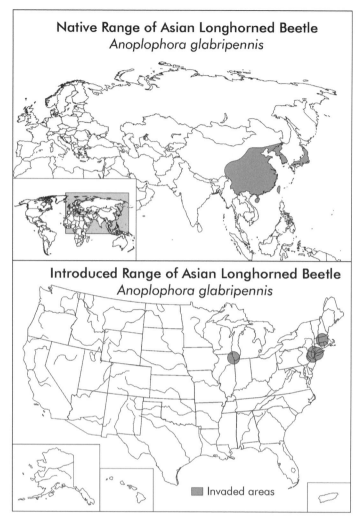

Native Range of Asian Longhorned Beetle
Anoplophora glabripennis

Introduced Range of Asian Longhorned Beetle
Anoplophora glabripennis

■ Invaded areas

Top: Asian longhorned beetles are native to eastern China, Korea, Japan, and Taiwan. Bottom: Asian longhorned beetles infest urban trees in and around Chicago, Illinois; the boroughs of New York City; New Jersey; and Worcester, Massachusetts.

Introduction History. Asian longhorned beetles were first reported in the United States in the Greenpoint section of Brooklyn, New York, in 1996, although they may have been present but undetected for 10 years before then. They probably arrived in wooden pallets or other solid-wood packing materials with pipes that were brought in from China in the late 1980s for a sewer project. From Brooklyn, they spread to Amityville on Long Island, Queens, and Manhattan. The introduction to Long Island was the result of a tree care company inadvertently disposing contaminated wood from Brooklyn at several locations. A separate introduction gave rise to the population in Chicago, first reported in 1998. Asian longhorned beetles showed up in Jersey City, New Jersey, in 2002 and had expanded their distribution to other towns (Woodbridge and Rahway) by 2004. They appeared on Staten and Prall's islands, New York, in 2007, and in Worcester, Massachusetts, in 2008. Live beetles continue to be discovered at ports and warehouses in various states.

Local dispersal from primary sites of infestation is most likely aided by the transport of infested firewood by vacationers, hunters, and others. Larvae hidden deep inside trees may be accidentally moved in live trees and cut timber as well. On their own, Asian longhorned beetles disperse very slowly. They were found exclusively on trees in cities and towns until 2008, when they were discovered in natural forests outside Worcester, Massachusetts.

Habitat. Asian longhorned beetles usually infest urban broadleaf deciduous trees, but since 2008, they have been found in natural hardwood forests.

Diet. Asian longhorned beetles differ from most other members of their family by eating the living wood of at least 18 species of broadleaf tree. (Most temperate-zone long-horned beetles have one preferred host and feed on dead and dying trees.) Maples (*Acer* spp.) are preferred, but Asian longhorned beetles have been reported to attack horse-chestnuts and buckeyes (*Aesculus* spp.), willows (*Salix* spp.), ash (*Fraxinus* spp.), birches

A. Adult Asian longhorned beetle. (Donald Duerr, USDA Forest Service, Bugwood.org). B. Asian longhorned beetle larvae. (Michael Bohne, Bugwood.org.)

(*Betula* spp.), and sycamores (*Platanus* spp.). Larvae eat healthy cambium, phloem, and xylem, forming long tunnels in the sapwood and heartwood of host trees.

Life History. In both New York and Chicago, adults reportedly emerge from pupae between July and October and bore their way through the bark of the host tree. This activity leaves telltale piles of frass and dime-size circular holes in the bark. The beetles usually fly only 150–250 ft. (50–70 m) away from the tree where they hatched. They feed on the bark of twigs on a new host tree for 2–3 days and then mate somewhere on the trunk or branches. The female chews a shallow, oval cavity into the cambium in which she lays a single egg; she will repeat this process 25–90 times. In 10–15 days, the eggs hatch into larvae that feed at first in the cambium but soon tunnel 4–12 in. (10–30 cm) deep into the tree. The larvae metamorphose into pupae and, about 18 days later, into adults inside the tunnels. Eggs, larvae, and pupae are able to overwinter if necessary. Adults remain active until late fall and then die. The life cycle is completed in 12–18 months.

Impacts. Large numbers of larvae infesting a tree weaken the host. Their tunneling girdles the tree, and dieback of the crown becomes noticeable. Weakened branches can break in strong winds. Ultimately, the tree dies.

The egg-laying cavities and exit holes joined to a gallery of tunnels inside the tree also open the host tree to infections and infestations of other kinds. Control efforts have resulted in the destruction of thousands of shade and ornamental trees in New York and Chicago. There is concern that if let loose in the broadleaf deciduous forests of New England, the Asian longhorned beetle could become a pest on the scale of gypsy moths (see Insects, Gypsy Moth) or chestnut blight (see Fungi, Chestnut Blight Fungus). Economic impacts in urban areas—where the beetles could rival Dutch elm disease (see Fungi, Dutch Elm Disease) in changing urban landscapes—involve diminished aesthetics and declining property values. Timber, maple syrup, nursery, and tourist industries are threatened in New England and elsewhere.

Management. The chief legal and effective way in the United States to deal with urban infestations is to cut down all trees known to host the beetle, chip and burn the wood, and grind out the stump. The trees that are removed are replaced with nonhost species. Infested areas are quarantined, and firewood, tree trimmings, and other tree by-products are prohibited from being transported out of the area. Contact insecticides are of little value, because much of the life cycle of the beetle is spent inside the tree where sprays cannot reach it. Systemic insecticides that can be taken up by the roots and distributed into stems and foliage have some potential, as does the application of parasitic fungi in bands on tree trunks to control adult beetles.

The Animal and Plant Health Inspection Service (APHIS) of the U.S. Department of Agriculture has inspectors at key ports of entry to check for beetle-infested cargo, targeting especially shipments from China with solid-wood packing materials. Since 1998, all such material is required to be treated prior to shipment to the United States, and compliance has been exceptionally high.

Federal quarantines were imposed in Chicago, New York, and New Jersey. In quarantine areas, only arborists certified to remove Asian longhorned beetles are allowed to cut down infested trees. No wood from the area can be used, moved, or disposed of without an inspection permit. New trees to be planted should be nonhost species. The quarantine zones in Hoboken and Jersey City, New Jersey, were deregulated in 2005; and the quarantine in Chicago was lifted in 2006. However, the quarantine area in the Middlesex/Union County area of New Jersey was expanded when new trees infested with the beetle were discovered. In July 2008, a new regulated area was established in Worcester, Massachusetts, that covers 62 sq. mi. (160 km^2). Visual surveys of trees and other means of early detection of and rapid response to new infestations are underway to attempt to prevent the spread of this pest into the natural forests of Massachusetts and surrounding states.

Selected References

"Asian longhorned beetle." University of Vermont, 2005. http://www.uvm.edu/albeetle/index.html.

Cavey, Joseph F. "*Anoplophora glabripennis*." National Information Center for State and Private Forestry, U.S. Forest Service, 2000. http://spfnic.fs.fed.us/exfor/data/pestreports.cfm?pestidval=53&lang display=english.

Muruetagoiena, Tamara. "Asian Longhorned Beetle (*Anoplophora glabripennis*)." Introduced Species Summary Project, Columbia University, 2004. http://www.columbia.edu/itc/cerc/danoff-burg/invasion_bio/inv_spp_summ/Anoplophora%20glabripennis.html.

■ | Asian Tiger Mosquito

Scientific name: *Aedes albopictus*
Synonym: *Stegomyia albopictus*
Order: Diptera
Family: Culicidae

Native Range. Asia. The Asian tiger mosquito occurs naturally from tropical Southeast Asia north through China, South Korea, and Japan, and on islands in the Pacific and Indian oceans, including Madagascar.

Distribution in the United States. Asian tiger mosquitoes are established in 26 states in the continental United States and in Hawai'i. Primarily a species of the Southeast, discontinuous outlying populations are recorded as far north as Minnesota and as far west as Nebraska, Kansas, Oklahoma, and Lubbock and Val Verde counties in Texas. On the East Coast, the northern limit is currently New Jersey, but that could change with a warming climate.

Description. The adult is a small mosquito with a black body and conspicuous silvery-white bands on the legs and feelers (palpus). A single white stripe runs along the back and head. Males are smaller than females and have feathery antennae and mouthparts modified for nectar-feeding. The female has unadorned feelers and a long proboscis adapted to biting and feeding on blood. As in all members of the genus *Aedes*, the abdomen narrows to a point. Adults are a little less than 0.25 in. (5 mm) long.

The wormlike larvae, or "wigglers," are easily disturbed by vibrations or a passing shadow and quickly descend to the bottom of the water container in which they live. They must periodically rise to the water surface to obtain oxygen through a somewhat inflated-looking breathing siphon. Fully grown wrigglers are about 0.25 in. (5 mm) long. The pupae are dark brown and curled into the shape of a comma. When disturbed, these "tumblers" roll end over end through the water.

Related or Similar Species. The yellow fever mosquito, *Aedes eagypti*, like the Asian tiger mosquito, is active by day and breeds in water held in containers. Its hind legs also bear white bands. In the field, the yellow fever mosquito is distinguished by the lyre-shaped pattern of white scales on its back and the white scales on the head of the female on a structure above its proboscis. In female Asian tiger

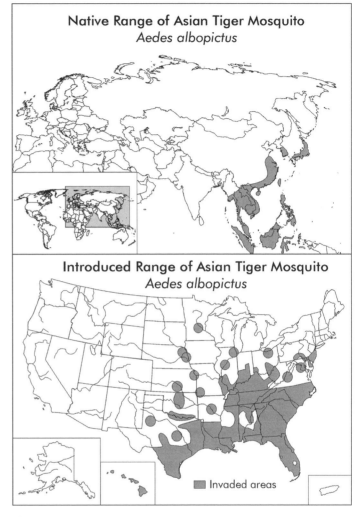

Top: The Asian tiger mosquito's native range extends from Southeast Asia to Japan and includes islands in the Pacific and Indian oceans. (Adapted from map by Gando, http://en.wikipedia.org/wiki/File:Albopictus _distribution_2007.png.) Bottom: Asian tiger mosquitos are currently established in 27 states. (Adapted from "Distribution of *Aedes albopictus* in the United States, by County, 2000." Centers for Disease Control and Prevention, Division of Vector-borne Infectious Diseases. www.cdc.gov/ ncidod/dvbid/arbor/albopic_new.htm.)

mosquitoes, this structure is black. The yellow fever mosquito originated in Africa and probably came to the New World on ships in the early days of trans-Atlantic trade and exploration.

Introduction History. Asian tiger mosquitoes probably arrived in Hawai'i soon after World War II. Their first verified occurrence in the continental United States, indeed in the New World, came in 1985 in Houston, Texas. It was accidentally transported from Japan in used tires imported for recapping. During the 1980s, importations of used tires increased, and Asian tires were preferred because of their high rubber content. They arrived in containers that were inadequately inspected at the ports of entry. The first

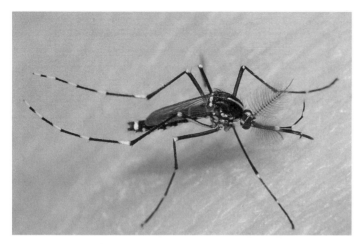

Asian tiger mosquito (male). (Susan Ellis, Bugwood.org.)

discovery of the mosquito in Maryland occurred in 1987 at a tire processing plant in Baltimore. The movement of used tires within the United States dispersed the insect to other eastern and midwestern states. Used tires are typically stored outdoors for long periods of time, and those that cannot be used for recapping often end up in illegal tire dumps. In both instances, they collect rainwater and become prime Asian tiger mosquito breeding grounds.

In California, Asian tiger mosquitoes were discovered in 2001 in containerized shipments of lucky bamboo (*Dracaena* sp.) from southern China. The popular ornamental plants had been transported in standing water so they would survive the long voyage.

The mosquitoes have been locally dispersed in moist vegetation and water containers, including cemetery flower pots. In several Florida counties, the mosquitoes were first found in cemeteries where fresh-cut flowers were placed at grave sites in plastic floral baskets. When the flowers died, they were discarded; however, the baskets were reused, often at different cemeteries. Eggs laid when flowers were fresh and standing in water in one cemetery would hatch later in another when water was again added.

Habitat. The Asian tiger mosquito likely evolved in forests where it could breed in tree-holes and in epiphytes and other places where water pooled. It still prefers the shade of densely vegetated areas, but exploits artificial containers in urban and suburban areas for breeding sites. It reproduces in bird baths, cemetery flower pots, empty soda cans, and other abandoned containers that hold water, but it is associated especially with old tires that collect water and organic-rich debris. The water only needs to be 0.25 in. (5–6 mm) deep. Northern range limits may be set by cold temperatures (daily mean temperatures in January of 32°F or 0°C) that prevent overwintering of eggs. Initially in the United States, the southern limit was established by summer day lengths shorter than 13.5 hours, but the species has overcome that controlling factor.

Diet. Asian tiger mosquitoes acquire energy by feeding on nectar. The females, however, need protein to produce eggs, and they feed also on blood. Humans are the chief host, but the females will feed on a variety of other mammals including dogs and cats, and on birds, reptiles, and amphibians. They generally feed at ground level, biting people around the ankles and knees.

The larvae are filter-feeders, consuming particulate organic debris and bacteria.

Life History. Mating occurs soon after adults emerge from pupae. The female mates only once and stores sperm for future egg production. A female feeds on blood for 4–5 days before laying eggs. She then deposits 40–150 eggs on the sides of containers just above water level. She will continue a weeklong cycle of feeding on blood and laying eggs for the remainder of her life. The eggs do not hatch until water level rises and covers them. Eggs may overwinter and not hatch until the following spring or summer, when they become flooded by water warmer than 60° F (15.5° C). The wriggling larvae develop through four instars and change into pupae about 10 days after hatching. Unlike many other insects, mosquito pupae are mobile, but they do not feed. Adults emerge from the pupae 10–14 days later. Adult mosquitoes live several days to a few weeks depending upon the weather. The lifespan is reduced during hot, dry spells. During their lives, adults rarely move more than a few hundred yards (or meters) from the container in which they hatched.

Impacts. The arrival of the Asian tiger mosquito raised major concerns of its potential as a vector for virus-based dengue fever. Dengue fever is a disease of the tropics. It once posed a sporadic health problem in southern states, but was eradicated in the United States in the 1940s. However, dengue fever is common in the Caribbean and tropical South and Central America, and could easily be reintroduced. A warming climate might enhance its ability to spread. It now appears that the Asian tiger mosquito is less likely than its close relative, the yellow fever mosquito, to transmit the dengue fever virus from human to human and cause major outbreaks of the disease. This is because the Asian tiger mosquito has multiple mammalian, avian, and reptilian hosts and is not restricted to a diet of human blood as is the yellow fever mosquito. If, after biting a human, the mosquito moves on to a bite another type of host, transmission of the virus is halted.

In laboratory studies, the Asian tiger mosquito has proved to be a competent vector for a number of other viruses, including eastern equine encephalitis, yellow fever, West Nile disease (see Microorganisms, West Nile Virus), LaCrosse encephalitis, St. Louis encephalitis, and western horse encephalitis. It is also a carrier of the nematode known as canine heartworm. Field populations have been found to carry eastern equine encephalitis virus and West Nile virus, but so far, no evidence exists that this mosquito has transmitted disease to any person.

The Asian tiger mosquito's greatest impact has been as an extreme nuisance. It is much more annoying than the yellow fever mosquito. An aggressive and persistent biter that feeds during the day, it tends to rest in shrubs in shady areas, where people also seek relief from the summer sun. The female interrupts her feeding to bite the same person several times and, because of her agility, is difficult to swat. She injects an anticoagulant from their salivary glands that causes the bite to swell and itch. Sometimes the itch may last for a week.

The Asian tiger mosquito has displaced the yellow fever mosquito from many areas in which the distribution areas of the two overlapped.

Management. The best defense against the Asian tiger mosquito is for individuals to remove potential breeding sites, such as old tires, empty buckets, or other containers that could fill with rainwater. Bird baths and wading pools should be emptied at least once a month. Roof gutters should be kept free of debris that would dam up water. The larvae can be killed with applications of the bacterium Bti (*Bacillus thuringiensis israelensis*). Adults are not readily trapped by contrivances that attract other mosquitoes, and they are resistant to many insecticides.

Biological controls include stocking predators of the larvae such as mosquitofish (*Gambusia* spp.; but see Fish, Mosquitofish) and developing a genetically engineered male that will pass a lethal gene to its female offspring.

Since 1988, the U.S. Public Health Service has required that all used tires imported into the United States from areas known to have Asian tiger mosquitoes be dried, cleaned, and fumigated.

Selected References

"The Asian Tiger Mosquito in Maryland." Maryland Department of Agriculture, n.d. http://www.mda.state.md.us/plants-pests/mosquito_control/_asian_tiger_mosquito_md.php.

Crans, Wayne J. "*Aedes albopictus* (insect)." ISSG Database, IUCN/SSC Invasive Species Specialist Group (ISSG), 2009. http://www.issg.org/database/species/ecology.asp?si=109&fr=1&sts.

Crans, Wayne J. "The Asian Tiger Mosquito in New Jersey." Fact Sheet. Rutgers Cooperative Research and Extension, New Jersey Agricultural Experiment Station, Rutgers, the State University of New Jersey, 1996. http://njaes.rutgers.edu/pubs/download-free.asp?strPubID=FS845.

O'Meara, G. F. "The Asian Tiger Mosquito in Florida." Entomology and Nematology Department, Cooperative Extension Service, Institute of Food and Agricultural Sciences, University of Florida, 2005. http://edis.ifas.ufl.edu/mg339.

Rios, Leslie, and James E. Maruniak. "Featured Creatures: Asian Tiger Mosquito." University of Florida, 2008. http://entnemdept.ufl.edu/creatures/aquatic/asian_tiger.htm.

■ | Brown Marmorated Stink Bug

Also known as: Yellow-brown stink bug; BMSB
Scientific name: *Halyomorpha halys*
Family: Pentatomidae

Native Range. East Asia: China, Japan, Korea, and Taiwan

Distribution in the United States. This rapidly spreading bug was reported in California, Delaware, Maryland, Missouri, New Jersey, New York, North Carolina, Oregon, Pennsylvania, Tennessee, Virginia, West Virginia, and the District of Columbia by 2010.

Description. Adults have the shield body shape typical of all stink bugs. The back is a marbled (marmorated) brown, and the underside a lighter tan color. Small round depressions on the head and first segment of the thorax (the pronotum) are coppery or bluish metallic in color. The edges of the pronotum are smooth, unlike the jagged margin of native stink bugs of the genus *Brochymena*. The next-to-last segment of the antennae bears pale bands, a distinguishing characteristic. The abdominal segments that extend beyond the wings have alternating bands of black and white; and the brown legs have faint light bands. Scent glands are located between the first and second pair of legs on the underside of the thorax. Adults are about 0.5 in. (12–17 mm) long and almost as wide.

Immatures or nymphs develop through five stages or instars. The first instars have yellowish-red abdomens with black bars and dark red eyes; they look like ticks and are about 0.09 in. (2.4 mm) long. Relatively inactive, they tend to remain clustered near the hatched egg mass. Successive instars become larger and more closely resemble adults. Their legs and antennae are black with white bands. The final, fifth instar has a whitish abdomen with red spots and is about 0.47 in. (12 mm) long. Knobs protrude in front of the scent glands on the underside, and spines occur on the legs, in front of the eyes, and on the sides of the thorax.

The pale green eggs are elliptical and measure about 0.04 in. (1 mm) in diameter. They are usually found on the underside of leaves in masses of 20 to 30.

Related or Similar Species. Brown marmorated stink bugs are easily confused with native species of stink bug, including the brown stink bug (*Euschitus servus*) and the green stink bug (*Acrosternum hilare*); with rough stink bugs (*Brochymena* spp.); and some leaf-footed or squash bugs (Family Coreidae). The Western conifer seed bug (*Leptoglossus occidentalis*) also invades buildings in the autumn in order to winter over, but it can be distinguished from the BMSB by the flattened structures on the lower segments of its hind legs.

Introduction History. This bug was accidentally introduced in packing materials to Allentown, Pennsylvania, where it was first collected in September 1998. It may have been present for a couple of years before it was discovered and identified. In 2000 and again in 2002, it was collected in New Jersey and, soon afterward, appeared in Delaware, Connecticut, and New York. In 2005, an established population was documented in Oregon. The stink bug has become established in many places on the East Coast of the United States and continues to expand its range. By 2009, it was established in Maryland, Virginia, and West Virginia and had been collected in California, Florida, Mississippi, Missouri, North Carolina, Ohio, Tennessee, and Washington, D.C. It is suspected that they are moving across the country in motor vehicles.

Habitat. Originally, populations were restricted to urban and suburban landscapes where they were associated with ornamental plants, garden crops, and fruit and shade trees. They

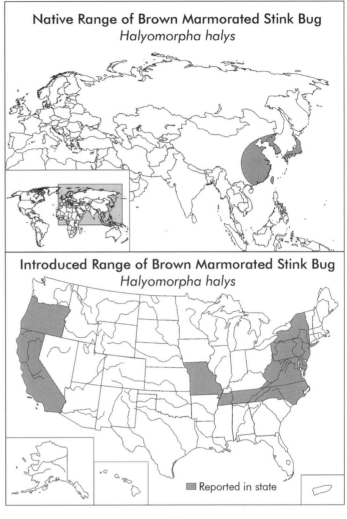

Top: The brown marmorated stink bug is native to East Asia. Bottom: This new invader was reported in 12 states and the District of Columbia by 2010.

have since expanded into agricultural areas, including orchards. Brown marmorated stink bugs seek winter shelter in buildings, entering in the fall through any small opening. They remain dormant much of the winter and leave houses and other buildings in the spring to reproduce and forage outside.

Diet. These bugs are sucking insects that feed on a wide variety of plants. They pierce fruit and leaves with specially adapted mouthparts and withdraw plant juices. Nymphs feed on fruits and seed pods. They are known to eat apples, peaches, pears, blackberries, tomatoes, green peppers, beans, sweet and field corn, and soybeans. Among ornamentals, they attack shade trees such as Norway maple (*Acer platanoides*) and ornamentals such as princess tree (*Paulownia tomentosa*), butterfly bush (*Buddleia* spp.), Rugosa roses (*Rosa rugosa*), and honeysuckle (*Lonicera* sp.)

Life History. Mating takes place in early summer, approximately two weeks after adults emerge from hibernation. Egg-laying begins in early summer and will continue through

The brown marmorated stink bug is rapidly expanding its range in the eastern United States. (David R. Lance, USDA APHIS PPQ, Bugwood.org.)

September in the Mid-Atlantic states. The female lays new egg masses on the undersides of leaves about once a week throughout the summer. The eggs hatch in four or five days and first instar nymphs usually stay near their egg mass until after the first molt. Each nymphal stage lasts about one week and is followed by a molt that produces the next instar. About two weeks after the final molt, the adults are sexually mature. Only one generation appears to be produced each year. However, since several generations are produced each year in the subtropical regions of its native range, the same may be expected when the stink bug becomes established in the Southeast. Adults live several years, and a single female may lay up to 400 eggs during her lifetime.

Impacts. Brown marmorated stink bugs are primarily nuisances in homes and other buildings, where they congregate in autumn and then overwinter. On warm days, they may rouse from hibernation and fly around, and in the spring, they fly to windows and doorways as they try to leave the premises. The bugs do no structural damage, do not lay eggs inside buildings, and are not known to carry any diseases. However, when crushed, they secrete a most unpleasant odor from their scent glands.

The stink bug can damage ornamental and garden plants and is becoming an agricultural pest in apple and peach orchards. Their feeding methods causes puckering, scarring, and deformation of fruits—a condition known as cat facing, which renders them unfit for sale as fresh market items. The fruits may still be processed, however, for juices, jams, sauces, and the like. Leaf-stippling will occur on leaves. The spread of the bug could threaten commercial apple, peach, and soybean crops. Damage to orchard crops has already been reported in Maryland and West Virginia.

Management. Preventing stink bugs from getting into buildings is the best line of defense. All potential cracks and crevices around windows and doors, chimneys, siding, fascia, and holes through which wiring passes should be caulked. Screens on windows and doors and on roof vents and weep holes should be kept in good repair. Insecticides applied around such openings in early fall when the bugs swarm and try to gain entry have short-term effects. Most degrade rapidly. A professional exterminator should be consulted for the best and safest results. With light infestations, individual bugs can be swept into a pail of water.

Once the insects are indoors, vacuuming individual stink bugs is the best remedy. Empty vacuum bags immediately to prevent a buildup of the stench injured, dying bugs exude. Do not spray insecticides in an attempt to kill bugs hiding in the house. The dead bugs will attract other pests such as carpet beetles, which will later feed on woolen items, dry goods, and other natural materials.

Spot treatment of garden plants with insecticides will reduce the damage from these insects.

Selected References

Day, Eric R., and Dini Miller. "Brown Marmotated Stink Bug, Homoptera: Penatomidae: *Halyomorpha halys*." Virginia Cooperative Extension, 2009. http://pubs.ext.vt.edu/2902/2902-1100/2902-1100.html.

Gyeltshen, Jamba, Gary Bernon, and Amanda Hodges. "Common Name: Brown Marmorated Stink Bug." Featured Creatures, Department of Entomology and Nematology, University of Florida, 2010. http://entnemdept.ufl.edu/creatures/veg/bean/brown_marmorated_stink_bug.htm.

Jacobs, Steve. "Brown Marmorated Stink Bug," Entomological Notes. College of Agricultural Sciences, Cooperative Extension, Pennsylvania State University, 2010. http://ento.psu.edu/extension/fact sheets/brown-marmorated-stink-bug.

MIPSP. "Brown Marmorated Stink Bug." *Massachusetts Introduced Pests Outreach Project*, 2008. http://www.massnrc.org/pests/pestFAQsheets/brownmarmoratedstinkbug.html.

■ | Common Bed Bug

Also known as: Bedbug, chinche, mahogany flat, redcoat, wall louse
Scientific name: *Cimex lectularius*
Order: Hemiptera
Family: Cimicidae

Native Range. Uncertain: possibly eastern Mediterranean. May have originated as a parasite of bats that transferred to humans in caves occupied by nomadic Stone Age peoples. Bed bugs apparently did not become a problem until people began to live in villages and cities.

Distribution in the United States. Throughout.

Description. Adult bed bugs are small, oval, wingless insects. Before they feed, bed bugs are flat and brown, but when engorged after a blood meal, they are swollen and dark red. Eyes are dark. The upper surface of the first segment of the thorax (the prothorax) forms a wide collar-like plate that curves slightly around the broad head. Antennae are prominent and have four segments. The abdomen has 11 segments. Before feeding, adult bed bugs are about 0.25 in. (4–9 mm) long and 0.06–0.1 in. (1.5–3 mm) wide.

Newly hatched nymphs are the size of poppy seeds and colorless. Later nymphal instars are smaller, paler versions of the adult. They become bright red after feeding. The cream-colored oval egg is about 0.04 in. (1 mm) long.

Bed bugs are rarely seen. Evidence of their presence develops in areas where they congregate and hide during the daytime. Such harborages will accumulate dark spots and staining from their feces, eggs, and eggshells, shed skins of maturing nymphs, and adult bugs. Red stains on bedding and mattresses occur when engorged bed bugs are accidentally crushed by their sleeping human hosts. Bed bugs do exude a distinct odor, variously described as resembling crushed coriander, rotting raspberries, or a sweet musty smell; but only when infestations are heavy can this be detected by people.

Bed bug bites are indicated by a patch of redness on the skin with a darker raised welt in the center. However, half of the people who are bitten experience no reaction; and among those who do show visible signs, the size and itchiness of the bite varies greatly. Bites usually occur on the exposed skin of the face, neck, arms, and hands and may be clustered or occur in a line. Identification of the presence of bed bugs from the bites alone is problematic.

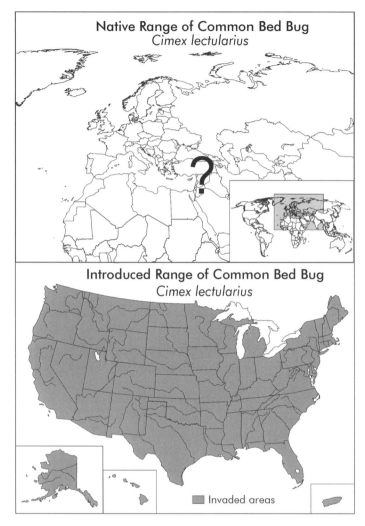

Top: The actual origins of the common bed bug remain unknown, but it may derive from the eastern Mediterranean. Bottom: Common bed bugs are found throughout the United States.

Related or Similar Species. Most other bed bug species prefer other mammals and birds as hosts. *C. lectularius* prefers humans, but will feed on bats, cats, dogs, and rodents as well as chickens and other birds. In tropical regions of the world, *C. hemipterus* is the more common parasite of humans. It has a noticeably narrower prothorax than *C. lectularius*. (The bed bug found in Hawai'i is *C. lectularius*.) Bed bugs could be confused with small ticks or cockroaches. Their bites can be mistaken for those of mosquitoes, fleas, and spiders.

Introduction History. The dispersal history of bed bugs is not well documented, but they became widespread in the temperate regions of the Northern Hemisphere during the colonial period and likely came to the United States in the 1700s when infestations on sailing ships out of Europe were notorious. They were known from Greece as early as 400 BC but were not mentioned in accounts from Germany until the eleventh century or from France until the thirteenth century. Although first noted in England in 1583, they did not become common until 1670. Presumably, sometime after that they made their way to what is now the United States. They had appeared in Jamaica, another British colony, by 1726.

In the early twentieth century, bed bugs were so common that they were deemed one of the top three household pests. The widespread use of DDT after World War II essentially eradicated bed bugs from the United States and other developed countries in the 1950s. However, bed bug populations quickly developed resistance to DDT, and the use of the pesticide was banned altogether in the United States in 1972 because of its deleterious effects on birds. In the 1990s, after an absence of nearly 50 years, bed bugs began a rapid increase, not only across the United States but in all the industrialized countries of the world. Once associated with poverty and underdevelopment, bed bugs now also occupy high-end hotels and residences. The resurgence may be a consequence of increases in international travel and immigration; the development of pesticide resistance, especially to pyrethroids; and

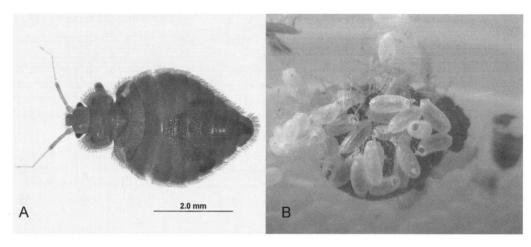

A. Adult bed bug. (James D. Young, USDA APHIS PPQ, Bugwood.org.) B. Bed bug eggs. (Mohammed El Damir, Pest Management, Bugwood.org.)

changing control practices for other vermin, such as cockroaches and ants, that no longer rely on broad-spectrum, persistent, and highly toxic pesticides, but focus on safer methods of bait-trapping. One study has traced the origin of the latest outbreak of bed bugs to poultry facilities in Arkansas, Delaware, and Texas. Workers could have unknowingly carried the pests home or elsewhere when they left the workplace. Another study suggests that recent infestations began in the Northeast and spread south and west from there. The last states to report major infestations were Hawai'i, Oregon, and Washington.

Habitat. These bugs occur where people sleep. All they need are a source of blood and a place to hide. They tend to hide close to where they feed. Their flat shape and small size allow them to congregate in tiny cracks and crevices. They prefer dry, rough surfaces such as wood and paper and avoid wet and hard surfaces such as stone, metal, and plaster. Favored sites include the seams of mattresses, box springs, bed frames, and cracks in molding. In hotels, they are often first found behind headboards. They may be found among items stored under beds; along and under the edges of wall-to-wall carpeting; behind picture frames, switch plates, and outlets; behind wallpaper; and in phones, smoke detectors, and televisions. In large infestations, they will spread to upholstered furniture and among clutter accumulated in closets.

While bed bugs were once associated in the popular mind with urban poverty and filth, this is not the case. Today's outbreaks commonly occur in fine hotels, vacation resorts, and cruise ships as well as in hospitals, college dormitories, and well-maintained apartment buildings and single-family residences.

Diet. Common bed bugs are obligate feeders on blood. Humans are their preferred hosts. They feed mainly at night while their hosts are asleep. They respond to warmth and carbon dioxide to locate their blood meal. Finding exposed skin, they pierce it with their mouthparts and inject their saliva, which contains an anticoagulant and numbing compound. A full meal takes 5–10 minutes. Typically during that time, a bed bug makes a row of three bites: "breakfast," "lunch," and "dinner." Adults may live several months without feeding.

Life History. After mating, a female lays 2–3 eggs per day in cracks and crevices. The female continues to lay throughout her lifetime, which may last 6–12 months, producing a total of 200–500 eggs. At room temperature (68°F or 20°C), nymphs hatch out in 1–2 weeks and immediately begin to feed. Bed bugs undergo incomplete metamorphosis in that the nymphs are simply tiny adults. Each of the five juvenile or nymphal states requires a blood meal in order to molt into the next stage. Full maturation takes from 9 to 18 weeks. Several generations can be produced in a year.

Impacts. Bed bugs are not known to transmit diseases to humans. In minor infestations, they are chiefly an annoyance and embarrassment. Their bites cause itchiness and inflammation in some people. Scratching the bites might result in secondary infections.

The occurrence of bed bugs can result in lawsuits against landlords and other property owners obligated to provide safe and habitable conditions for tenants.

Management. Eradication, especially of major infestations in multiple-unit buildings, is difficult and requires the assistance of experienced, professional pest control companies in administrating an integrated pest management program. Although the media has recently tried to blame EPA and its ban of persistent pesticides for the current resurgence in bed bugs, many populations had already developed resistance to DDT and other organophosphates in the 1950s and 1960s as well as to the pyrethroids used in their place. Today, few over-the-counter insecticides registered for use in homes are effective against bed bugs, but licensed pest control operators do have some options available. Furthermore, some companies are using specially trained dogs to help locate infestations.

Several nonchemical treatments exist that help control minor infestations. Careful and repeated vacuuming of places where bed bugs like to hide can physically remove bugs from sites that a high suction wand can reach. Heat (temperatures above 120°F or 49°C) kills eggs, nymphs, and adults. Clothes and bedding can be washed in hot water. Backpacks, toys, or shoes and the like—dry or wet—can treated in a clothes dryer set at medium-to-high heat for 20 minutes. Larger items such as suitcases can be put in plastic bags and set in the sun or in a closed car parked in the sun for a day. Freezing temperatures will also kill bed bugs, but items must be left outside in winter or in a freezer for several days for the cold to be effective.

Preventing an infestation is paramount in the fight against bed bugs. Bed bugs are transported from place to place on luggage, clothing, boxes, secondhand furniture and mattresses. They spread through buildings by crawling out entry ways and through holes in walls and ceilings. Caulking cracks and crevices and sealing or removing loose wallpaper will help deter them, but careful inspection of items brought into the house is also important. This is especially true for purchasers of secondhand beds and sofas and frequent travelers. In hotels, keep suitcases off the floor and on hard surfaces such as luggage stands and table tops. Put clothing in disposable plastic bags and directly into a washing machine and/or dryer upon arrival home. Remove clutter from floors and monitor or eliminate any other potential bed bug harborages. Keeping sheets and blankets from contacting the floor by tucking them in and placing the legs of a bed frame in a small dish of mineral oil will keep bed bugs from crawling up into a bed. Usually a combination of techniques is required, as is the cooperation of all inhabitants of a building. Frequent inspection of one's surroundings will at the very least allow early intervention by a professional.

Bed bug bites may be avoided by wearing pajamas that cover arms and legs. Insect repellents have not proven very effective deterrents, although mosquito netting impregnated with permethrin may help—until bed bugs develop resistance to that pesticide.

Today's rapid increase in bed bug populations is making them a major public health concern and instigating more research on ways to control and eradicate this long-neglected insect. New products such as the insecticide chlorofenpyr and the insect growth regulator hydroprene, though slow acting, seem to be effective. Still, no chemical means of control by itself can eliminate an infestation entirely.

Anyone with a suspected infestation should contact a licensed pest control operator for confirmation and notify the public health department. Do *not* apply any insecticide or pesticide directly to a mattress or other surface that comes into direct contact with people or pets.

Selected References

Anderson A., and K. Leffler. "Bed Bug Infestation in the News: A Picture of an Emerging Public Health Problem in the United States." *Journal of Environmental Health* 70(9): 24–27, 2008.

Brooks, Shawn E. "Bed Bug, *Cimex lectularius* Linneaus (Insecta: Hemiptera: Cimicidae)." Document EENY-140 (IN297), Featured Creatures, Entomology and Nematology Department, Florida Cooperative Extension Service, Institute of Food and Agricultural Sciences, University of Florida, 2009. https://edis.ifas.ufl.edu/in297.

Centers for Disease Control and Prevention, U.S. Environmental Protection Agency. "Joint Statement on Bed Bug Control in the United States from the U.S. Centers for Disease Control and Prevention (CDC) and the U.S. Environmental Protection Agency (EPA)," 2010. http://www.cdc.gov/nceh/ehs/publications/Bed_Bugs_CDC-EPA_Statement.htm.

Pollack, Richard, and Gary Alpert. "Bed Bugs." Harvard School of Public Health, 2005. http://www.hsph.harvard.edu/bedbugs/.

Potter, Michael F. "Bed Bugs." ENTFACT-636, Department of Entomology, University of Kentucky College of Agriculture, 2010. http://www.ca.uky.edu/entomology/entfacts/ef636.asp.

Staab, Tina. "The History of Bed Bugs." eHow.com, 2009. http://www.ehow.com/about_5376204_history-bed-bugs.html#ixzz0y7tJ0BxF.

■|Emerald Ash Borer

Also known as: EAB
Scientific name: *Agrilus planipennis*
Order: Coleoptera
Family: Buprestidae

Native Range. Emerald ash borers are native to northeastern China and adjacent areas of Mongolia and Russia, Korea, Japan, and Taiwan, where forests are composed of broadleaf trees closely related to those in the eastern United States.

Distribution in the United States. This beetle currently occurs in Illinois, Indiana, Kentucky, Maryland, Michigan, Minnesota, Missouri, New York, Ohio, Pennsylvania, Virginia, West Virginia, and Wisconsin.

Description. Adult emerald ash borers are small, metallic-green beetles. The slender, elongate bodies are about 0.5 in. (7.5–13.5 mm) long, males being somewhat smaller than females. When open, the wing covers expose a metallic-purplish-red abdomen. The segment just behind the head, to which the first pair of legs is attached, is wider than the head. Larvae are whitish with a brown head. Their abdomens have 10 segments, are flattened dorsoventrally, and end with a pair of brown pincers. Late stage larvae are about 1.0 in. (26–32 mm) long.

Adults are active and seen in the open only from mid-May to September and usually only in the afternoons of warm, sunny days, making infestations difficult to detect. Early signs and symptoms that emerald ash borers have invaded a tree include jagged orange scars in the bark made by woodpeckers on the upper trunk and branches and a top-down thinning of the canopy and yellowing of leaves. Bark may split vertically where larval feeding galleries have been excavated beneath it. If the bark of infested ash trees is cut away, serpentine tunnels filled with

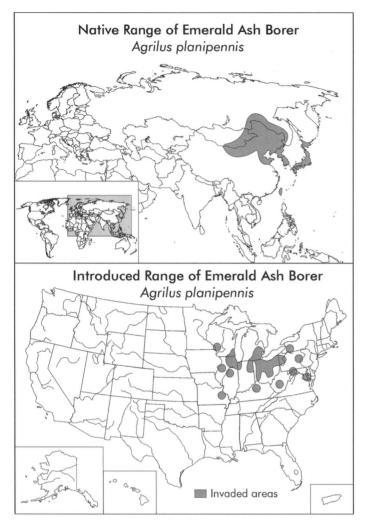

Native Range of Emerald Ash Borer
Agrilus planipennis

Introduced Range of Emerald Ash Borer
Agrilus planipennis

■ Invaded areas

Top: Emerald ash borers are native to the broadleaf forests of northeastern China, Korea, Japan, and Taiwan. (Adapted from "Native range of emerald ash borer in Asia." USDA Forest Service, Northern Research Station. http://www.nrs.fs.fed.us/disturbance/invasive_species/eab/local-resources/images/native_range.gif.) Bottom: The emerald ash borer currently affects ash trees in cities, suburbs, and forests in 13 states. (Adapted from map by Cooperative Ash Borer Project, USDA Forest Service, 2010. http://www.emeraldashborer.info/files/MultiState_EABpos.pdf.)

a fine sawdust-like frass are the obvious signs that the beetles are present. With heavy infestations, as trees begin to die, they resprout from the roots.

Related or Similar Species. A number of ash borers are native to the United States but are not similar in color, body shape, or shape of the exit hole. However, several metallic wood-boring beetles native to American forests are likely to be confused with emerald ash borers. Among them are the bronze birch-borer (*Agrilus anxius*) and the two-lined chestnut borer (*Agrilus bilineatus*), both of similar size and shape but not green; and the six-spotted tiger beetle (*Cinindela sexguttata*) and the caterpillar hunter (*Calosoma scrutator*), both of which are green but have distinctly different body shapes.

Introduction History. The emerald ash borer was first confirmed in the United States in 2002 in Canton, Michigan. It probably had arrived at least 15 years earlier in wood packing materials coming from China. From Michigan, the beetle migrated to Ohio (2003) and Minnesota (2009). In 2003, it was carried illegally to Prince George, Maryland, in nursery stock originating in Michigan and sold in Maryland and Fairfax County, Virginia. The emerald ash borer continues to be spread to other parts of infested states and to new states in live trees, green lumber, firewood, wood chips, and debris. It had arrived in Indiana by 2004; Upper Peninsula, Michigan, by 2005; Cook County, Illinois, by 2006; several counties in Pennsylvania by 2007, and at New River Gorge in Fayette County, West Virginia, by 2007. It was confirmed in Missouri and Wisconsin in 2008 and in Kentucky, Minnesota, and New York in 2009.

Habitat. Emerald ash borers require ash trees. They survive in urban and suburban parks, yards, and roadways, woodlots, and natural broadleaf deciduous forests.

A. Adult emerald ash borer. (Pennsylvania Department of Conservation and Natural Resources–Forestry Archive, Bugwood.org.) B. Emerald ash borer larva. (Pennsylvania Department of Conservation and Natural Resources–Forestry Archive, Bugwood.org.) C. Exit holes made by emerald ash borer. (Joseph O'Brien, USDA Forest Service, Bugwood.org.) D. Twisting tunnels or galleries made by emerald ash borer larvae. (Art Wagner, USDA APHIS PPQ, Bugwood.org.)

Diet. Emerald ash borers feed exclusively on trees of the genus *Fraxinus*. Adult beetles eat along the edges of leaves. Larvae eat phloem and xylem in the outer sapwood. They tend to prefer the upper trunks and branches of large trees. In the United States, green ash (*F. pennsylvanica*) and black ash (*F. nigra*) are more vulnerable to attack than white ash (*F. americana*) and blue ash (*F. quadrangulata*), but all ash species are at risk.

Life History. Females deposit their eggs one by one in crevices in the bark or under flaps of bark on the trunk and branches of ash trees. Seven to 10 days later, the eggs hatch, and the larvae bore through the bark into the phloem to begin feeding. They feed for several weeks, creating ever wider, s-shaped tunnels in the outer sapwood as they grow. A single feeding gallery will range in length from 4 to 12 in. (20–50 cm). Feeding ends in the fall, and the prepupal larvae overwinter in shallow chambers about 0.5 in. (1 cm) deep in the outer sapwood or bark. The pupal stage begins in late April or May, and adults begin to emerge in late May, with peak emergence occurring from early to mid-June. Adults are able to fly immediately upon emerging and, although strong fliers, usually move less than 0.5 mi. (0.8 km) to find a mate and a new host tree. They will feed on the foliage for a few days while they mature, and then they mate. Females begin egg-laying after another week or two of feeding. Adult males survive about one month after emergence, females for about two months.

Research in Michigan indicates that the life cycle is longer in healthy trees than stressed ones. In newly infested trees, many larvae overwinter in the earliest stage of development and feed during a second summer. In stressed trees, nearly all larvae overwinter in the late prepupal stage, and then pupate and emerge as adults the next summer.

Impacts. Emerald ash borers kill all trees that they infest. Adult beetles do relatively little damage when they feed on the foliage. It is the larvae that are so destructive: they feed on phloem and interrupt the flow of nutrients through the tree, so the tree weakens and dies. The deaths of ash trees affect city and suburban landscapes, where they are planted as shade trees and ornamentals in parks and yards. In forests, their deaths mean a loss of browse, shelter, and seed that may be consumed by birds, small mammals, and insects. Although many dead or dying trees produce sprouts at the base, these too are attacked once they attain diameters of about 1 in. (0.5 cm). The demise of white ash across large swaths of eastern North America could result in the loss of forest biodiversity and have widespread impacts on habitat and watershed quality. In economic terms, trees valuable in the timber industry are being lost, and livelihoods based upon the sale of wood products such as lumber, mulch, and firewood are threatened. A special concern exists among some Native American groups who view black ash as a cultural resource because they use it in their basket making.

Management. Containment rather than eradication is the main goal of management. The federal government (APHIS) and infected states set up quarantine zones in efforts to prevent the further spread of the emerald ash borer. Within each quarantine area, the movement of firewood, green lumber, ash nursery stock, and ash debris as well as all wood chips is restricted. As of March 2010, federal quarantine areas cover the entire states of Illinois, Indiana, and Ohio; all of the Lower Peninsula of Michigan; and parts of Kentucky, Maryland, Minnesota, New York, Pennsylvania, Virginia, West Virginia, and Wisconsin. Interstate movement of all firewood is prohibited, since it is difficult to distinguish ash from other hardwoods. Surveys of areas surrounding infected counties are conducted frequently to detect new infestations. Monitoring tools such as prism purple panel traps, baited with a lure and a nontoxic glue on the outer surfaces to catch any adult beetles attracted to the trap, are used.

Experiments to see if a parasitic wasp can help control emerald ash borers are focusing on three tiny stingless wasps that lay their eggs in beetle larvae or eggs. In their native range in China, emerald ash borers are forest pests but do not have devastating effects, in part because Chinese ashes have evolved resistance to their attacks. Asian trees survive in Michigan, a state with a large infestation. It may be possible to hybridize American ashes and Chinese species and select for resistance as has been done with the American chestnut (*Castanea dentata*; see Fungi, Chestnut Blight Fungus) and American elm (*Ulmus americana*; see Fungi, Dutch Elm Disease).

Selected References

Bauer, Leah S., Therese M. Poland, and Deborah L. Miller. "Emerald Ash Borer." Forest Disturbance Processes. U.S. Forest Service, Northern Research Station, 2010. http://nrs.fs.fed.us/disturbance/invasive_species/eab/biology_ecology/planipennis/.

"Emerald Ash Borer." Pennsylvania Department of Conservation and Natural Resources, 2009. http://www.dcnr.state.pa.us/forestry/fpm_invasives_EAB.aspx.

"Emerald Ash Borer." Wikpedia, 2010. http://en.wikipedia.org/wiki/Emerald_ash_borer.

"Emerald Ash Borer Information." Wisconsin Department of Agriculture, Trade and Consumer Protection, n.d. http://datcp.wi.gov/Environment/Emerald_Ash_Borer/index.aspx.

McCullough, Deborah G., Noel F. Schneeberger, and Steven A. Katovich. "Emerald Ash Borer. Pest Alert." USDA Forest Service, Northeastern Area State and Private Forestry, 2008. http://www.na.fs.fed.us/spfo/pubs/pest_al/eab/eab.pdf.

■|Formosan Subterranean Termite

Also known as: FST
Scientific name: *Coptotermes formosanus*
Order: Isoptera
Family: Rhinotermitidae

Native Range. The Formosan subterranean termite is probably native to southern China. It was first described from the island of Taiwan (Formosa) and hence its name. It may have been transported to Japan before 1600.

Distribution in the United States. As of 2010, it was established in 10 states: Alabama, California, Florida, Georgia, Hawai'i, Louisiana, Mississippi, North Carolina, South Carolina, Tennessee, and Texas. Isolated colonies have been reported in Arizona, Arkansas, New Mexico, and Virginia.

Description. Like other termites, the Formosan subterranean termite is a social insect that lives in colonies composed of three castes: workers, soldiers, and the reproductives (alates or swarmers, queen, king, and immature alates). The majority are white to off-white workers that resemble most other termites. Identification of the species relies upon distinguishing soldiers and alates. Soldiers have large teardrop-shaped heads that are orange-brown. Their mandibles are black and sickle-shaped and, when crossed, form an X. Their yellowish-white bodies are about 0.25 in. (6.4 mm) long. Soldiers are aggressive and, when attacked, produce drops of a milky-white, glue-like fluid from a pore (fontanelle) on the front of the head. Soldiers comprise 10–15 percent of the colony.

Alates are the winged reproductives; they swarm at night. Since they are attracted to light, they are often found at windows or around light fixtures. They are yellowish brown and have two pairs of wings covered with tiny hairs, visible under the low-power magnification of a hand lens. The wings are clear and of equal length. Two thick veins occur on the leading edge of each wing. Total length is about 0.5 in. (12–15 mm).

Since these termites spend much of their lives underground, one is more apt to see signs of their presence rather than the insects themselves. A good indication that Formosan subterranean termites have invaded is the occurrence of shelter tubes made of mud on hard surfaces such as on tree bark, up the side of foundations, and along concrete slabs. Nests made of carton may become exposed on door frames, ceilings, stairs, and near the base of trees. These nests may also be constructed in elevated locations without a connection to the ground. Damage to windows, doors, and floors as well as utility poles, fences, and landscape timbers are other signs that large termite colonies are nearby.

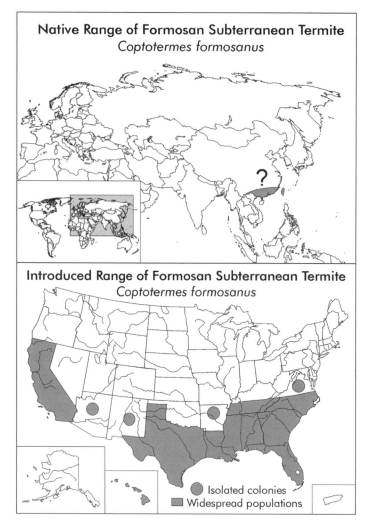

Native Range of Formosan Subterranean Termite
Coptotermes formosanus

Introduced Range of Formosan Subterranean Termite
Coptotermes formosanus

● Isolated colonies
■ Widespread populations

Top: The Formosan subterranean termite is believed to be native to southern China. It was first described from the island of Formosa, now Taiwan. Bottom: The Formosan subterranean termite was established in 10 states by 2010 and reported in 4 others. (Adapted from "Just the Facts ... Formosan Subterranean Termite," 2006.)

Related or Similar Species. No other termites in the United States build nests of carton, a mixture of termite excrement, chewed wood, saliva, and soil. The feeding galleries of Formosan subterranean termites are essentially free of soil, whereas native subterranean termites fill their galleries with soil and fecal material. Infested timbers often have layers of moist soil in areas where termites are active. The alates of native subterranean termites (*Reticulitermes* spp.) are smaller, their wings have no hairs, and they swarm in daylight. Among native subterranean termites, the proportion of soldiers to workers is usually only 1–2 percent, and the soldiers do not emit a white substance from the fontanelle. Native soldiers' heads are oblong, and the mandibles do not cross.

Introduction History. The Formosan subterranean termite may have come to Hawai'i as early as 1869, but it was not officially reported there until 1913. Its first record in the continental United States is officially 1965, when it was discovered at a shipyard in Houston, Texas. It probably arrived in crated and palleted military supplies on ships returning from the Pacific after World War II. Indeed, the termite seems to have spread around the world on ships; the earliest sites of infestations are at ports. In 1966, it was reported in Louisiana (New Orleans and Lake Charles). The first official report of the termite in South Carolina (Charleston) was made in 1967, but a search of earlier collections revealed its presence by 1957. First reports of the Formosan subterranean termite in Florida date to 1980–1983; Alabama, 1985–1987; Tennessee (Memphis), 1985; North Carolina, 1990; California (San Diego County), 1991; and Georgia (Atlanta), 1993. The termites are poor flyers and do not spread rapidly on their own. After the initial introductions, the termites were likely accidentally transported in contaminated building and plant materials brought in from previously infested areas. They are commonly associated with used railroad ties, a popular landscape timber. In general they have remained confined to the southeastern

United States at latitudes south of 32°30′ N. However, the availability of centrally heated buildings may allow them to extend their range farther north.

Habitat. Natural forests, planted stands of trees, and urban areas. Formosan subterranean termites live underground in moist areas with moderate temperatures. Their nest will be below frost level, but above the water table. They also build aerial nests on structures where they can find moisture. Such sites include boats, porches, flat roofs where water pools, rooftops with vegetation, and gutters. They will exploit live and dead trees as nesting sites. Often

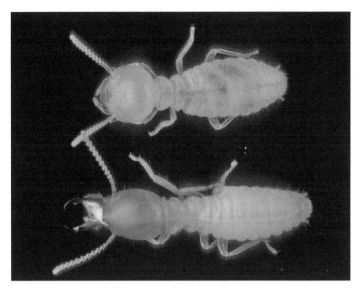

Formosan subterranean termite worker (top) and soldier (bottom). (Gerald J. Lenhard, Louisiana State University, Bugwood.org.)

the nest is outdoors, and a man-made wooden structure is used only as food. The nest is generally accessed from the ground. Mud galleries or shelter tubes connect nests to food sources.

Diet. These termites consume cellulose from wood, cardboard, or paper. Bacteria and other microorganisms in the gut digest the cellulose, releasing nutrients and energy for the termite. They usually attack wood in contact with the ground and hollow out galleries by feeding between the growth rings. Although they will chew through metal sheeting, asphalt, some plastics, foam insulation, and plaster, they do not eat these materials. It is often reported that they can hew through concrete, but they actually need preexisting cracks, which they then enlarge. The termites feed over a foraging territory that may be several thousand square feet in area.

Life History. Major swarms of Formosan subterranean termites begin in late spring and continue through the summer. They occur on humid, still evenings, usually at dusk. A single colony may release 70,000 or more winged reproductives. They fly only a short distance (60–150 ft. or 20–50 m) and then shed their wings. The female searches for a nesting site in moist crevices near a good supply of wood and the male follows her. Together, they hollow out a royal chamber. The pair, now king and queen, mate, and in a few days, the queen produces 15–30 eggs. The eggs hatch in 2–4 weeks and are cared for by the king and queen until they become third instars. These young termites will care for the next batch of larvae, which will hatch from the second laying of the queen. The queen will continue to lay 2,000 eggs a day. It may take 3–5 years before the colony is large enough to produce alates or to cause noticeable damage to trees and structures. Mature colonies will have as many as 10 million workers, soldiers, a primary queen, and several secondary reproductives. Workers forage and take care of eggs and larvae and feed the larvae, soldiers, and reproductives. Soldiers defend the colony. The secondary reproductives take over egg production if the queen or king dies.

Impacts. Termites feed on wood used in construction as well as on living trees. They hollow out the structure they are feeding upon, leaving a papery-thin covering behind. This activity weakens and destroys beams, floors, sills, and the like and kills trees, including those of aesthetic and commercial value. While individual Formosan subterranean termites do not eat more wood than native species, the colonies are so much larger that more damage results more quickly. Reportedly, in Hawai'i where a house was built on top of a termite nest, the house was almost completely destroyed in two years. In Hawai'i, it is the single most economically important insect pest in the state. Before Hurricane Katrina, the Formosan subterranean termite cost New Orleans $300 million a year for control, repairs, and replacement of utility poles due to this species. After hurricanes Rita and Katrina, the highest concentrations of this pest in the continental United States were in the flooded cities of New Orleans and Lake Charles, Louisiana. Cleanup of soggy debris and downed trees only spread the termite, which was also attracted to brown rot fungi growing on wet wood and wallboard. The IUCN has nominated this insect to be among "100 of the 'World's Worst' invaders."

Management. Preventive measures include the use of pressure-treated wood wherever timbers come in contact with the ground. It is also important to prevent the build-up of moisture from leaky pipes, lawn irrigation sprinklers, clogged gutters, air conditioning condensate, and rainwater. Chemical barriers can be placed outside a structure both during and after construction. Baits that use a chitin synthesis inhibitor are effective in eliminating an entire colony. General groundskeeping is also a good preventive practice. Remove dead trees and stumps and scrap wood piles from the property and reduce or do away with the use of mulch near a building's foundation. Once a population has become well established, it may impossible to eradicate it.

Selected References

Carlson, Elizabeth. "The Formosan Subterranean Termite." BugwoodWiki. Center for Invasive Species and Ecosystem Health, University of Georgia, 2008. http://wiki.bugwood.org/The_Formosan _Subterranean_Termite.

Hu, Xing Ping. "Formosan Subterranean Termites." Alabama Cooperative Extension System, Alabama A&M University and Auburn University, 2003. http://www.aces.edu/pubs/docs/A/ANR-1035/.

"Just the Facts . . . Formosan Subterranean Termites." U.S. Army Center for Health Promotion and Preventive Medicine, Entomological Sciences Program, 2006. http://phc.amedd.army.mil/PHC %20Resource%20Library/FormosansubterraneantermitesJan2010.pdf.

National Biological Information Infrastructure (NBII) and IUCN/SSC Invasive Species Specialist Group (ISSG). "*Coptotermes formosanus* (Insect)." ISSG Global Invasive Species Database, 2006. http:// www.issg.org/database/species/ecology.asp?si=61&fr=1&sts=sss.

Su, Non-Yao, and Rudolf H. Scheffrahn. "Featured Creatures: Formosan Subterranean Termite." Entomology Department, University of Florida, 2000. http://www.entnemdept.ufl.edu/creatures/ urban/termites/formosan_termite.htm.

■ | Glassy-Winged Sharpshooter

Also known as: GWSS
Scientific name: *Homalodisca vitripennis*
Synonym: *Homalodisca coagulata*
Order: Hemiptera
Family: Cicadomorpha

Native Range. Glassy-winged sharpshooters are native to the southeastern United States, where it occurs from eastern Texas to southern North Carolina.

Distribution in the United States. This leafhopper is a native transplant to California.

Description. The glassy-winged sharpshooter is a relatively large leafhopper, measuring from 0.4–0.5 in. (1.1–1.4 cm) long. Females are slightly larger than males. The back is generally brown; small ivory to yellow spots dot the head and thorax. The underside of the abdomen is white; face and legs are yellow-orange. Wings are transparent with red veins. Some females have a large white spot on the middle of each wing composed of a powdery material (brochosomes) that is secreted by the insect. Nymphs look like adults except that they are gray and wingless.

Related or Similar Species. The glassy-winged sharpshooter can be told from almost all other sharpshooters by its large size. It is most similar to the native smoke tree sharpshooter (*Homalodisca liturata*), a close relative. The smoke tree has wavy markings on its body instead of the spots characteristic of the glassy-winged sharpshooter.

Introduction History. Glassy-winged sharpshooters were first collected near Irvine, California, in 1989, although it was mistakenly identified as the

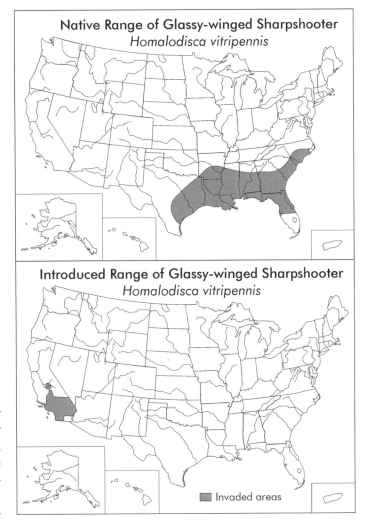

native smoke tree sharpshooter at the time and not recognized as an introduced species until 1994, when it was properly identified in Ventura County. It had probably arrived as egg masses on plants imported from the southeastern United States. It first became abundant in commercial groves of citrus and avocado and on some woody ornamentals such as crape myrtle. Its presence was visible when "leafhopper rain" evaporated and whitened leaves. Toward the middle of the 1990s, the glassy-winged sharpshooter moved inland in Riverside and San Diego counties. By the end of the 1990s, large populations occurred in southern Kern County in citrus groves and vineyards. It is expected that it will become a permanent resident in suitable habitats in the Central Valley and throughout northern California.

Top: The glassy-winged sharpshooter is native to the southeastern United States. (Adapted from Conklin and Mizell 2009.) Bottom: Glassy-winged sharpshooters infest several counties in southern California and threaten agricultural areas is the Central Valley and northern parts of the state. (Adapted from "Glassy-Winged Sharpshooter in California." California Department of Food and Agriculture, 2010. http://www.cdfa.ca.gov/pdcp/Maps/GWSS_Distribution2010.jpg.)

Adult glassy-winged sharpshooter. (Russ Ottens, University of Georgia, Bugwood.org.)

Habitat. In California, it prefers riparian woodlands in coastal and foothill areas. In its native range, it occupies the forest edge. It also feeds and reproduces on woody ornamental trees, vines, and annuals. Crape myrtle and sumac seem to be preferred in its native range; in California, it inhabits eucalypts and coast live oaks, and it is also found on grapevines and citrus, avocado, and macadamia trees.

Diet. Adult glassy-winged sharpshooters suck the sap from the xylem—the tissues that distribute water and dissolved nutrients from the roots—of a wide variety of host plants. Feeding times are synchronized with peak nutrient content in the host plant. The insects insert mouthparts that serve as straws into the xylem. Nutrients in xylem fluid are very diluted, so large volumes of fluid must be processed. The result is that glassy-winged sharpshooters produce large amounts of watery wastes ("leafhopper rain") that can coat the canopies of infected trees with a white residue as the excreta evaporates. Adult glassy-winged sharpshooters are able to feed on older wood than other sharpshooters because of their larger mouthparts, and this also allows them to feed on dormant trees and vines during the winter. They tend to feed closer to the base of grapevines and other plants than do other sharpshooters. Young nymphs feed on the stems of the plant on which they hatch.

Life History. In California, mating occurs in spring and summer. Two generations are produced each year. Initial egg-laying continues from late February through May. Each female produces 10–12 eggs at a time and deposits them underneath the lower epidermis of leaves of selected host plants. As she lays the eggs, she covers them with white scrapings of brochosomes from her wings. Two weeks later, the eggs hatch into small nymphs that will undergo five molting stages before they are mature. The first generation matures and, from June through September, lays eggs for the second generation of the year. Adults live about two months between generations. The adults from the second generation will produce young the following year. Populations peak in summer and begin to decline in late August. With the approach of winter, adults move into forested areas, where they go into semi-hibernation in the leaf litter until warmer temperatures return and they can begin to mate.

Impacts. The glassy-winged sharpshooter, throughout its life cycle, is a vector for plant diseases associated with strains of the bacterium *Xylella fastidiosa.* As such, it is a major threat to California's billion-dollar grape industry, where it is already implicated in an increase in Pierce's disease. Pierce's disease clogs the xylem and destroys a plant's ability to draw water and nutrients from the soil to its leaves. At first, the tips of leaves turn brown and die, and in 1–2 years, the entire infected vine dies. Pierce's disease itself is not new to California and has probably been there for at least 100 years. The problem is that the glassy-winged sharp-shooter is a more effective transmitter of the bacterium than native sharpshooters because it is more of a generalist in terms of hosts, has greater mobility (up to 0.25 mi. per day), and has larger mouthparts. While other species feed on the parts of the vine that get pruned away, the glassy-winged sharpshooter feeds at the basal stems, and the infection can become systemic or chronic, spread to other vines, and wipe out whole vineyards where table grapes, wine, or raisins are produced.

Different strains of the *X. fastidiosa* bacterium cause diseases in other valuable plant species, and the glassy-winged sharpshooter could become a vector for almond leaf scorch, phony peach disease, alfalfa dwarf, and citrus variegated chlorisis, which affects orange trees. It is already involved in the spread of oleander leaf scorch. Oleanders are widely planted as ornamental shrubs in southern California, particularly in freeway medians, but also in parks and suburban yards. The cost to the California Department of Transportation should it have to remove diseased oleanders and replant medians with other ornamentals would be well over $50 million. The wholesale nursery trade is also economically affected, as ornamental plants require inspection and treatment with insecticides before transport to other parts of California or neighboring states.

Management. Management is aimed at containing the invasion and preventing outbreaks of Pierce's disease and related plant diseases in new areas. Better early detection methods are needed. Short-term strategies to slow the spread of the disease include the use of systemic insecticides and behavior modifiers that would disrupt the transmission of the bacteria. Biological control of the glassy-winged sharpshooter may be possible by using a small wasp from Texas and northern Mexico, *Gonatocerus truguttatus,* that is an egg parasite of the glassy-winged sharpshooter; or by inserting genes conferring resistance to *X. fastidiosa* into host plants or rootstocks. Other bacteria or fungi may be found that are restricted to xylem tissues and interfere with *X. fastidiosa*'s reproduction or transmission. One candidate is the fungus *Hirsutella* sp. that affects glassy-winged sharpshooters in their native range.

Selected References

Cimino, Andria. "Glassy-Winged Sharpshooter, Xylophagous Leafhopper (*Homalodisca coagulata*)." Introduced Species Summary Project, Columbia University, 2002. http://www.columbia.edu/itc/cerc/danoff-burg/invasion_bio/inv_spp_summ/Homalodisca_coagulata.html.

Conklin, Tracy, and Russell F. Mizell III. "Featured Creatures: Glassy-Winged Sharpshooter." University of Florida, Institute of Food and Agricultural Sciences, 2009. http://entomology.ifas.ufl.edu/creatures/fruit/glassywinged_sharpshooter.htm.

Garrison, Rosser W. "New Agricultural Pest for Southern California: Glassy-Winged Sharpshooter (*Homalodisca coagulata*)." County of Los Angeles, Agricultural Commissioner/Weights and Measures Department, 2001. http://acwm.co.la.ca.us/pdf/GWSSeng_pdf.pdf.

Hoddle, Mark S. "The Glassy-Winged Sharpshooter." Center for Invasive Species Research, University of California, Riverside, 2003. http://cisr.ucr.edu/glassy_winged_sharpshooter.html.

National Biological Information Infrastructure and IUCN/SSC Invasive Species Specialist Group. "*Homalodisca vitripennis* (Insect)." ISSG Database, 2006. http://www.issg.org/database/species/ecology.asp?si=240&fr=1&sts.

Pierce's Disease Research and Emergency Response Task Force. "Glassy-Winged Sharpshooter (*Homalodisca coagulata*)." University of California, Agriculture and Natural Resources, n.d. http://news.ucanr.org/speeches/glassywinged.html.

■ | Gypsy Moth

Also known as: European gypsy moth
Scientific name: *Lymantria dispar*
Order: Lepidoptera
Family: Lymantridae

Native Range. The gypsy moth comes from the temperate broadleaf forests of Eurasia and northern Africa. European and Asian subspecies have evolved. The gypsy moth that is now invasive across the eastern United States originated in Europe. The first egg masses were brought from France.

Distribution in the United States. Gypsy moths are established in many northeastern states and continue to expand their distribution area westward and southward. Currently they occur from northern North Carolina to Maine and west through Ohio and northern Indiana into Wisconsin.

Description. Gypsy moths are sexually dimorphic. The female is white with dark chevrons across the forewings. She has a wing span of about 2 in. (80 mm) but is flightless. Her body is covered with hairs, and her antennae are slender. The male is darker and smaller. His wings are dark brown with black banding, and he has feathery antennae. A strong flier, the male has a wingspan of about 1.5 in. (60 mm). The abdomen of the male is narrower than that of the female.

Eggs are deposited in oval masses that are covered with soft, buff-colored hairs from the female's abdomen. The velvety masses are about 1.5 in. (60 mm) long and 0.75 in. (30 mm) wide.

Newly hatched larvae are about 0.125 in. (3 mm) long and initially tan, but become black within four hours. Younger caterpillars (first to fourth instars) are brown to black and have long body hairs. Late instars are black with 11 pairs of bumps (tubercles) along the back. The forward five pairs are blue; the rear six pairs are red. Each tubercle has a tuft of yellow or brown hairs. A single yellow line runs the length of the back of younger caterpillars from the head to the last segment. Additional yellow lines adorn the head of fourth through sixth instars. True legs are dark red. At maximum growth, the caterpillars are 2–3 in. (80–120 mm) long.

Pupae are dark red-brown and teardrop-shaped, rounded in the front, and tapered toward the rear. Each pupa attaches to a substrate by means of a few strands of silk. Male pupae are 0.75–1.5 in. (30–60 mm) long; female pupae are up to 2.5 in. (100 mm) long.

Related or Similar Species. The Asian gypsy moth is a subspecies of *Lymantria dispar* that is occasionally intercepted at western ports. The adult moths look very similar to the European variant that has invaded the northeastern United States, but the female is a strong flier. Asian caterpillars vary more in color than European ones. It is safe to assume, for now, that any flying white moth is not a gypsy moth.

Gypsy moth caterpillars might be mistaken for some native tent caterpillars; however, gypsy moths never make tents or webs. The eastern tent caterpillar (*Malacosoma americanum*) has a white line down its back and light blue and black spots on the sides. The forest tent caterpillar (*Malacosoma disstria*) has a line of white blotches the length of its back, and light blue stripes on the sides. The fall web worm (*Hyphantria cunea*) is greenish or yellow with long white hairs and has a black stripe down the back and a yellow stripe on each side.

Introduction History. The introduction of the gypsy moth can be traced to a single person and a specific address. A Frenchman and amateur entomologist, Etienne Leopold Trouvelet, had come to Medford, Massachusetts, with his family in 1852. He was interested in identifying American silkworms for possible use in producing silk. For unknown reasons, in 1868 or 1869, after a visit to France, he brought some gypsy moth egg masses back to his house at 27 Myrtle Street. It seems he let the eggs develop on trees in the backyard. When some caterpillars escaped, he notified local authorities, but no effort was made to eradicate them. In 1882, a gypsy moth

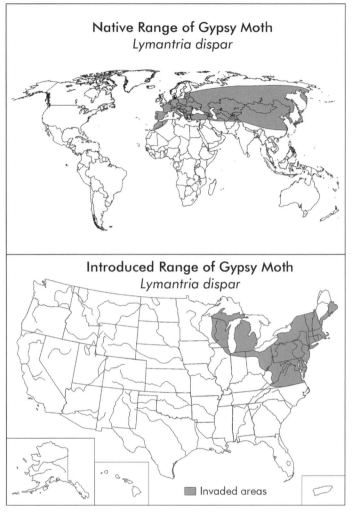

Top: The gypsy moth's native range coincides with the temperate broadleaf forests of Eurasia and North Africa. (Adapted from "Gypsy Moth around the World." USDA Forest Service. http://www.fs.fed.us/ne/morgantown/4557/gmoth/world/.) Bottom: The gypsy moth continues to spread southward and westward from established populations in the Northeast and Great Lakes region. (Adapted from "Gypsy Moth Quarantine Map." USDA Forest Service, 2007. http://www.fs.fed.us/ne/morgantown/4557/gmoth/atlas/q2007.gif.)

outbreak on Myrtle Street marked the first evidence of an emerging problem. The gypsy moth population continued to grow, and in 1889, the first program to eradicate the moth began. The Massachusetts State Board of Agriculture implemented manual removal of egg masses from trees and structures, applied pesticides, and burned infested forests, all to no avail. Eradication efforts ceased in 1900. The historical rate of spread between 1900 and 1915 is estimated to have averaged about 6 mi./year (9.5 km/year). By 1934, gypsy moths had spread north through Vermont, New Hampshire, and Maine and south into Connecticut and Rhode Island. Expansion rates had slowed to less than 2 mi./year

A. Gypsy moth caterpillar. (John H. Ghent, USDA Forest Service, Bugwood.org). B. Adult moths, male (left) and female (right). (USDA Forest Service Archive, USDA Forest Service, Bugwood.org.)

(2.8 km/year) between 1916 and 1965, so that by 1965, only eastern New York State, including Long Island, was added to the distribution area. After 1965, the rate of spread again increased, as did the range of the gypsy moth. At an estimated spread rate of nearly 13 mi./year (21 km/year), the gypsy moth moved south through Pennsylvania and New Jersey into Maryland, West Virginia, and Virginia and west across northern Ohio, into Michigan, and eventually Wisconsin and northeastern Minnesota. Nothing suggests that this spread has stopped.

Newly hatched (first instar) gypsy moth caterpillars sometimes move down from the tree canopy on silken threads that break in strong winds, carrying the larvae out of the canopy. Updrafts can carry them as much as 12 mi. (nearly 20 km) from the tree where they hatched. However, most dispersal to new areas is probably achieved by unintentional human transport. Gypsy moth egg cases and larvae can hitchhike on firewood, nursery stock, vehicles, and outdoor equipment. The typical pattern of spread is one of isolated populations appearing ahead of the expanding boundary of the distribution area and then coalescing to form a continuous, new area of infestation.

Habitat. Temperate broadleaf deciduous forests are preferred habitat. Populations become densest and resulting defoliation most intense on dry ridges and other areas with shallow soils or excessively drained soils. Coniferous forests, urban and suburban landscapes, and agricultural environments also have habitat value for gypsy moths.

Diet. Gypsy moth caterpillars feed on foliage. Adult moths do not have mouthparts and thus do not eat at all. Host plants are extremely varied; over 500 different types can be consumed. Suitability of plant species for forage depends on the developmental stage of larvae. In general, trees under stress are more vulnerable to attack than healthy specimens. In the northeastern United States, oaks are the most favored hosts, but maples, hickories, and many other hardwoods are also eaten. Trees that are avoided include American holly, ash (*Fraxinus* spp.), black walnut (*Juglans nigra*), butternut (*Juglans cinerea*), flowering dogwood (*Cornus florida*), mountain laurel (*Kalmia latifolia*), balsam fir (*Abies balsamea*), and arborvitae (*Thuja occidentalis*).

Life History. Moths emerge from pupae in summer, males usually 1–2 days ahead of females. A few hours after emergence, females release a pheromone that attracts males, which fly as much as 0.5 mi. (0.8 km) upwind to mate with them. Females deposit eggs within 24 hours of mating. Egg masses are placed on tree trunks and the undersides of branches as well as in crevices in the bark and under loose bark. They also may be laid on camping gear and in vehicles or anywhere else to which females can crawl. The eggs are covered with hairs from the female's abdomen that probably provide some protection from predators and cold temperatures. Although development from embryo to larva is completed within a month, larvae enter diapause and overwinter inside the egg. Eggs hatch the following spring at about the time leaf buds open on oak trees.

Most larvae in a given egg mass hatch within seven days of each other. In cool or rainy weather, they stay near the remains of the egg mass, but once the sun shines, they move upward toward the light into the canopy of the tree. Larvae feed first on new leaves and, when not feeding, stay attached by silk threads on the undersides of leaves. The caterpillars undergo molts about once a week to accommodate their growth. Males have four molts; females five. The larval period lasts about 40 days, after which time each caterpillar finds a pupation site, wraps itself in a thin silk net, and rests for 1–2 days before becoming a pupa. The pupa erupts from the larval skin and becomes dark reddish-brown; it stays in the silk net for the next two weeks while development into an adult moth is completed. When the adult moth emerges, it takes a few hours to expand its wings. The males then fly away in search of females. Adult moths live about a week.

Gypsy moth populations are cyclical. A low-level, *innocuous phase* may last several years, to be followed by *a release phase* of 1–2 years' duration that sees an increase in population density of several orders of magnitude. The next phase is the *outbreak phase*, when population density is great enough that defoliation is evident over a considerable area of forest. Outbreaks rarely last more than one or two years before starvation and disease cause the population to crash in what is known as the *decline phase*. The foliage of infested trees may respond by increasing levels of toxic chemicals and decreasing the nutritive value of leaves, contributing to the die-off. Naturally occurring pathogens such as the fungus Em (*Entomophaga maimaiga*)—originally introduced from Asia when Massachusetts was attempting to stamp out gypsy moths in the early 1900s—and NPV (nucleopolyhedrosis virus) commonly contribute to population declines. Population cycles are synchronous over large distances, but they proceed aperiodically.

Impacts. Gypsy moths, especially during outbreaks, defoliate hardwood and sometimes softwood trees. This can weaken the tree, making it more susceptible to drought, diseases, and other pests such as the shoestring fungus (*Armillaria mellea*) and the two-lined chestnut borer (*Agrilus bilineatus*). Defoliated forests open the forest floor to sunlight, causing accelerated drying of leaf litter and increased forest-fire risk. Most trees can recover from a single defoliation event, but several defoliations in successive years can kill them. In the Northeast, an estimated 20 percent of forest trees die after heavy infestations. How this will affect species composition and ecological dominance is as yet unknown, but in time, ecologically dominant oaks could be replaced by less palatable tree species, which would have ripple effects throughout the forest community. The loss of trees and the unsightly mess of hundreds of caterpillars and a constant rain of their feces from the treetops during an outbreak can negatively affect not only timber production, but also tourism and recreation. Other economic costs are incurred in inspection and other control measures. Automobile and train wrecks on roads and tracks slippery with caterpillars have been reported during some severe outbreaks.

Management. Gypsy moth control is aimed at suppressing existing populations, eradicating pioneer populations, and slowing the spread of this invasive moth. Homeowners can control small local infestations by manually scraping and destroying egg masses from trees in their yards or attaching sticky barrier bands to tree trunks to prevent the migration of caterpillars. More widespread suppression of populations is achieved by spraying areas with Bt (*Bacillus thuringiensis*), a bacterium that produces a chemical toxic to moth and butterfly larvae.

The National Gypsy Moth Slow the Spread (STS) program, a joint local, state, and federal initiative, was implemented in 1999 to monitor recently established and still-isolated low-level populations in zones between infested areas and non-infested areas. Along the front line of range expansion, gypsy moth population monitoring is conducted with pheromone traps to capture males, and several types of barriers are placed on tree trunks to collect larvae. Direct counts of egg masses are also employed. Once new infestations are identified, Integrated Pest Management (IPM) and other controls are initiated to eradicate or suppress the population. APHIS has established quarantine areas in much of the infested region. Nursery stock, firewood, vehicles, outdoor equipment, and household items moving out of these areas into non-infested ones must be inspected by certified inspectors.

Selected References

Baczynski, Tracy. "European Gypsy Moth (*Lymantria dispar*)." Introduced Species Summary Project, Columbia University, 2002. http://www.columbia.edu/itc/cerc/danoff-burg/invasion_bio/inv_spp_summ/Lymantria_dispar.htm.

"Gypsy Moth." Animal and Plant Health Inspection Service, U.S. Department of Agriculture, 2010. http://www.aphis.usda.gov/plant_health/plant_pest_info/gypsy_moth/index.shtml.

"Gypsy Moth Pest Profile." California Department of Food and Agriculture, State of California, 2010. http://www.cdfa.ca.gov/phpps/pdep/target_pest_disease_profiles/gypsy_moth_profile.html.

"Integrated Pest Management Manual: Gypsy Moth." National Park Service, U.S. Department of the Interior, 2010. http://www.nature.nps.gov/biology/ipm/manual/gypsymth.cfm.

Liebold, Sandy. "Gypsy Moth in North America." Forestry Sciences Laboratory, Forest Service, Northeastern Research Station, U.S. Department of Agriculture, 2003. http://www.fs.fed.us/ne/morgantown/4557/gmoth/.

■ | Hemlock Woolly Adelgid

Also known as: Woolly
Scientific name: *Adelges tsugae*
Order: Homoptera
Family: Adelgidae

Native Range. Hemlock woolly adelgids are native to Asia. They are known to be from Japan, Taiwan, southwestern China, and India. Genetic analysis has determined that those in the eastern United States came from southern Japan. In their native range, these adelgids utilize both hemlocks and spruces and undergo both asexual and sexual reproduction. Asian hemlocks appear to have developed resistance to adelgid attacks.

Distribution in the United States. The hemlock woolly adelgid is invasive in 16 eastern states from southeastern Maine to northern Georgia and west to eastern Kentucky, Tennessee, and West Virginia. Its distribution area continues to expand; and it may soon become established in Ohio, where isolated collections have occurred. In the western United States, where it is not a pest, it occurs from northern California to southeastern Alaska.

Description. The hemlock woolly adelgid is a tiny, aphid-like insect less than 1/16 in. (1.5 mm) long. They are oval and blackish gray in color. Newly hatched nymphs (first instars) are about the same size, but reddish brown. The insect probably would not be noticed were it not for the fact that nymphs covers themselves in white wax-like filaments and adults secrete a white "woolly" ovisac in which to lay their eggs. The woolly or cottony casings are about 1/8 in. (3 mm) in diameter and quite visible at the base of needles on the undersides of the outermost tips of hemlock branches from late fall to early summer.

Introduction History. The hemlock woolly adelgid was first identified on the West Coast in 1924. First reports on the East Coast came from Richmond, Virginia, in 1951. It may have been accidentally imported on nursery stock from Japan. Today, it continues to spread on ornamental hemlocks via the nursery trade, but early-stage larvae and eggs can be transported by birds and mammals moving through hemlock forests. The wind can disperse infested twigs as well as the mobile crawlers. Also, humans can unintentionally transport eggs, nymphs, and adults on live ornamental trees and with debris from dead and dying hemlocks. The hemlock woolly adelgid appeared in Shenandoah National Park in Virginia in the 1980s, moved southward on the Blue Ridge Parkway in

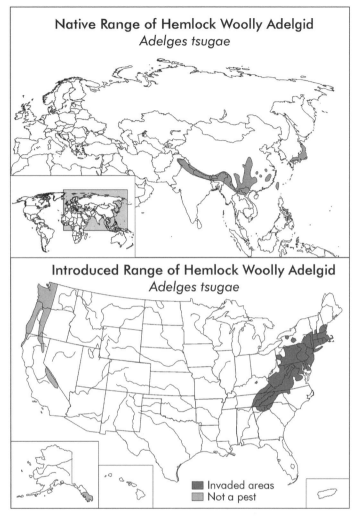

Top: The hemlock woolly adelgid is native to coniferous forests from India to Japan. (Adapted from Reardon, R., et al. "Biological Control of Hemlock Woolly Adelgid." U.S. Department of Agriculture, Forest Service. Forest Health Technology Enterprise Team. Burlington, VT. FHTET-2004-04. http://wiki.bugwood.org/Archive:HWA/Introduction.) Bottom: The hemlock woolly adelgid is currently invasive in 16 eastern states, and it continues to expand its range. In the west it is not considered a pest. (Adapted from "Counties with Established HWA Populations 2009." USDA Forest Service. Http://na/fs/fed/us/fnp/hwa/maps/2009.pdf.)

the 1990s, and was in Great Smoky Mountains National Park by 2002. In 2007, it was spreading at a rate of nearly 10 mi./year (15.6 km/year) in the southern part of its range and 5.0 mi./year (8.13 km/year) in the northern part. By some estimates, it could kill all the hemlocks in the southern Appalachians by 2020.

Habitat. In the United States, these aphid relatives are restricted to hemlocks, although in their native range, certain spruces are hosts during the sexually reproducing stages of the life cycle. In the eastern United States, this adelgid infests both eastern hemlock (*Tsuga*

A. Woolly casings on a hemlock twig infested with hemlock woolly adelgids. (Connecticut Agricultural Experiment Station Archive, Connecticut Agricultural Experiment Station, Bugwood.org). B. The aphid-like adult adelgid. (Michael Montgomery, USDA Forest Service, Bugwood.org.)

canadensis) and Carolina hemlock (*T. caroliniana*). In the West, the woolly hemlock adelgid occurs on both western hemlock (*T. heterophylla*) and mountain hemlock (*T. mertensia*), but these trees are resistant, and the adelgid is not considered a pest.

Diet. Nymphs and adults suck sap from the twigs of hemlocks. They feed at the base of new needles by extracting starch-rich fluids from the tissues in the xylem that manufacture and store the plant's food. At the same time, they inject toxins in their saliva into the plant.

Life History. Hemlock woolly adelgids have both sexual and asexual (parthenogenic) stages in the life cycle, but U.S. populations only multiply asexually; all are females. Six stages of development define the life cycle: egg, four instars of nymphs, and adult. Two generations are produced each year. In March, females that had overwintered lay 20–75 eggs and cover them in cottony material excreted by the adult. Larvae, known as crawlers, emerge in April and May and move to feeding places at the bases of hemlock needles. They may hitchhike to nearby hemlocks on birds and mammals or be blown there by the wind. When the crawlers settle, they become immobile nymphs. By June, they have matured into wingless and winged adults. The winged females fly off looking for spruce trees, but since no suitable host spruces exist in the United States, most of them presumably die without reproducing. If there were such spruces, these females would reproduce sexually. The wingless females reproduce asexually and lay 100–300 eggs that will become the second generation of the year. The larvae hatch in early July, settle at the bases of hemlock needles, and then go into dormancy through the hot summer months. Nymphs resume development in October and November and feed throughout the winter, maturing into adult females early the next spring.

Impacts. Hemlock woolly adelgids weaken and kill eastern and Carolina hemlocks, including those in forests and those cultivars used as ornamental trees and hedges. To date, the most severe impacts have occurred in parts of Virginia, New Jersey, Pennsylvania, and Connecticut. In the Northeast, hemlock decline and death takes 4–10 years; in the southern Appalachians, the pace is more rapid, and trees may die in only 3–6 years. Infestations often start in large, mature hemlocks, which may be 150 ft. (45 m) tall and more than 500 years old. Infested trees can be identified first by their grayish-green needles that contrast with the glossy green of healthy hemlocks. Defoliation follows, progressing from the lower branches upward and ultimately causing the death of the tree. The rate and extent of hemlock decline are accelerated by other stress factors, such as drought, poor soil conditions, and other pests and diseases. Hemlocks form important habitat for numerous birds from warblers to turkeys, and for small mammals such as snowshoe hares and rabbits; they produce a deep shade along mountain streams that maintains the cool water temperatures needed by trout and other fish. The loss of the hemlock canopy increases light penetration to the forest floor, drying and warming riparian and riverine habitats in the summer and reducing shelter in upland habitats in winter. Research suggests that nitrogen cycles are altered when hemlock stands perish. The demise of hemlocks could have cascading effects that would alter forest ecosystems in much of the eastern United States.

Management. Few environmentally sound methods for controlling hemlock woolly adelgids in forests are available. Natural predators and pathogens exist, but fail to lower population levels enough to prevent tree deaths. Biological control by Asian beetles, such as the ladybird beetle (*Pseudoscymnus tsugae*), that limit their attacks to adelgids and control them in Japan, hold some promise; several states have released them experimentally. DNA research suggests that Asian hemlocks have evolved a degree of resistance to adelgid infestations. Crossing eastern hemlocks with Asian hemlocks has successfully maintained the appearance of the eastern hemlock but produced a tree that discourages adelgid settlement and slows the growth of those nymphs that do feed upon it.

Isolated trees in yards or parks can be treated with insecticidal soaps, or adelgids can be physically removed with strong sprays of water; but these treatments must be repeated often. Removal of infected branches will slow the decline of the tree. In Great Smoky Mountains National Park, treatments with systemic insecticides that are applied at the roots or injected into the trunk of trees in campgrounds have reversed some of the impacts of the infestation and remained effective for several years.

Preventing or slowing the spread of the adelgid can be aided by cultural control methods such as not moving live plants, logs, firewood, or bark chips from infested areas. Selective removal of heavily infested trees reduces that likelihood that wind, birds, and other wildlife will disperse eggs and nymphs. When replanting landscape trees, the use of adelgid-resistant hemlock species, such as *T. diversifolia* and *T. sieboldii* from Japan or western and mountain hemlocks from the Pacific Northwest, instead of eastern hemlock will limit the availability of suitable host trees and may help control adelgid populations.

Selected References

Chowdhury, Shahrina. "Hemlock Woolly Adelgid (*Adelges tsugae*)." Introduced Species Summary Project, Columbia University, 2002. http://www.columbia.edu/itc/cerc/danoff-burg/invasion_bio/inv_spp_summ/Adelges_tsugae.html.

"Hemlock Woolly Adelgid." Pest Alert, State and Private Forestry NA-PR-09-05. U.S. Department of Agriculture, Forest Service, Northeastern Area, 2005. http://na.fs.fed.us/spfo/pubs/pest_al/hemlock/hwa05.htm.

McClure, Mark S. "Hemlock Woolly Adelgid, *Adelges tsugae* (Annand)." Connecticut Agricultural Experiment Station, 1998. http://www.ct.gov/caes/cwp/view.asp?a=2815&q=376706.

Pennsylvania Department of Conservation and Natural Resources. "Hemlock Woolly Adelgid." Commonwealth of Pennsylvania, 2009. http://www.dcnr.state.pa.us/forestry/woollyadelgid/index.aspx.

■|Japanese Beetle

Scientific name: *Popillia japonica*
Order: Coleoptera
Family: Scarabaeidae

Native Range. Japan.

Distribution in the United States. The Japanese beetle is established in all states east of the Mississippi River except Florida. Noncontinuous infestations occur west of the Mississippi in Arkansas, Colorado, Iowa, Kansas, Minnesota, Missouri, Nebraska, and Oklahoma. Far-western states are protected by quarantine, and any beetles arriving have, so far, been eradicated.

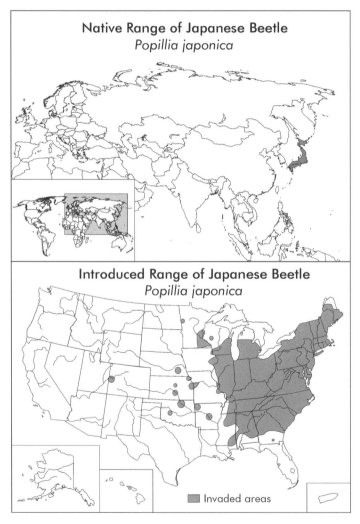

Top: The Japanese beetle is native to Japan. Bottom: In the United States, the Japanese beetle occurs in all states east of the Mississippi River except Florida. Isolated populations occur as far west as Colorado. (Adapted from APHIS, 2004.)

Description. The Japanese beetle is an oval-shaped, metallic-green insect about 0.5 in. (10–13 mm) long and 0.25 in. (6–7 mm) wide. The wing covers are bronze-colored and extend almost the full length of the abdomen. Five small white tufts of hair line each side of the body and two others lie just posterior to the wing covers. Males are usually slightly smaller than females. Japanese beetles travel and feed in groups.

Eggs are spherical and almost translucent. They swell to a diameter of about 0.08 in. (2 mm) before hatching. The grub or larvae is white with a reddish-brown head. It has three pairs of legs and lies curled in the form of a C. Full-grown larvae are about 1.0 in. (26 mm) long. The pupa looks like a cream-colored or tan adult with the legs folded close to the body.

Related or Similar Species. Japanese beetle larvae look like those of most beetles and can be distinguished by a V-shaped row of spines beneath the abdomen.

Introduction History. The Japanese beetle was first discovered in a nursery in Riverton,

A. Adult Japanese beetle. (David Cappaert, Michigan State University, Bugwood.org.) B. Japanese beetle larvae (grubs) are commonly found in the soil beneath lawns. (David Cappaert, Michigan State University, Bugwood.org.)

New Jersey, in 1916. It is believed that grubs contaminated shipments of iris bulbs sometime before 1912, when inspections of imports began in the United States. In the next 60 years, it spread throughout 22 states east of the Mississippi River. The climate is favorable to the insect, large expanses of turf in lawns and golf courses provide excellent habitat for reproduction, and abundant foliage is available to support adults. The beetle can be spread during any stage of its life cycle in plant materials, soil, and sod. It has now moved into states west of the Mississippi, but for the most part, populations have remained isolated. Hitchhikers on aircraft are continually intercepted at airports in western states, where it could become a major pest in orchards and truck farms.

Habitat. Open woods, meadows, farms, gardens, and lawns.

Diet. Adult Japanese beetles eat the leaves, flowers, and fruits of hundreds of different plants. Their hosts include trees, shrubs, vines, and perennial and annual herbs, including crops and ornamentals. Preference is for young leaves, which they skeletonize by consuming the green tissue between the veins. Grubs feed primarily on the roots of grasses, but will also consume the roots of annual fruit and vegetable plants.

Life History. Adults begin to emerge from pupae in early summer and congregate on plants to feed and mate. Each female leaves the plant upon which she is feeding again and again to deposit 1–4 eggs at a time. In the course of the summer, she will lay 40–60 eggs total. She deposits her eggs in soil at depths of 3–4 in. (1.2–1.6 cm), and preferably beneath turf. The larvae hatch out about two weeks later and begin to feed on plant roots and grow. With the approach of cold weather, the grubs move deeper into the soil to overwinter. When soil temperatures warm in the spring, the grubs migrate back up toward the surface and resume feeding. They pupate and, in 8–20 days emerge as the next generation of adults. The life cycle usually takes a year. Ten months of the cycle are spent in the larval stage.

Impacts. Leaves of heavily infested ornamental, truck, and garden plants will turn brown and die. Soft fruits such as grapes and berries may be eaten completely. Grapes injured by beetles become vulnerable to attack by native green June beetles (*Cotinus mitida*), which are unable to bite into intact grapes themselves. Corn is damaged when Japanese beetles eat the silk and prevent the formation of kernels. The larvae damage turf, turning patches of lawns brown when numbers are high.

Management. Biological controls can be effective in controlling grubs and hence adult beetles. Applications of milky spore (the bacterium *Bacillus popillae*) to turf can reduce populations for decades if properly used. Nematodes (*Heterorhabditis* spp.) also work well in destroying grubs. Commercially available traps lure Japanese beetles away from plants, but may also attract beetles into a yard. Landscape plantings of non-palatable plants such as forsythia, holly, juniper, arborvitae, boxwood, spruce, and yew may deter beetles from massing in suburban properties.

Selected References

APHIS. "Managing the Japanese Beetle: A Homeowner's Handbook." Animal and Plant Health Inspection Service (APHIS), U.S. Department of Agriculture, 2004. http://www.aphis.usda.gov/lpa/pubs/pub_phjbeetle04.pdf.

Bilberry, S. "*Popillia japonica*." Animal Diversity Web, 2001. http://animaldiversity.ummz.umich.edu/site/accounts/information/Popillia_japonica.html.

Day, Eric, Pete Schultz, Doug Pfeiffer, and Rod Youngman. "Japanese Beetle." Virginia Cooperative Extension, Virginia Tech, and Virginia State University, 2009. http://pubs.ext.vt.edu/2902/2902-1101/2902-1101.html.

"Japanese Beetle Pest Profile." California Department of Food and Agriculture, 2009. http://www.cdfa.ca.gov/phpps/pdep/target_pest_disease_profiles/japanese_beetle_profile.html.

"Japanese Beetles." *Forest Insect and Disease Newsletter*. Minnesota Department of Natural Resources, 2002. http://www.dnr.state.mn.us/fid/november02/japanese.html.

■ | Multicolored Asian Lady Beetle

Also known as: Japanese ladybug, Asian lady beetle, Halloween lady beetle
Scientific name: *Harmonia axyridis*
Order: Coleoptera
Family: Coccinellidae

Native Range. Asia. The native range stretches from the Altai Mountains east to the Pacific Ocean and Japan and from southern Siberia south to southern China. The distribution area includes Kazakhstan, Uzbekistan, Russia, Mongolia, China, Korea, Taiwan, and Japan. This lady beetle has at various times been imported to the United States from Japan, Korea, and Russia.

Distribution in the United States. Well-established populations exist in many parts of the Midwest, Northeast, South, and Northwest.

Description. Like most other lady beetles, the adult multicolored Asian lady beetle has a domed, oval shape. Populations in the United States contain individuals in a mix of colors, from tan or pale yellow-orange to bright red-orange. They may or may not have black spots on the wing covers. Those that do have up to 10 spots on each wing cover. The middle body segment (pronotum) is white to straw-yellow and usually marked with a distinct black M. Adults are relatively large, measuring about 0.3 in. (7–8 mm) long.

Larvae are elongate and flattened; their bodies are covered with flexible spines. Late (fourth) instars are bluish black with a yellow-orange patch on each side of the abdomen. Eggs are bright yellow.

Related or Similar Species.
Many native lady beetles occur
in the United States. The most
common is the convergent lady
beetle (*Hippodamia convergens*).
Adults are somewhat elongate
in shape compared to the multicolored Asian lady beetle.
Convergent lady beetles have
black spots on red wing covers.
Behind the head, white lines
converge on a black back-
ground. They range in size from
0.16 to 0.28 in. (4–7 mm).
Native lady beetles do not overwinter indoors.

Another introduced lady
beetle, the seven-spotted lady
beetle (*Coccinella septempunctata*), from Europe, is established in some northeastern
and north central states. It is
about the same size as the
multicolored Asian lady beetle
but has a white spot on either
side of its black head. The wing
covers are red or orange with 1–
5 spots on each one. It overwinters in sheltered areas outdoors
near the fields in which they
feed.

Introduction History. Multicolored Asian lady beetles have
been deliberately introduced to
the United States numerous
times as agents for the biological control of aphids and

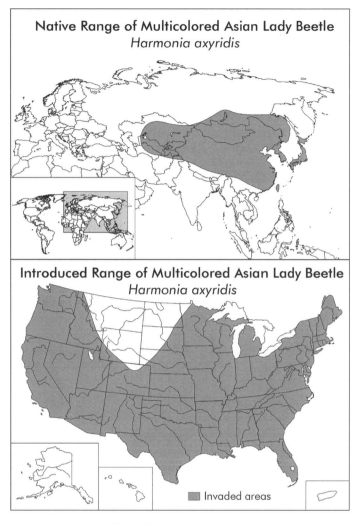

Top: The multicolored lady beetle occurs across a broad swath of Asia
from Kazakhstan to Japan. Bottom: This alien lady beetle is invasive in
most of the lower 48 states. (Adapted from "Multicoloured Asian Lady
Beetle." Project UFO. http://www.projectufo.ca/drupal/multicoloured
_asian_lady_beetle.)

aphid-like insects. They were first released by the U.S. Department of Agriculture in
California in 1916 and again in the mid-1960s to control pecan aphids. From 1978 to
1982, additional releases were made in Connecticut, Delaware, Georgia, Louisiana, Maine,
Maryland, Mississippi, Ohio, Pennsylvania, and Washington. Interestingly, none of these
efforts seem to have given rise to established populations.

In 1988, a population was discovered north of New Orleans in Louisiana. This may have
derived from an accidental introduction. From this point of origin, multicolored lady beetles
spread quickly through southern and midwestern states. By 1994, they were reported in
Alabama, Florida, Georgia, North Carolina, South Carolina, Ohio, and Minnesota. They
were also established in the Northeast by that time, perhaps as a result of other, independent

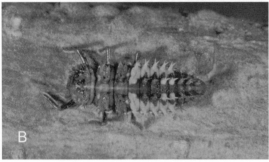

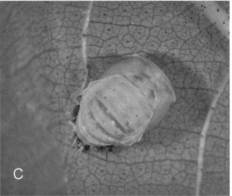

A. Adult multicolored Asian lady beetle. (Gerald J. Lenhard, Louisiana State University, Bugwood.org.) B. Larva. (Gerald J. Lenhard, Louisiana State University, Bugwood.org.) C. Pupa. (John Ruberson, University of Georgia, Bugwood.org)

introductions rather than dispersal from the South or Midwest. It is known, for example, that they arrived at ports in South Carolina and Delaware as accidental hitchhikers on imported nursery stock.

Habitat. Multicolored Asian lady beetles live in cultivated fields, orchards, vineyards, nurseries, and gardens. They usually overwinter in buildings.

Diet. Aphids and similar soft-bodied insects such as scales and psyllids comprise the main food of both adult and larval multicolored Asian lady beetles. An adult can devour 90–270 aphids a day, and a single larva may consume 600–1,200 while it develops. These lady beetles also feed on the larvae of butterflies and of other beetles as well as on injured fruits such as grapes.

Life History. Mating takes place in early spring after adults emerge from overwintering sites. Females lay clumps of roughly 20 eggs on the undersides of leaves. The eggs hatch in 3–5 days, and the fast-moving larvae forage on the host plant for aphids and scale insects. The larvae will molt four times as they grow larger. After the fourth molt, they enter an immobile pupa stage. The adult beetle emerges from the pupa 5 or 6 days later. The complete life cycle from egg to adult takes 15–25 days. Before the first killing frost of autumn, as prey become less abundant, adults seek shelter indoors. They seem to be attracted by bright reflective surfaces such as the south-facing sides of light-colored buildings. They enter homes and other buildings in huge numbers though cracks and other poorly sealed openings and occupy cool places inside walls, floors, and attics. There they become dormant. Winter aggregations may number in the tens of thousands. On warm, sunny days, they wake up and move to the interior of the building, where they seek the light at windows. With the warming temperatures of springtime, they exit their winter shelters and begin to mate. They do no physical damage to structures and do not reproduce indoors. They may return to the same overwintering sites every year. Adults can live 2–3 years.

Impacts. The introduction of the multicolored Asian lady beetle as a biological control agent on shrubs, trees, and a variety of crops has been largely successful. Its consumption of large numbers of aphids, scales, and thrips has reduced the need for chemical pesticides. In California, it has nearly eradicated pecan aphids, and elsewhere significantly reduced soybean aphids after their recent introduction from China.

Multicolored Asian lady beetles are a minor, but perhaps increasing, agricultural pest. They appear to displace native lady beetles and other beneficial insects through competition for food and predation. This lady beetle may also depress numbers of the exotic seven-spotted lady beetle (*Coccinella septempunctata*). In autumn, the beetles congregate and feed on ripening fruits such as pears, apples, and grapes—especially if the fruits have already been damaged by birds or other insects. This is particularly troublesome in vineyards, where it is difficult to remove them from the grapes being harvested for winemaking. Lady beetles get crushed along with the grapes and taint the flavor of the wine.

The multicolored Asian lady beetle may be a threat to eggs and caterpillars of the monarch butterfly (*Danaus plexippus*). Its real infamy, however, derives from its status as an annoying overwintering invader of homes. Numbers can be so great that homeowners can hear them moving in the walls. If a lady beetle is frightened or crushed, it exudes an unpleasant odor and secretes, via reflex bleeding, a yellowish liquid from its leg joints that will stain drapes, carpets, and other light-colored surfaces. They are more a nuisance than a threat, however. They do not destroy wood or textiles or otherwise damage human property. They do not carry diseases. They do not sting, although they can bite. Sensitive people may have allergic reactions to their presence.

Management. Preventing multicolored Asian beetles from entering homes is the best way to deal with them. All possible entries should sealed; door and window screens, as well as those in all vents, should be kept in a state of repair; and door sweeps should be installed on exterior doors. Once inside, it is difficult to remove them. The most recommended method is by vacuuming them and employing a means to discard the catch before survivors escape, and before the smell of dead beetles becomes too strong. Swatting them or sweeping them up can lead to reflex bleeding and bad odors. Do not use bug bombs. They may have little effect and will only attract other scavenger insects.

Selected References

"Biological Control: A Guide to Natural Enemies in North America: *Harmonia axyridis*." Cornell University, College of Agriculture and Life Sciences, Department of Entomology, n.d. http://www.nysaes.cornell.edu/ent/biocontrol/predators/Harmonia.html.

Hahn, Jeffrey. "Multicolored Asian Lady Beetles." University of Minnesota Extension Service, 2004. http://www.extension.umn.edu/distribution/housingandclothing/m1176.html.

Jones, Susan C., and Joe Boggs. "Multicolored Asian Lady Beetle." Ohio State University Extension Fact Sheet, HSE-1030-01.Ohio State University, n.d. http://ohioline.osu.edu/hse-fact/1030.html.

Koch, R. L. "The Multicolored Asian Lady Beetle, *Harmonia axyridis*: A Review of Its Biology, Uses in Biological Control, and Non-Target Impacts." *Journal of Insect Science* 3: 32, 2003. Published online, http://www.ncbi.nlm.nih.gov/pmc/articles/PMC524671/.

■ | Red Imported Fire Ant

Also known as: RIFA
Scientific name: *Solenopsis invicta*
Order: Hymenoptera
Family: Formicidae

Native Range. South America. Red imported fire ants come from a narrow strip extending from Porto Velho, Brazil, northwestward to Santa Fe, Argentina. Genetic studies suggest northeastern Argentina as the most likely source area of fire ants in the United States.

Distribution in the United States. Imported red fire ants can be found across the southeastern United States. They occur from southeastern Virginia and eastern and southern North Carolina across the Gulf states and eastern Texas. Isolated populations occur in parts of Maryland, Tennessee, Arkansas, Oklahoma, New Mexico, and California. They are likely to spread through southwestern Texas to southern Arizona and from southern California north along the Pacific coastal region into northern California and Oregon. The ants also occur in Puerto Rico and the U.S. Virgin Islands.

Description. Red imported fire ant colonies produce three types of workers, all sterile females and distinguished from other ants by their very aggressive behavior. The smallest, the so-called "minor workers," are about 0.1 in. (1.6 mm) long and the largest, comprising about 35 percent of a mature colony, the "major workers," are about 0.25 in. (6 mm) long. Intermediate in size are the many "media workers." These ants are reddish to dark brown. The waist (pedicel) has two segments, and all workers possess a stinger at the end of their black bulbous abdomens (gaster). Winged females and males, known as "reproductives," are both about 0.34 in. (8.8 mm) long. Each colony will have one or more queens, also about 0.34 in. (8.5 mm) long.

Red imported fire ant mounds occur in open areas and are rarely more than 18 in. (45 cm) in diameter or 16 in. (40 cm) high. They are hard and have no visible entrances. Those built in clay soils are typically symmetrical domes, while those constructed in sandy soils are irregularly shaped. However, in some circumstances, no mounds are evident. In urban and suburban areas, they may nest under concrete slabs, in the walls of buildings, or in electrical equipment.

Related or Similar Species. Among the many fire ant species that live in the United States, three other members of the genus *Solenopsis* are also pests. Separating one from another is difficult and requires a sample size of 40 or more workers for proper identification even by experts. The black imported fire ant (*S. richteri*) is limited in its distribution, found only in a small part of northern Mississippi and Alabama. It is displaced by and may actually be a subspecies or regional variant of the red imported fire ant. The southern fire ant (*S. xyloni*) is native to the Southeast. Nests are often built under stones or boards or at the bases of plants. The nest usually appears as loose soil with many craters spaced over an area of 2–4 sq. ft (0.18–0.37 m^2). The tropical fire ant (*S. geminata*) is another native species of the southeastern United States. Tropical fire ant workers have square heads that are large in proportion to their bodies. Their mounds are often built around clumps of vegetation.

Introduction History. Imported red fire ants were introduced to the United States through the port of Mobile, Alabama, in the mid-1930s. Genetic studies trace the origins to 9–20 queens and (presumably) their workers. A secondary introduction may have occurred 60 miles to the west of Mobile. From these pioneers, populations spread to other southeastern states. They arrived in Virginia in 1989, and were discovered in Los Angeles, Orange and Riverside counties, California, in 1998. In single-queen colonies, growth is outward at a rate no more than 40 m per year. In multi-queen colonies, dispersal of winged reproductives covers greater distances. Most new colonies form within a mile of the reproductives' birthplace, but some may be 10 miles or more distant. Nonetheless, most of the range expansion in the United States is probably the result of human actions. Ants are spread when mated females are transported in sod, hay, the root balls of ornamental plants, and on earth-moving equipment.

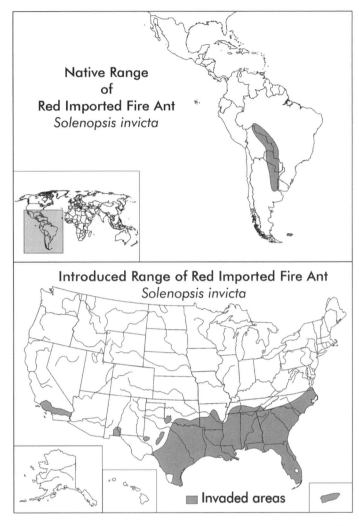

Top: The red imported fire ant is native to South America. Bottom: The red imported fire is invasive in the Gulf States and east Texas and expected to expand its range in coming years. It has also been introduced into Puerto Rico. (Adapted from USDA Agricultural Research Service map, 2007. http://www.ars.usda.gov/fireant/Imported.htm.)

Habitat. Red imported fire ants are most often associated with disturbed habitats such as agricultural fields, pastures, lawns, and other open, sunny sites. Rarely do they occur in natural forests or similar well-shaded locations. These ants are limited by cold temperatures and climates receiving less than 20 in. (510 mm) of precipitation a year. In dry regions, however, they can exist by colonizing areas near permanent sources of water or areas regularly irrigated. Red imported fire ants can regulate their microclimate to a degree by moving their broods to higher and lower levels of their mounds, which may extend as much as 4 ft. (1.2 m) below the ground.

Diet. Red imported fire ants are omnivores. They are voracious predators of other arthropods, but also consume emerging seedlings of field and truck crops such as soybeans, eggplant, cucumbers, corn, okra, and a host of others. They are known to chew the bark of

1.0 mm

The waist of the red imported fire ant has two distinct segments. All workers have a stinger at the end of the abdomen. (Eli Sarnat.)

citrus trees and eat fruits. In homes, they prefer oily and greasy foods high in protein. These ants also tend aphids and consume honeydew. Worker ants can only swallow liquids. Solids are cut to a manageable size and carried back to the nest to be fed to larvae. The larvae digest the food and regurgitate a protein-rich brew that feeds the workers and queen. The workers also regurgitate what they have swallowed so that their nest mates can lick or suck it up.

Life History. Biological patterns for fire ants must be described in terms of both the individual and the colony, since the latter is the essential social unit. The individual's life cycle begins as an egg. A grub-like legless larva hatches in 8–10 days and begins the major growth stage of the individual. It periodically sheds its skin as it grows larger, and becomes a pupa with legs in 6–12 days. The pupal phase lasts another 9–16 days, after which time adult sterile workers emerge and begin foraging or reproductives emerge, take flight, and mate some 300–800 ft. (90–240 m) up in the air. The complete transition from egg to adult takes 22–38 days.

A colony begins life when a mated female descends to the earth, breaks off her wings, and digs a founding nest 2–5 in. (5–12 cm) straight down into the ground. She seals herself off and lays 12–14 worker eggs. It takes about a month for these eggs to develop into adults. During this period, the queen does not feed, but she nourishes the larvae with sterile eggs, regurgitated oils, and salivary secretions. These workers will be the smallest in the colony cycle and are called "minims." They open the nest, forage for food, rear a new group of workers, and feed the queen. The queen becomes essentially an egg-making machine and can produce 200 eggs a day. As the colony grows, more foragers (10–20% of total workers) mean more food, and a larger proportion of larvae become majors. At the end of its second year, a colony matures and produces winged reproductives. The colony at that time averages 27,000 workers. Very large mature mounds may house 240,000 workers. Minor workers live 1–2 months, media workers 2–3 months, and major workers 3–6 months. Queens live 2–6 years.

Impacts. Red imported fire ants are mainly feared for their bites. Their venom creates a burning pain in the victim and, within 24 hours, raises itching pustules on the skin. These aggressive ants swarm over anything disturbing their nests, so bites are multiple. For sensitive people, bites can result in anaphylactic shock and, rarely, death. Increasingly, red imported fire ants are invading homes, outdoor electric meters and air conditioners, traffic control boxes, and airport runway lights, increasing opportunities for human-ant interaction, interfering with switching mechanisms, and causing short circuits and shutdowns. Infestations of red imported fire ants also pepper lawns and parks with unsightly mounds.

Fire Ants Leave the United States for Distant Shores

Since 2000, red imported fire ants have invaded China, Taiwan, and Australia. The southern United States appears to be the source of these invasions. Recent genetic studies indicate two or more introductions to each area have occurred. The ants that arrived in Taiwan stem directly from populations in California; however, California received its fire ants from southern U.S. states. American populations carry distinct combinations of genes (haplotypes) that are extremely rare in South America but common in the United States, Asia, and Australia and allow the invasion history of the ants to be tracked. Hitchhikers in cargo, fire ants appear to be following the world's major trade routes to new locales.

Source: Ascunce, Marina S., Chin-Cheng Yang, Jane Oakey, Luis Calcaterra, Wen-Jer Wu, Cheng-Jen Shih, Jérôme Goudet, Kenneth Ross, and DeWayne Shoemaker. "Global Invasion History of the Fire Ant *Solenopsis invicta*." *Science* 331: 1066–68, 2011.

In agricultural fields, mounds can damage farm equipment. Ants feed on young shoots, protect aphids that feed on plants, and can interfere with root growth; many crops suffer significant damage where ant colonies occur. On the other hand, red imported fire ants are major predators of pests such as boll weevils, sugarcane borer, and ticks. It is credited with reducing the range of the lone star tick, which targets livestock.

In more natural settings, the red imported fire ant is a strong competitor of native ants, especially the tropical fire ant; and displaces the invasive Argentine ant (*Linepithema humile*). It has been implicated in the decline of the bobwhite (*Colinus virginianus*) in the southeastern United States and has negatively impacted other ground-nesting birds and reptiles through bites, predation, and competition for space and food.

Management. Poisonous baits have proven most effective in reducing or eliminating colonies, since red imported fire ants carry food back to their nests. Boiling water poured on individual mounds may help reduce populations, but area flooding only causes the ants to link together in rafts and disperse. Biological controls hold promise, particularly the use of the nematode *Neoaplectana carpocapsae* and parasitic phorid flies that lay their eggs in ants. Spread of the ant is being slowed with the establishment of quarantine areas.

Ant-proofing structures can prevent the entry of ants to buildings. Caulking and sealing all cracks and crevices keep ants out, since they usually nest outdoors and enter homes only to feed. Generally sanitizing outdoor areas through the frequent emptying of trash cans and dumpsters also discourages foraging ants.

Selected References

Apperson, Charles, and Michael Waldvogel. "Red Imported Fire Ant in North Carolina." Insect Note_ENT/rsc-35, Department of Entomology, North Carolina Cooperative Extension, n.d. http://www.ces.ncsu.edu/depts/ent/notes/Urban/ifa.htm.

Collins, Laura, and Rudolf H. Scheffrahn. "Red Imported Fire Ant." Featured Creatures, Department of Entomology and Nematology, University of Florida, Institute of Food and Agricultural Sciences, 2008. http://entnemdept.ufl.edu/creatures/urban/ants/red_imported_fire_ant.htm.

"Integrated Pest Management Manual: Fire Ants." National Park Service, U.S. Department of the Interior, 2010. http://www.nature.nps.gov/biology/ipm/manual/fireants.cfm.

IUCN/SSC Invasive Species Specialist Group. "*Solenopsis invicta* (Insect)." ISSG Database, 2006. http://www.issg.org/database/species/distribution.asp?si=77&fr=1&sts=&lang=EN.

"Red Imported Fire Ant Pest Profile." California Department of Food and Agriculture, 2010. http://www.cdfa.ca.gov/phpps/pdep/target_pest_disease_profiles/rifa_profile.html.

■ Vertebrates

■ Fish

■ | Alewife

Also known as: Big eye herring, freshwater herring, gray herring, kyack, sawbelly, white herring, branch herring, river herring, glut herring, mulhaden, golden shad
Scientific name: *Alosa pseudoharengus*
Synonym: *Pomolobus pseudoharengus*
Family: Clupeidae

Native Range. Native to the Atlantic Ocean from Red Bay, Labrador, to South Carolina; spawning in estuaries and Atlantic Slope rivers of the eastern United States as much as 100 miles inland.

Distribution in the United States. As a native transplant, it is established in all five Great Lakes and also reported in streams, landlocked lakes, and reservoirs in Colorado (headwaters of the Colorado River basin), Georgia (Savannah River), Kentucky (Ohio River), Indiana (Bass Lake), Maine (Belgrade Lakes), Nebraska (Merritt Reservoir, Ainsworth Canal), New York (Lake Otsego, Cayuga, Upper Saranac, Big Moose, Woodhull, Saratoga, and Seneca lakes; St. Regis headwaters, mountain lakes in Adirondacks), Pennsylvania (Delaware Gap National Recreation Area), Tennessee (Dale Hollow Reservoir, Watauga Reservoir), Vermont (Lake St. Catherine), Virginia (Claytor Lake, John W. Flanagan Reservoir, Lake Chesdin, Leesville Reservoir, Smith Mountain Lake), West Virginia (Bluestone Reservoir, New and Kanawha rivers), Wisconsin (Kangaroo Lake, Pigeon River, Pigeon Lake, East Twin River, Sheyboygan River, Green Bay, St. Louis River estuary, Sauk Creek, and Milwaukee River).

Description. The transplanted, landlocked alewife is smaller than the native anadromous alewife and typically reaches a length of 6–10 in. (15–25 cm) and weight of less than 4 oz. (0.11 kg). The body is laterally compressed and relatively deep. It has a silvery color with a darker greenish sheen on the back. A distinctive black dot occurs on the body behind the eye. Scales merge along the belly to form scutes that create a serrated edge, the reason for one of the fish's common names, sawbelly. The tail or caudal fin is forked. Alewives move in large schools.

Introduction History. Alewives were first reported from Lake Ontario in 1873. Some have suggested they are actually native to the lake, having migrated up the St. Lawrence River from the Atlantic Ocean at some earlier time in geologic history. Another possible route for entry to the Great Lakes was the Erie Canal, opened in 1825 to connect Lake Erie with the Hudson River and the Atlantic Ocean. The alewife was first recorded from Lake Erie in 1831. The fish slowly dispersed upstream, perhaps aided by the Welland Canal, which connects Lakes Ontario and Erie. Populations were established in Lake Huron by 1933, Lake Erie by 1940, Lake Michigan by 1949, and Lake Superior by 1954. Genetic information suggests that alewives arrived via the Erie Canal and are not descendants of a population native

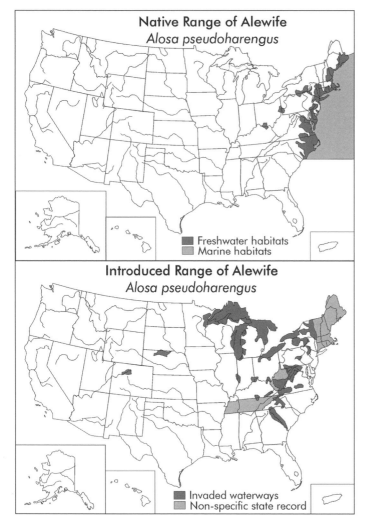

Native Range of Alewife
Alosa pseudoharengus

■ Freshwater habitats
■ Marine habitats

Introduced Range of Alewife
Alosa pseudoharengus

■ Invaded waterways
■ Non-specific state record

Top: The alewife is native to the western Atlantic Ocean, from Labrador south to South Carolina. In its native range it is anadromous. Bottom: As a native transplant, the alewife is established in all of the Great Lakes and in streams, lakes, and reservoirs in several states. (Both maps adapted from Fuller, Maynard, and Raikow 2009.)

to Lake Ontario. The fish is now abundant in Lake Huron and dominant in Lake Michigan. Its numbers are held in check in Lake Superior by cold water and in Lake Erie by water too shallow to provide many over-wintering refuges.

Invasion of the Great Lakes was aided by prior overfishing of its major predators, Atlantic salmon and lake trout. The introduction of the sea lamprey (see Fish, Sea Lamprey) may also have reduced the population sizes of competitors. Equally important, alewife physiology was such that they could thrive in the landlocked freshwater habitats to which they were introduced without the need to return to the sea after spawning, part of the natural life cycle of populations in their native range.

In water bodies other than the Great Lakes, all introductions were intentional, through both legal (e.g., in Virginia and West Virginia) and illegal (e.g., Lake Otsego and Adirondack lakes in New York and Lake St. Catherine in Vermont) stocking of streams and lakes with forage fish.

Habitat. Temperate lakes, reservoirs, and rivers. They prefer the deeper waters of lakes by day and move into shallower waters near shore at night. Alewives spawn in shallow waters and tributary streams in the spring. They overwinter in the deepest parts of lakes, but extremely cold winters may cause die-offs.

Diet. Primarily a filter-feeder, the alewife consumes zooplankters such as copepods. However, they are generalists and also feed on insect larvae, fish eggs, and small fish fry.

Life History. Alewives spawn in the spring during the night. Each female produces 12,000–50,000 eggs, each about 0.9 mm in diameter; they are broadcast as the males release sperm. Eggs are deposited over all types of substrate. No parental care is invested in eggs or larvae; the adults leave the spawning grounds as soon as spawning is completed. The larvae hatch after 3–6 days, depending upon water temperature; and 3–5 days after that they begin to feed. Larvae transform into juvenile fish slowly. Landlocked fishes mature in 1–3 years, faster than native anadromous members of the species. They live fewer than 10 years. Mass die-offs occur periodically, usually in the spring.

Alewives are prey for many predators, including native lake trout (*Salvelinus namaycush*), eels (Anguillidae), bigmouth bass (*Micropterus salmoides*), and whitefish (*Coregonus* spp.). Herons and other fish-eating birds take alewives, as do semi-aquatic mammals such as otter and mink.

Impacts. Alewives transformed the Great Lakes ecosystem. In some parts of the lakes, they are now a keystone spe-

Transplanted alewives inhabit landlocked lakes, reservoirs, and streams. (Eric Engbretson.)

cies. In Lake Michigan, 70–90 percent of the fish, by weight, are alewives. They outcompeted native zooplankton-feeders and are blamed for the demise of the lake trout in Lake Michigan, once a mainstay of the lake's commercial fishery. Their preying on fish fry is implicated in declines of emerald shiner (*Notropis atherinoides*), yellow perch (*Perca flavescens*), deepwater sculpin (*Myoxocephalus thompsoni*) and burbot (*Lota lota*). In smaller bodies of water, they compete with yellow perch, rainbow smelt (*Osmerus mordax*), and young bass, all of which survive on zooplankton. Alewives may also interfere with reproduction in landlocked Atlantic salmon (*Salmo salar*) and lake trout in areas where alewives are the main prey species: Alewives have a high amount of the enzyme thiaminase, so that the eggs of their predators become deficient in thiamine, resulting in high mortality rates among the fry.

Alewives are so well established in the Great Lakes ecosystem that their removal at this point would be disruptive. Instead, they are seen as a boon to efforts to reestablish an important native game fish in the lakes, the Atlantic salmon, for which they serve as an important forage species. Pacific Chinook salmon (*Oncorhynchus tshawytscha*) and coho salmon (*Oncorhynchus kisutch*) have been introduced to Lake Michigan to help control alewives and to provide a sport fishery.

Mass die-offs, such as occurred during the 1960s in the Great Lakes, polluted the beaches with rotting fish and were both nuisances and health hazards.

In other waters, where alewives are more recent invaders, they have the potential to alter the food web and reduce biodiversity.

Management. Population reduction is the only option available in the Great Lakes, where the alewife is a permanent addition to the fish fauna, and eradication might actually cause more damage to the ecosystem. Reestablishment of the Atlantic salmon and the stocking of other predatory fish such as Pacific salmon and brown trout (*Salmo trutta*: see Fish, Brown Trout), itself an exotic species, may help control alewife numbers. Elsewhere, culling extant populations and the prevention of new introductions are essential to reduce threats to native ecosystems. Legislation and enforcement of laws that make transport of alewives illegal are important. Construction of net barriers to prevent downstream dispersal out of lakes has been suggested. In some instances, fishing of alewives and collecting them when they congregate on spawning or overwintering grounds have been recommended. Alewives are sometimes used for bait and for pet food.

Selected References

Bean, Tim. "Alewife (*Alosa pseudoharengus*)." Introduced Species Summary Project, Columbia University, 2002. http://www.columbia.edu/itc/cerc/danoff-burg/invasion_bio/inv_spp_summ/alewife.html.

Fuller, Pam, Erynn Maynard, and David Raikow. "*Alosa pseudoharengus*." USGS Nonindigenous Aquatic Species Database, Gainesville, FL, 2009. http://nas.er.usgs.gov/queries/FactSheet.asp?speciesID=490.

National Biological Information Infrastructure (NBII) and IUCN/SSC Invasive Species Specialist Group (ISSG). "*Alosa pseudoharengus* (Fish)." ISSG Global Invasive Species Database, 2005. http://www.issg.org/database/species/ecology.asp?si=625&fr=1&sts=&lang=EN.

Tobias, V., and W. Fink. "*Alosa pseudoharengus*." Animal Diversity Web, University of Michigan Museum of Zoology, 2004. http://animaldiversity.ummz.umich.edu/site/accounts/information/Alosa_pseudoharengus.html.

■| Asian Swamp Eel

Also known as: Rice eel, rice-paddy eel, belut, white ricefield eel, yellow eel.
Scientific name: *Monopterus albus*
Synonym: *Fluta alba*
Family: Synbranchidae

Native Range. Indian subcontinent, Southeast Asia, East Asia. It is also native to Central and South America and may be native to Australia.

Distribution in the United States. Florida (North Miami area, Little Manatee River and Bullfrog Creek drainages near Tampa, and a canal system near Homestead, South Miami–Dade County); Georgia (Chatahoochee Nature Center, Roswell, Fulton County; Chattahoochee National Recreation Area, Gwinnett County); Hawai'i (O'ahu), and New Jersey (vicinity Silver Lake, Gibbsboro, Camden County).

Description. Asian swamp eels have elongated bodies and a compressed, tapering tail. They lack scales and fins and are covered with mucous. A single V-shaped gill opens on the underside just behind the head. The nose is blunt, and the eyes are small and dark. The upper lip is thick and covers part of the lower lip. Teeth are bristle-like (villiform). Body color ranges from olive green to brown; some are spotted with flecks of gold, yellow, or black. Total length is about 3 ft. (1 m); adults weigh about 1 lb. (0.5 kg).

Related or Similar Species. Asian swamp eels are not true eels, but they may be mistaken for native American eels (*Anguilla rostrata*). True eels have scales and fins and are anadromous. Asian swamp eels also resemble sea lampreys (see Fish, Sea Lamprey), which have obvious dorsal and caudal fins and seven gill openings on each side of the head. Lampreys lack jaws and have oval-shaped mouths. In Florida, Asian swamp eels might be confused with two large native aquatic amphibians, the two-toed amphiuma (*Amphiuma means*) and the greater siren (*Siren lacertina*). The former have four tiny legs and the latter have small front legs and bushy external gills.

Introduction History. Asian swamp eels were in Hawai'i by 1900. They were likely deliberately introduced for food. Swamp eels are a common food fish in China and other parts of Asia, where they are sold live. It is likely that immigrants brought them to Hawai'i. The first reports of swamp eels occurring in Florida and Georgia stem from the 1990s; these may have been accidentally released from fish farms or intentionally "freed" by aquaria owners. In Florida, two populations (North Miami and Tampa) can be genetically traced to China, and the third (South Miami–Dade County) to Southeast Asia. The Georgia population stems from Japan or Korea, which may be why it is cold-tolerant.

Habitat. Swamp eels are freshwater creatures and prefer shallow (less than 10 ft. or 3 m deep) sluggish, even stagnant waters. They are found in ponds, reservoirs, wetlands, streams, canals, and ditches.

Swamp eels can survive relatively wide temperature ranges and tolerate cold, even freezing, temperatures. They also are able to withstand low oxygen levels in water since they can "breathe" through their skin. During periods of drought, they can remain burrowed in damp mud for weeks without eating. Their adaptability to a wide range of ecological conditions extends to brackish and saline water, and they can even crawl across land if moist enough. During the day, Asian swamp

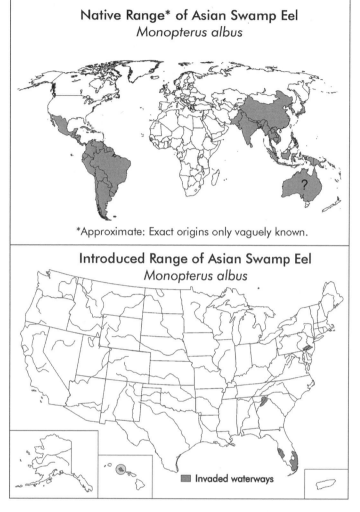

eels burrow in wet mud or hide in crevices or beneath dense vegetation.

Top: The so-called Asian swamp eel is native not only to Asia, but to Central and South America as well. It may also be native to Australia. Bottom: The Asian swamp eel is established in several widely separated locations in the United States, where it has apparently been introduced as a food fish. (Adapted from Nico and Fuller 2009.)

Diet. Nocturnal predators, swamp eels consume a variety of animals, including oligochaete worms, aquatic insects, amphipods, crayfish, tadpoles, frogs, turtle eggs, fish eggs, and other fish. They also eat detritus.

Life History. Asian swamp eels go through their complete life cycle in freshwater. Reproduction has been reported throughout the year. Eggs are laid in floating bubble nests near the mouths of burrows and are guarded by males. All young hatch as females, but some mature females are transsexual and transform into males after a yearlong nonfunctional stage. The males are larger than the females.

Impacts. To date, negative impacts on native ecosystems are more of a threat than a reality. Their broad range of tolerances, diet, and ability to move over land give Asian swamp

A. Adult Asian swamp eel. (U.S. Geological Survey.) B. Close-up of head. (U.S. Geological Survey.)

eels the potential to be aggressive invasive organisms. Furthermore, they have no known predators in the United States. Their burrowing habits and nocturnal activity periods could allow them to spread without being detected. They might reduce populations of native prey species as well as native predators such as large fish, frogs, turtles, and wading birds that compete for the same food source. A major concern is that they will invade Everglades National Park. The Homestead, Florida, collection site is only 0.5 mi. (0.8 km) from the park.

Management. Because of their ability to breathe air, control by usual fish eradication methods of poisoning water is difficult. Electrical barriers may prevent movement to new areas. Removal of vegetation could help. Electro-fishing devices are the only way to capture and detect Asian swamp eels for research and monitoring. The best practice is prevention: Do not stock these fish; do not release pet aquarium eels (or any other organisms) into local waters; and do not transport them as bait, forage, or aquarium pets.

Selected References

Bricking E. M. "Asian Swamp Eel." Introduced Species Summary Project, Columbia University, 2002. http://www.columbia.edu/itc/cerc/danoff-burg/invasion_bio/inv_spp_summ/Monopterus _albus.html.

Hamilton, H. *Frequently Asked Questions about the Asian Swamp Eel.* Florida Integrated Science Center, Gainesville, FL: USGS, 2006. http://fl.biology.usgs.gov/Nonindigenous_Species/Swamp_eel_FAQs/ swamp_eel_faqs.html.

National Biological Information Infrastructure (NBII) and IUCN/SSC Invasive Species Specialist Group (ISSG). "Ecology of *Monopterus albus*." ISSG Global Invasive Species Database, 2005. http:// www.issg.org/database/species/ecology.asp?si=446&fr=1&sts.

Nico, Leo, and Pam Fuller. "*Monopterus albus*." USGS Nonindigenous Aquatic Species Database, Gainesville, FL, 2009. Revised November 17, 2008. http://nas.er.usgs.gov/queries/FactSheet.asp ?speciesID=974.

Rotham, Carly J. "Asian Swamp Eel Threatens N.J. Wildlife," New Jersey Real-Time News, 2008. http:// www.nj.com/news/index.ssf/2008/09/asian_swamp_eel_threatens_nati.html.

■|Bighead Carp

Scientific name: *Hypophthalmichthys nobilis*
Synonym: *Aristichthys nobilis*
Family: Cyprinidae

Native Range. China, where it is native to lowland rivers of the North China Plain and South China, including the Yangtze (Chiang Jiang), Pearl, and West Xi Kiang.

Distribution in the United States. Established in the middle and lower Mississippi and Missouri rivers, in Illinois and Missouri. It may also be established in the Tallapossa Drainage in Sougahatchee Creek and Yates Reservoir, Alabama; in the Big Muddy, Cache, and Kaskaskia rivers in Illinois; and in a backwater outlet of the Black River, Louisiana. It has been reported in Arizona, Arkansas, California, Colorado, Florida, Hawai'i, Indiana, Iowa, Kansas, Kentucky, Louisiana, Mississippi, Nebraska, Oklahoma, South Dakota, Tennessee, Texas, Virginia, and West Virginia.

Description. Bighead carp are deep-bodied, laterally compressed fish with, as their name implies, very large heads. A complete lateral line arcs ventrally in the anterior part of the body. Scales on the body are tiny; the head and opercle are scaleless. Body color is dark gray on top and off-white on sides and belly. Mature specimens have dark grayish blotches on the top of the body. Young up to eight weeks old are silvery. On the underside, a distinct, smooth keel runs from near the base of the pectoral fins to the vent. The large upturned mouth has bony, rigid lips without barbels; the lower jaw protrudes slightly beyond the upper one. There are no teeth in the jaws. The eyes are close to the mouth and lie on the body midline. Fins of small individuals lack spines, but large specimens have a heavy, stiff,

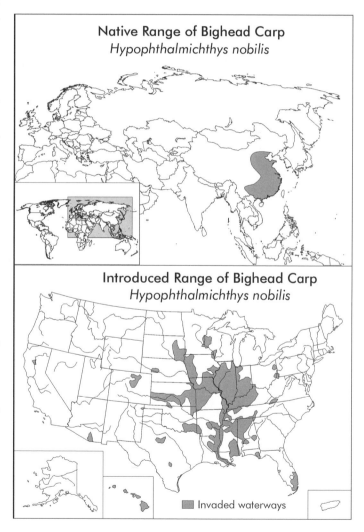

Top: The bighead carp is native to the lowland rivers of eastern China. Bottom: Bighead carp are invasive in the Mississippi-Missouri River systems and are reported in a number of other waterways. (Adapted from Nico and Fuller 2009.)

Bighead carp can exceed 4 ft. in length and weigh more than 100 lb. (U.S. Geological Survey.)

non-serrated spine at the origin of the dorsal fin, which has 8–9 soft rays. A slightly stiffened spine appears at the origin of the anal fin, which is hooked and has 13–14 soft rays. Pectoral fins on large males have sharp, non-serrated ridges along several of the anterior rays. The gill rakers are long and closely spaced, but are not fused together. The pharyngeal tooth count is 4–4. Large individuals may be more than 4 ft. (1.2 m) long and weigh over 100 lbs. (45 kg). Males are larger than females.

Related or Similar Species. Silver carp (*Hypophthalmichthys molitrix*), another invasive Asian carp (see Fish, Silver Carp), is a close relative and occurs in many of the same places as bigheads. The silver carp has a longer keel that runs from the base of the anal fin to the isthmus at the base of the gills. Its body color is greenish dorsally and silvery below the midline; it lacks the blotches of mature bigheads. The anal fin has 12–13 rays, and pectoral fins have 15–18 rays and a hard spine with serrated rear margin. Gill rakers are fused into a plate and covered with a net-like, porous matrix with which they strain the smaller phytoplankton from the water. Silver carp are somewhat smaller than bighead carp, attaining a length of 3 ft. (1.0 m) and a weight of 60 lbs. (27 kg). Silver carp are known for leaping high into the air when disturbed. Juvenile bigheads can be confused with gizzard shad (*Dorosoma cepedianum*; see Fish, Gizzard Shad), often used as bait fish.

Introduction History. Both bighead and silver carp were imported into Arkansas in 1972 by a catfish farmer interested in their potential for improving water quality in his fish ponds by their ability to consume huge amounts of algae. The fish spread to other aquaculture operations for the same purpose. Carp first appeared in the open waters of the Ohio and Mississippi rivers in the early 1980s; they probably escaped from culture ponds. Thousands of bighead carp made their way into the Osage River, Missouri, when a fish hatchery near Lake of the Ozarks was inundated by floodwaters in April 1994. Fish that had escaped into the Missouri River multiplied and spread into the lower Kansas River in Kansas after 1990.

Hojo-e

Bighead carp are a food fish for some Asian Americans and are usually purchased live at markets in large cities such as Chicago and New York. Cultural practices among Chinese, Vietnamese, and other Asian immigrant groups could provide a potential pathway for the fish to enter the Great Lakes. As part of a Buddhist ceremony known as *hojo-e*, releasing captive animals secures merits for the afterlife and lengthens the life of the practitioner. The ritual commonly takes place at Buddhist temples under the guidance of monks. In the United States, goldfish, turtles, and birds are usually released, but the practice may explain the appearance of Asian carp in public ponds and lagoons in the Great Lakes region. Chicago and New York now both have regulations requiring that Asian carp be killed before they are sold.

Source: Higbee, E., S. Fellow, and K. G. Shwayder. "The Live Food Fish Industry: New Challenges in Preventing the Introduction and Spread of Aquatic Invasive Species." Great Lakes Panel on Aquatic Nuisance Species, *ANS Update* 10(2): Fall–Winter 2004. Online at http://www.glc.org/ans/ansupdate/pdf/2004/ANSUpdateFW.pdf.

Illegal introductions by commercial fish farmers in the late 1980s are responsible for carp in Grand River, Oklahoma; California; and Cherry Creek Reservoir, Colorado.

Habitat. Bighead carp are native to subtropical and temperate freshwater habitats, preferring large rivers and the lakes connected to them. They usually are found in the upper or euphotic layer of the water column, where food is most abundant. They may migrate up streams from lakes to spawn, since they require a current for their eggs to float and develop properly.

Diet. Bighead carp filter relatively large particles (10–100 μm) from the water. As plankton feeders, their diet changes as they grow. Larvae consume mostly small phytoplankters such as protococcaceans and cyanobacteria; but large individuals specialize on larger particles, including zooplankters such as cladocerans and midge larvae and algae such as diatoms and colonial phytoplankters. They are opportunistic feeders, however, and what they eat depends upon the type of suspended materials that are most abundant. They have been said to consume their weight in plankton each day. They also feed on detritus.

Life History. Males and females mature in 2–4 years in warmer, subtropical waters and 5–7 years in cooler, temperate waters. They are prolific breeders, with females producing up to a million eggs during their lifetimes. Eggs have diameters near 0.2 in. (4.5–5 mm). Spawning takes place in the spring as water levels in rivers rise. Adults may migrate 100 miles or more upstream to breed. Eggs and larvae then float downstream to the lower reaches of rivers and to lakes. Intolerant of brackish water, carp spend their entire lives in freshwater.

Impacts. As voracious plankton and detritus feeders, bighead carp can outcompete native filter-feeding organisms such as mussels, fish larvae, and some adult fishes. The adult fish most likely to suffer from carp invasions are paddlefish (*Polydon spathula*), bigmouth buffalofish (*Ictiobus cyprinellus*) and gizzard shad (*Dorosoma petense*). Bighead carp have the potential of disrupting entire aquatic food webs. Their introduction into the Great Lakes system is particularly feared for this reason.

Management. The main focus of management is preventing the introduction of bighead carp into the Great Lakes. They already inhabit the Illinois River, which connects to Lake Michigan via the Chicago Sanitary and Ship Canal. By 2011, the U.S. Fish and Wildlife Service, the EPA, the U.S. Army Corps of Engineers, the State of Illinois, the International Joint Commission, and the Great Lakes Fishery Commission had completed a multimillion-dollar series of three electric dispersal barriers on the canal.

Selected References

"Asian Carp: Key to Identification." U.S. Fish and Wildlife Service, 2002. http://www.fws.gov/Midwest/fisheries/library/broch-asiancarpkey.pdf.

Bighead and Silver Carp (*Hypophthalmichthys nobilis* and *H. molitrix*). Wisconsin Department of Natural Resources, 2004. http://dnr.wi.gov/invasives/fact/asian_carp.htm.

Nico, Leo, and Pam Fuller. "*Hypophthalmichthys nobilis.*" USGS Nonindigenous Aquatic Species Database, Gainesville, FL, 2009. Revised November 10, 2010. http://nas.er.usgs.gov/queries/FactSheet.asp?speciesID=551.

U.S. Army Corps of Engineers, Chicago Sanitary and Ship Canal Aquatic Nuisance Species Dispersal Barrier System, 2009. http://www.lrc.usace.army.mil/AsianCarp/BarriersFactSheet.pdf.

■ | Brown Trout

Also known as: German trout
Scientific name: *Salmo trutta*
Synonyms: *Salmo fario, Fario argenteus*
Family: Salmonidae

Native Range. Eurasia. In Europe, this anadromous fish is native to Atlantic, Baltic, and Black Sea and Caspian Sea drainages. In western Asia, it is reported as native to Afghanistan, Armenia, and Turkey. It is also considered native in parts of North Africa. The fish has been introduced into areas where it is not native on every continent except Antarctica. Those imported into the United States came from Germany.

Distribution in the United States. The brown trout has been introduced into almost all 50 states and the Commonwealth of Puerto Rico. The only states without brown trout are Alaska, Louisiana, and Mississippi. The fish may not be breeding in most states, but it is continually restocked for recreational fishing.

Description. Brown trout are so named because of their brown to golden-brown color. The sides are yellow or silvery and bellies white to yellow. Red spots with blue halos and black spots adorn the sides of stream-dwelling browns, but are faint on lake-dwelling individuals. The lateral line is iridescent when light hits it from the right angles. There are two dorsal fins, the rear one a small fatty (or adipose) fin with a reddish color. The anal fin has 9–10 rays. The tail is square. Adults in the United States can be 13–16 in. (33–40 cm) long and weigh up to 10 lbs. (4.5 kg).

Related or Similar Species. Atlantic salmon (*Salmo salar*) are close relatives. They have no red on the adipose fin and the tail is slightly forked. Rainbow trout (*Onchorhynchus mykiss*) have lines of black spots on the tail (see Fish, Rainbow Trout, for a fuller description).

Introduction History. Brown trout were first imported from German hatcheries into the United States in 1883, when a shipment of 80,000 eggs landed at Cold Spring Harbor, New York. These eggs were distributed by the U.S. Fish Commission to the Caledonia Fish hatchery in New York and the Norville Hatchery in Michigan. In 1886, Pennsylvania began stocking brown trout in streams where native brook trout populations had been extirpated or reduced by practices such as logging, farming, dam construction, and industrial discharges, all of which had led to warmer water temperatures and siltation. Since then, brown trout have been stocked by state and local agencies across the country for sport fishing.

Habitat. Although anadromous in its native range, brown trout in the United States are freshwater fish and prefer streams and lakes. They hide during the day in shallow beds of aquatic vegetation, in shallow rock-strewn areas, under submerged logs, or in deep pools. They are most active at dawn and dusk and when the water temperature is near 55°F (12.8°C). They generally prefer water temperatures between 65° and 75°F (18–24° C) and tolerate warmer water temperatures than do native brook trout.

Diet. Smaller individuals feed on insects such as mayflies, caddisflies, and midges that the stream carries to them. Larger brown trout have a broader diet that includes large aquatic insects, mayflies, caddisflies,

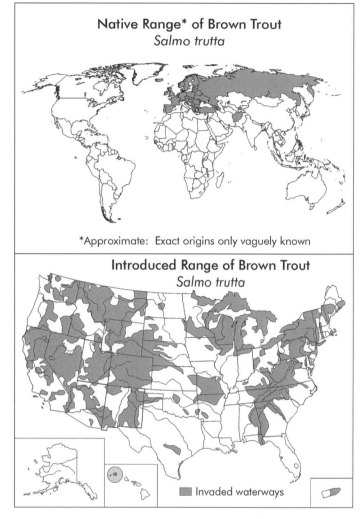

Top: The brown trout is native to Europe and western Asia. It has been widely introduced throughout Eurasia, making delineation of the actual native range difficult. Bottom: Brown trout have been introduced for sport-fishing in all states except Alaska, Louisiana, and Mississippi. (Adapted from Fuller, P. "Salmo trutta." USGS Nonindigenous Aquatic Species Database, Gainesville, FL, 2009. http://nas.er.usgs.gov/queries/speciesmap.aspx?speciesID=931.)

crustaceans, snails, amphibians, and small fish. They actively hunt their prey primarily at night. Both small and large trout also consume detritus washed from the shore.

Life History. Brown trout spawn from October to December in shallow headwater streams, ideally where the water is about 1 ft. (30 cm) deep, the current is about 7 in./second (18 cm/second), and substrate particles are small (diameters of ca. 0.5 in. or 1.27 cm). The female hollows out a nest, or redd, in which to release her eggs. As she releases her eggs, the male releases milt to fertilize them. The female covers the fertilized eggs with sand and fine gravel, and both adults leave. The larvae stay in the redd for 2–3 weeks until they are about 0.1 in. (25 mm)

Red spots with blue halos and black spots appear on brown trout but are faint on lake-dwelling individuals. (Tramper/Shutterstock.)

long. They then move downstream or into lakes for the first two years of their life. Brown trout begin to establish territories when they are juveniles. They mature at 3–4 years of age.

Many populations in the United States are maintained through restocking efforts rather than natural reproduction.

Impacts. Brown trout may compete with native fish, especially other salmonids (trout and salmon) for food. They grow larger and more rapidly than native species. They reportedly have also reduced populations of native trout through predation. As aggressive defenders of territories, they can displace native fishes from prime habitat. They have been implicated in population reductions of brook trout (*Salvelinus fontinalis*) and Modoc sucker (*Catostomus microps*), an endangered fish in California. Although it rarely happens, brown trout can hybridize with native trout. Brown trout has been nominated as one of "100 of the 'World's Worst' invasive alien species" by the IUCN and Global Invasive Species Programme.

Management. Brown trout are regularly stocked into most waters. Sport fishing often reduces numbers so that annual restocking is necessary. In New York, programs aimed at improving water quality for native brook trout become control methods for brown trout. In California, attempts are being made to eliminate brown trout where they may compete with golden trout (*Oncorhynchus mykiss aguabonita*).

Selected References

"Brown Trout in Pennsylvania." Pennsylvania Council of Trout Unlimited, n.d. http://www.patrout.org/Documents/Reference/brown.pdf.

Idema, A. "Salmo trutta." Animal Diversity Web, Museum of Zoology, University of Michigan, 1999. http://animaldiversity.ummz.umich.edu/site/accounts/information/Salmo_trutta.html.

Lauterbach, Sandra. "Brown Trout (*Salmo trutta*)." Introduced Species Summary Project, Columbia University, 2006. http://www.columbia.edu/itc/cerc/danoff-burg/invasion_bio/inv_spp_summ/Salmo_trutta.htm.

National Biological Information Infrastructure (NBII) and IUCN/SSC Invasive Species Specialty Group (ISSG). "*Salmo trutta* (fish)." ISSG Global Invasive Species Database, 2006. http://www.issg.org/database/species/ecology.asp?si=78&fr=1&sts=sss&lang=EN.

■ | Gizzard Shad

Also known as: Hickory shad, mud shad, nanny shad, skipjack, winter shad
Scientific name: *Dorosoma cepedianum*
Family: Clupeidae

Native Range. Native to eastern and central North America in the Mississippi, Atlantic and Gulf coast drainage systems. Probably native in the Arkansas, South Platte, and Republican drainage systems in eastern Colorado. Possibly native to the St. Lawrence River and the Great Lakes.

Distribution in the United States. The gizzard shad has been introduced to reservoirs and natural bodies of water both within and peripheral to its natural range. It is nonnative but established in the Colorado and Salt rivers and their connected reservoirs in Arizona; in the upper Colorado River drainage system in Arizona, Colorado, and Utah; and in Wyoming. Within its native range, its distribution has been extended by stocking in Colorado, Illinois, Indiana, Kansas, Kentucky, Minnesota, Nebraska, Pennsylvania, Utah, and Virginia. They may or may not be native to the Great Lakes, where they are found in all lakes except Lake Superior. Relatively recent natural range expansion probably explains their occurrence in Connecticut (Connecticut River), Maine

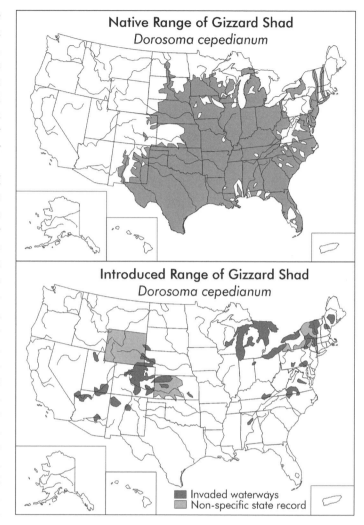

Top: The gizzard shad is native to the eastern and central United States. Bottom: Gizzard shad have been introduced to rivers, lakes, and reservoirs within and beyond the borders of its native range. (Both maps adapted from Fuller 2009.)

(lower Saco and Kennebec rivers), Massachusetts (Connecticut and Merrimack rivers), and Vermont (Lake Champlain and Connecticut River).

Description. This is a deep-bodied, moderately compressed herring that reaches lengths of 22 in. (56 cm) and weights of up to 7 lbs. (3.2 kg). It has a dark-blue or gray back, silvery sides that may reflect various colors, and a white belly. A dark, purplish blotch occurs high on the side behind the opercle in young and small specimens; it is faint or absent in older, larger individuals. It has no lateral line. The small mouth has a deep notch in the center of the upper jaw, which protrudes slightly beyond the lower jaw. The snout is blunt, and the eye large. The short, soft-rayed dorsal fin is centered on the back and has 10–12 rays; the last ray is a long filament that is distinctive for the species. The tail (caudal fin) is deeply forked. Scute-like scales form a distinct saw-toothed keel along the belly. Gill rakers are fine and number more than 400. The stomach is thick-walled and acts like a gizzard, giving the fish its common name.

Related or Similar Species. Threadfin shad (*D. petense*) is a smaller fish, rarely more than 8 in. (20 cm) long. The ray extending from its dorsal fin is much longer than that of gizzard shad and almost reaches the tail. The mouth especially distinguishes the two: in the threadfin, the lower jaw extends beyond the upper jaw—the reverse of the situation in gizzard shad. Fins of threadfin shad tend to be yellowish, whereas those of gizzard shad are grayish.

Introduction History. In most cases, gizzard shad were intentionally introduced to ponds, lakes, and reservoirs as forage fish, especially in recreational sport fisheries. From their sites of entry, however, they have dispersed rapidly into connecting bodies of water, often assisted by man-made facilities. Gizzard shad in the Upper Colorado River basin of Arizona, Colorado, and Utah may trace their origins to an accidental introduction in a contaminated stocking of largemouth bass from a hatchery in Texas into Morgan Lake, New Mexico, near the San Juan River around 1996. Gizzards were first reported in the San Juan arm of Lake Powell in 2002 and are now found throughout the lake. From Lake Powell, they dispersed downstream to Lake Mead and, by 2008, were found in the Grand Canyon stretch of the Colorado River. They also dispersed upstream into the headwaters of the Colorado: in 2006 they were collected in the Gunnison and Middle Green rivers. Gizzard shad were stocked in Lake Havasu in the lower Colorado River drainage and, from there, have dispersed upstream as far as Davis Dam and downstream to the Mexican border and into the Salton Sea.

In Connecticut, Massachusetts, and Vermont, gizzard shad have expanded their range upstream in the Connecticut River. They were first recorded at the mouth of the river around 1980, part of what seems to be a natural range expansion northward along coastal rivers, which has also brought them into the Merrimack River in Massachusetts and lower Saco and Kennebec rivers in Maine. By 1986, they had appeared at Holyoke Dam and now occur in the mainstem of the Connecticut River as far north as the Bellows Falls dam. Fish "ladders" built to let American shad (*Alosa sapidissima*) and Atlantic salmon (*Salmo salar*) circumvent the dams on their annual spawning runs have aided this movement. Gizzards entered Lake Champlain, Vermont, through the Hudson Barge Canal.

Some believe gizzard shad may be native to the Great Lakes and St. Lawrence River, but it is also possible that they entered at Green Bay, Wisconsin, from the Mississippi River via the Fox-Wisconsin Canal or via the Chicago River Canal. Entry into Lake Erie was facilitated by the Ohio Canal.

Gizzards in Wyoming stem from intentional introductions made in Nebraska.

Habitat. Freshwater lakes and reservoirs, slow-moving rivers, and pools of smaller streams are preferred habitat. Gizzard shad can also be found in the brackish waters of estuaries. Native populations in the Mid-Atlantic states are anadromous. They often feed in schools over mucky and sandy bottoms.

Diet. Gizzard shad are filter-feeders and, as adults, strain algae and detritus from bottom sands and muds. They also ingest sand to aid in grinding food in their gizzard-like stomachs. Young shad capture zooplankters (copepods and cladocerans) from the water column. Shad feed in large schools and characteristically leap from the water and flip onto their sides, a behavior that gave them the common name of skipjack.

Life History. Gizzard shads spawn near shore during spring nights as water temperatures rise near 70°F (21°C). Spawning may continue for several weeks, after which adults return to deeper water. A single female may produce 300,000 or more eggs. Eggs and milt are broadcast near the surface in shallow water, and fertilized eggs settle down through the water and stick on roots, plants, and debris. They hatch in 2–3 days. Larvae grow rapidly and, during their first year, young shad may reach lengths of 4–7 in.—too large to be prey for all but the largest predatory fish. Most are sexually mature at age 2. Life span is 2–3 years. Adults experience die-offs in winter and after spawning.

The gizzard shad, a member of the herring family, can attain a length of 22 in. and weigh as much as 7 lb. (Eric Engbretson.)

Impacts. As larvae, gizzard shad may compete with the young of other, more desirable fish for zooplankton. Not only do shad populations grow rapidly, but they also spawn earlier than some other fish and thus may deplete the food resource enough to negatively affect growth and survival of sport and other fishes. They are usually stocked as a forage fish, but their rapid growth rate soon makes them too large to be taken by most predatory fishes. However, while they are small, they can be important prey for striped bass, largemouth bass, crappie, walleye, and other sport fishes. The closely related threadfin shad, also introduced to reservoirs as forage and baitfish, do not grow so large and so are favored by sport fisheries managers, but they too may be outcompeted by gizzard shad. In the Colorado River, gizzard shad are viewed as one more threat to endangered endemic fishes such as Colorado River pike minnow (*Ptychocheilus lucius*), bonnytail chub (*Gila elegans*), humpback chub (*Gila cypha*), and razorback sucker (*Xyrauchen texanus*) as well as to a number of other sensitive species.

Management. Little to none.

Selected References

Finney, Sam T., and Mark H. Fuller. "Gizzard Shad (*Dorosoma cepedianum*) Expansion and Reproduction in the Upper Colorado River Basin." *Western North American Naturalist* 68(4): 524–25, 2008.

"Fish Facts—Gizzard Shad." U.S. Fish and Wildlife Service, Connecticut River Coordinator's Office, n.d. http://www.fws.gov/r5crc/Fish/zc_doce.html.

Fuller, Pam. *Dorosoma cepedianum*. USGS Nonindigenous Aquatic Species Database, Gainesville, FL, 2009. Revised March 10, 2008. http://nas.er.usgs.gov/queries/FactSheet.asp?speciesID=492.

"Gizzard Shad *Dorosoma cepedianum*," Utah Division of Wildlife Resources, n.d. http:// wildlife.utah.gov/pdf/AIS_plans_2010/AIS_12nGizzardShad-Dan-final.pdf.

"Shad Species Used in Striped Bass Fishing." Ben Sanders' ArkansasStripers.com, n.d. http:// www.arkansasstripers.com/shad_species.htm.

Steiner, Linda "Herrings," Chapter 10 in *Pennsylvania Fishes*. Pennsylvania Fish and Boat Commission, 2000. http://www.fish.state.pa.us/pafish/fishhtms/chap10.htm.

■|Grass Carp

Also known as: White amur
Scientific name: *Ctenopharyngodon idella*
Family: Cyprinidae

Native Range. Native to eastern Asia, from the Amur River, Russia, to the West River in southern China. They have been widely distributed throughout the world as a food fish. The first imported into the United States reportedly came from Taiwan and Malaysia.

Distribution in the United States. Grass carp have been widely used to control aquatic vegetation and have been reported from 45 states. States without records of grass carp are Alaska, Maine, Montana, Rhode Island, and Vermont. The fish is established in parts of the Mississippi River drainage, including the Mississippi River itself, in Arkansas, Kentucky, Illinois, Louisiana, Montana, and Tennessee; the Illinois River in Missouri, the lower Missouri River in Montana; the Ohio River in Illinois; and the Trinity River in Texas.

Description. The grass carp is a large fish, often attaining a total length greater than 3 ft. (1.0 m) and a weight of 40 lbs. (18 kg) or more. It has a thick, oblong body with a rounded belly and broad head. The anal fin is set close to the caudal fin, which is forked. The dorsal fin is short, with 7–8 rays and set over the pelvic fins. All fins are soft-rayed and usually green-gray to dull silver. The lateral line is complete and slightly down-curved. Eyes are positioned at or slightly above midline of body. Body color is silver to pale gray, darker on the back and brassy on the sides. The large scales on the back and sides have distinct dark edges that give a characteristic cross-hatched pattern to the body. The pharyngeal teeth are long and serrated, with deep, parallel grooves. Gill rakers are short, unfused, and widely spaced.

Related or Similar Species. Black carp (*Mylopharyngodon piceus*), another introduced Asian carp but one of much more restricted distribution in the United States, is darker in color, although not black. Its pharyngeal teeth are smooth and resemble human molars. It is a bottom-dwelling species that feeds on snails and other mollusks. All native cyprinids (minnows) have anal fins positioned more anteriorly than grass carp.

Introduction History. Grass carp first entered the United States in 1963, when they were imported by the aquaculture program at Auburn University, Alabama, and the Fish Farming Experimental Station in Stuttgart, Arkansas. Fish escaping from the latter facility in 1966 were the first released into the White River, but later intentional stockings in Arkansas took place in lakes and reservoirs with free passage to river systems. Grass carp were reported in the Illinois section of the Mississippi River in 1971. Soon thereafter, grass carp were being commonly reported in the Mississippi and Missouri rivers. The fish has been widely dispersed since then through legal and illegal interstate transport and release by private individuals and organizations, escapes from farm ponds and aquaculture facilities, and stockings by federal, state, and local government agencies, among them the U.S. Fish and Wildlife Service, the Tennessee Valley Authority, and state fish and game agencies in Arkansas, Delaware, Florida, Iowa, New Mexico, and Texas.

Habitat. Grass carp prefer shallow, quiet waters in lakes, pools, and the backwaters of large rivers that have abundant aquatic vegetation. They undertake long spawning migrations into faster-moving rivers.

Diet. A herbivore, the grass carp feeds on algae, submerged and floating aquatic macrophytes, and even overhanging terrestrial plants. Although they are known to consume hundreds of different kinds of aquatic plants, they show distinct preferences. Young fish select soft, succulent species such as filamentous algae, pondweeds, elodea, duckweeds, chara, and the soft, growing tips of submerged plants. As they grow older and larger, they add more fibrous, less succulent plants and tend to avoid filamentous algae. Adults will eat such invasive plants as waterhyacinth (see Volume 2 Aquatic Plants, Waterhyacinth, *Eichhornia crassipes*), hydrilla (see Volume 2 Aquatic Plants, Hydrilla, *Hydrilla verticillata*), and Eurasian watermilfoil (see Volume 2, Aquatic Plants, Eurasian watermilfoil, *Myriophyllum spicatum*), especially if more palatable species are not present. They may also take insects and other invertebrates and detritus, particularly if it is attached to aquatic plants.

Food consumption rates vary with water temperature, age and size of fish, and plant species available. Optimal consumption occurs when water temperatures are between 70° and 86°F (21–30°C). Carp feed only intermittently when temperatures fall into the 30s (below 0°C). Essentially a warm-water fish, they are dormant during the cold winters of temperate regions.

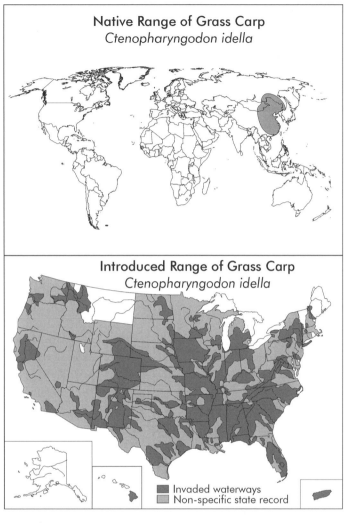

Top: The grass carp is native to eastern Asia. Bottom: Grass carp, introduced to control aquatic plants, have been reported from 45 states. (Adapted from Nico, Fuller, and Schofield 2009.)

Life History. Grass carp make long annual migrations to spawn in fast-moving large rivers. They often congregate in areas of turbulent water such as in rapids or at falls or the base of dams. Spawning occurs in the spring as water temperatures rise into the 60s and 70s (15–low 20s°C). The eggs must stay afloat in order to develop, which they do as they drift downstream, perhaps as much as 110 miles. Eggs hatch in 2–3 days, and larvae shelter in the vegetated areas of floodplains and lakes. They may winter in deep pools in rivers. Grass carp become sexually mature between two and five years of age in the subtropics and four to seven years or longer in cooler regions. Females usually mature a year later than males. Females produce 500,000–700,000 eggs, giving carp enormous reproductive potential.

Impacts. Grass carp are stocked to remove aquatic weeds, many of which are also classified as invasive. This may be advantageous for some game fish, but overconsumption by

The large, black-edged scales on the back and sides of grass carp give them a distinctive appearance. (Oleg_Z/Shutterstock.)

carp can also modify habitats and alter the species composition and trophic structure of ecosystems by eliminating the aquatic vegetation that provides food, breeding sites, and cover for native invertebrates, fishes, and birds. They may destroy spawning substrates, disturb sediments and increase water turbidity, increase nutrient levels in the water (their short gut makes their digestive system very inefficient, so large volumes of waste are produced), promote algal blooms, and decrease oxygen levels. Reportedly, they have contributed to reductions in populations of bluegill (*Lepomis macrochirus*), sunfish, smelt, and pike. Grass carp also have been implicated in the declining numbers of certain waterfowl such as Gadwall (*Anas stepera*), American Wigeon (*Anas americana*), and American Coot (*Fulica americana*) in areas to which the fish has been introduced.

Carp may transmit several parasites and diseases to native fish. They are probably the source of the Asian tapeworm (*Bothriocephalus opsarichthydis*), which has infected some native cyprinids, including the endangered woundfin (*Plagopterus argentissimus*), a small minnow endemic to the Virgin River, a tributary of the Colorado River.

Although considered an invasive species, grass carp is still stocked and even recommended as a biocontrol method for noxious weeds (see sidebar on Triploid Grass Carp).

Management. All states have restrictions on the importation of grass carp for aquatic weed control, but most allow stocking with proper permitting. Because of the reproductive potential and propensity to disperse, and because of environmental concerns, sterile triploid fish (see sidebar on Triploid Grass Carp) are widely used today for weed control. Fish screens may be required at all inlets and outlets to a water body to prevent grass carp from migrating out of lakes and into rivers. Carp may or may not have the desired effect of removing aquatic weeds. They seem to be an all-or-nothing proposition and should be stocked only where complete elimination of submerged vegetation is desired or can be tolerated.

Triploid Grass Carp

Scientists have sought ways to reduce the likelihood of reproduction in grass carp and thereby control their spread into open waters. Crossing female grass carp with male bighead carp produced 100 percent triploid hybrids that were sterile. Triploid means each fish has three copies of each chromosome instead of the normal two (diploid). Later, it was discovered that shocking the eggs (in fish hatcheries) with cold or heat under high hydrostatic pressure also produced triploids, and this became the standard method of producing sterile fish. Since triploids have larger red blood cells and nuclei than diploids, the two are readily separated in a simple blood test. The U.S. Fish and Wildlife Service now has an established testing and inspection protocol to screen all fish leaving hatcheries and to guarantee that only triploids are imported into those states that do not permit the introduction of diploid grass carp.

Selected References

"Grass Carp, *Ctenopharyngodon idella* (Valenciennes 1844)." Florida Integrated Science Center, USGS, 2005. http://fl.biology.usgs.gov/Carp_ID/html/ctenopharyngodon_idella.html l.

National Biological Information Infrastructure (NBII) and IUCN/SSC Invasive Species Specialist Group (ISSG). "*Ctenopharyngodon idella* (Fish)." ISSG Global Invasive Species Database, 2005. http://www.issg.org/database/species/ecology.asp?si=369&fr=1&sts=sss&lang=EN.

Nico, L. G., P. L. Fuller, and P. J. Schofield. "*Ctenopharyngodon idella*." USGS Nonindigenous Aquatic Species Database, Gainesville, FL, 2006. http://nas.er.usgs.gov/queries/FactSheet.asp?speciesID=514.

■ | Lionfish

Also known as: Red lionfish, turkeyfish, devil firefish, zebrafish
Scientific name: *Pterois volitans/P. miles*
Family: Scorpaenidae (Scorpionfishes)

Native Range. The Indo-Pacific Ocean region, from southern Japan to the Philippines to Micronesia and the east coast of Australia through western Polynesia to the Marquesas Island and Oeno in the Pitcairn islands. The red lionfish (*P. volitans*) is mostly a Pacific Ocean fish; its range extends eastward through Oceania; the devil firefish (*P. miles*) occurs primarily in the Indian Ocean, from Sumatra northwestward into the Red Sea. Both fishes are found off Sumatra. The devil firefish has passed through the Suez Canal into the Mediterranean Sea.

Whether or not these are two true species or simply regional variants of the same species has not been conclusively determined. Genetic studies indicate that both now occur as exotic taxa in the Atlantic Ocean off the U.S. coast, but the Atlantic population consists primarily of *P. volitans*. Their likely source area is the Philippines.

Distribution in the United States. Apparently established and reproducing from the Florida Keys to Cape Hatteras, North Carolina. Divers have reported sightings of juveniles as far north as Long Island, New York, New Jersey, and Rhode Island; but it is assumed they were carried north in the Gulf Stream as the young of the year and will either migrate south or die in winter. Winter bottom-water temperatures may create a northern limit to their distribution area as it does for other tropical fishes whose young are frequently encountered off the coast of the northeastern United States in summer.

Description. Lionfish are unmistakable. The red lionfish has long, separated dorsal fin spines. The appearance of the dorsal spines together with the elaborate fan-like pectoral fins resulted in the common name, turkeyfish. The body is white to cream-colored with red-to-maroon vertical stripes that gave it another of its common names, zebrafish. The stripes have a regular, alternating pattern between wide and narrow bands and sometimes converge in a V on the sides. Fleshy tabs or tentacles occur over the eyes and above the mouth.

The dorsal, anal, and pelvic spines of lionfish are highly venomous. A pair of venom glands is found on each spine, which is covered by a sheath. When a spine is pushed into flesh, the sheath presses down on the glands, and venom is released into a groove that runs along the spine, delivering the poison to the wound. The resulting sting is extremely painful, but not fatal to humans.

The largest lionfish collected off the East Coast of the United States was about 17 in. (43 cm) long and weighed about 2.5 lbs. (1.1 kg).

Related or Similar Species. The devil firefish and the red lionfish are very similar in appearance. They are distinguished from each other by the number of rays in the dorsal

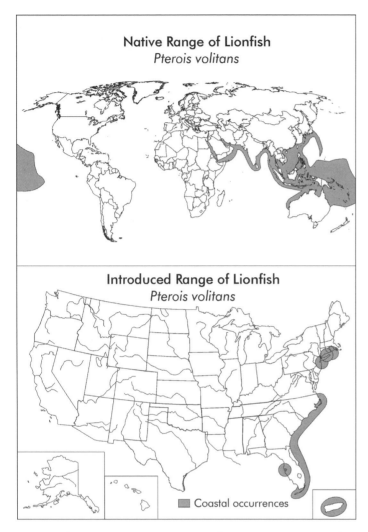

Native Range of Lionfish
Pterois volitans

Introduced Range of Lionfish
Pterois volitans

■ Coastal occurrences

Top: Lionfish are native to the Indo-Pacific region. Bottom: Lionfish appear to be established along the Atlantic coast of the United States from the Florida Keys north to Cape Hatteras, North Carolina. Juveniles have been reported as far north as Rhode Island. (Adapted from Schofield, Morris, Langston, and Fuller 2009.)

and anal fins and by genetic assays. The lionfish usually has 11 dorsal-fin rays and 7 anal fin rays, whereas the devil firefish has 2 and 6, respectively. The two fishes are very closely related and may actually be the same species. A 2007 study showed that the composition of the Atlantic population was 97 percent lionfish and 7 percent devil firefish.

Introduction History. The first known occurrence of lionfish in the Atlantic off the United States was a specimen collected near Dania, Florida, in October 1985. Later, six fish escaped into Biscayne Bay from a large private aquarium that was destroyed during Hurricane Andrew in 1992. These fish were seen alive in the bay a few days later. Sporadic or undocumented sightings of red lionfish were reported from Florida's east coast between 1993 and 2002. In February and March 2002, three specimens were taken off northeast Florida near St. Augustine, Jacksonville, and Amelia Island.

Lionfish were first recorded off Georgia in 2001, when a single adult was collected. The same year, two juveniles were taken off Long Island, New York.

Small groups of lionfish were observed during research dives in submersibles off South Carolina and North Carolina in 2002. This was the first, though circumstantial, evidence that the Atlantic population was reproducing.

Although water ballast is a possible route of entry for the fish, it is highly unlikely. The red lionfish is a very popular marine aquarium species, and the aquarium trade is the most probable pathway by which the fish came to the Atlantic Ocean. The color patterns of Atlantic lionfish are quite similar to those from the Philippines, where many are collected for the aquarium trade. Either accidentally or deliberately, whether they had grown too large or for whatever reason, these (and other) aquaria pets are being dumped into the open waters of the ocean.

The wild population is growing rapidly. That juvenile fish smaller than those sold for aquaria are showing up off the North Carolina coast and elsewhere is a strong indication that the species is reproducing. This is the first time a western Pacific fish has become established in the Atlantic Ocean off the United States.

Habitat. Lionfish are tropical marine fish associated with the continental shelf biome. They inhabit reefs at depths of 30 to 575 ft. (9–175 m), where they can hide in crevices by day. Along the southeastern coast of the United States, they have been sighted on reefs and other hard substrates, including shipwrecks. In their native distribution areas, they are found in lagoons and turbid inshore waters, including harbors, in addition to offshore coral reefs and rocky outcrops.

The lionfish, with its long dorsal spines and fan-like pectoral fins, is unmistakable. (Albert Kok.)

Diet. Lionfish are carnivores that prey on crustaceans and small fish, including young lionfish and juveniles of some commercially important species such as grouper and snapper. They focus increasingly on fish as they age. Nocturnal hunters, lionfish move into deeper waters at night to forage. They are ambush predators and sweep up and trap their prey in their extended pectoral fins. They then quickly pounce and ingest it.

Life History. Lionfish are solitary and territorial most of the year, only forming small groups during mating, when males join with several females. Males use their spines and fins in competitive visual displays. Females release a pair of mucous-encapsulated clusters of 2,000–15,000 eggs that are externally fertilized by the males. The egg mass floats, and as microbes decompose the mucous, eggs are released into the water column. Early life-history stages are poorly known. Presumably, lionfish larvae have a pelagic stage during which they are dispersed by ocean currents. The larval stage lasts an estimated 25–40 days. Age at sexual maturity and average lifespan are unknown. There are few known predators of lionfish, although off the Bahamas, lionfish have been found in the stomachs of native groupers.

Impacts. Recent (2008) studies in the Bahamas have shown that lionfish may have negative impacts on native Atlantic reef-fishes. Reefs with lionfish had 79 percent fewer juvenile reef fish compared to reefs lacking lionfish. Prey items found in lionfish stomachs included fairy basslet (*Gramma loreto*), bridled cardinal fish (*Apogon aurolineatus*), white grunt (*Haemulon plumierii*), bicolor damselfish (*Stegastes pertitus*), several wrasses, striped parrotfish (*Scarus iserti*), and dusky blenny (*Malacoctenus gilli*). Juvenile spiny lobster (*Panulirus argus*) is possibly a food item. These findings suggest that lionfish affect recruitment of coral fishes and may thereby compete with native piscivores. They could potentially reduce the number of ecologically important species such as parrotfish and other herbivores that prevent macroalgae from taking over the reef.

Management. Very little can be done in a marine environment to control invasive species. Environmental changes occurring on the southeastern-U.S. continental shelf favor the continued expansion of lionfish in these waters. Many important native reef fish predators have

---

Tournament Held to Help Control Lionfish Population in the Florida Keys National Marine Sanctuary

In September 2010, the Reef Environmental Education Foundation (REEF) and the Florida Keys National Marine Sanctuary sponsored a lionfish hunt with cash and other prizes for teams of divers that would catch invasive lionfish. The 27 participating teams collected a total of 534 lionfish during the weekend tournament, the winning team catching 111 in one day. The largest was 11 in. (27.9 cm) long, but most were too small to filet, so the planned follow-up feast was a failure. Nonetheless, sanctuary superintendent Sean Morton said, "The sanctuary is thrilled by the response from the dive community. The volume of fish caught during this single-day event demonstrates that dedicated diver removal efforts can be effective at helping keep this invasive [species] at bay."

Source: Frink, Stephen. "The Tipping Point(s)." *Alert Diver*, Fall 2010: 8.

---

been overfished, but even if they were at past densities, it is unknown if native predators could eventually control lionfish numbers. None of them have prior experience with a prey fish with venomous spines. The fish fauna in general has become more tropical during the last decades of the twentieth century, so it appears conditions increasingly favor the breeding and recruitment of lionfish.

Selected References

Hare, J. A., and P. E. Whitfield. "An Integrated Assessment of the Introduction of Lionfish (*Pterois volitans/miles* Complex) to the Western Atlantic Ocean." NOAA Technical Memorandum NOS NCCOS 2, 21 pp., 2003. http://coastalscience.noaa.gov/documents/lionfish_ia.pdf.

"Invasive Lionfish Threaten Native Fish and the Environment in U.S. Atlantic Coastal Waters." National Ocean Service, NOAA, 2009. http://oceanservice.noaa.gov/facts/lionfish.html.

Masterson, J. "*Pterois volitans*." Smithsonian Marine Station at Fort Pierce, 2007. http://www.sms.si.edu/IRLSpec/Pterois_volitans.htm.

Schofield, P. J., J. A. Morris Jr., J. N. Langston, and P. L. Fuller. "*Pterois volitans/miles*." USGS Nonindigenous Aquatic Species Database, Gainesville, FL, 2009. http://nas.er.usgs.gov/queries/FactSheet.asp?speciesID=963.

■| Mosquitofish

Scientific names: *Gambusia affinis* (western mosquitofish)
Gambusia holbrooki (eastern mosquitofish)
Family: Poeciliidae

Native Range. The western mosquitofish is native to the south-central United States, from Illinois and Indiana south into Mexico, west to New Mexico, and east into the Mobile River drainage system. Also native in the Chattahoochee and Savanna rivers. Apparently, most introductions to other states (and worldwide) stem from a few populations from Georgia, Illinois, Tennessee, and Texas.

The eastern mosquitofish is native to the Atlantic and Gulf coast drainages from southern New Jersey south and west to southern Alabama.

Distribution in the United States. The western mosquitofish has been widely introduced into most western states. The eastern mosquitofish has likely been transplanted outside its native range only in eastern states. It is known to have been introduced into New Jersey and Tennessee as well as Alabama. Sometimes they were moved to other regions within the same state to which they are native. Such is the case in Virginia, for example. The two mosquitofish species are so similar in appearance and so closely related, and early records often so imprecise, that it is difficult to always be sure which species was transplanted where.

Description. These two little fishes are very similar in appearance. They have arched backs and deep bellies. The head is large, its upper surface flattened. The eyes are very large relative to body size. The small mouth is upturned and protrusible to allow feeding from the surface. Dorsal and anal fins are rounded; small black spots appear on the dorsal fin. The western mosquitofish usually has six dorsal rays, and the eastern mosquitofish seven.

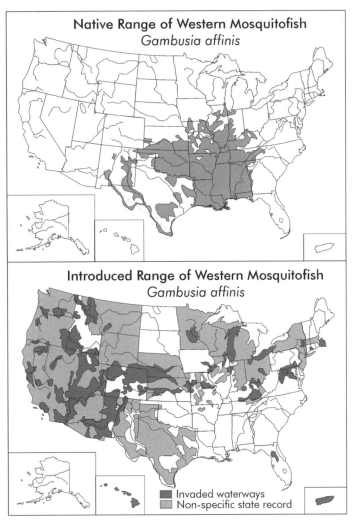

Top: The western mosquitofish is native to water bodies in the south central United States. Bottom: *Gambusia affinis* has been introduced into most western states. (Both maps adapted from Nico, Fuller, and Jacobs 2009.)

The head and much of the body are covered with large scales outlined by dark pigments to give a characteristic diamond pattern on the body. The back is greenish or brown, the sides gray-blue, and the belly silvery-white. In mature females, a black patch appears on the rear of the abdomen above and forward of the vent. Anal, pelvic, and pectoral fins are a pale, translucent amber.

Mosquitofish are sexually dimorphic; the females are much larger than the males. Adult females may be as much as 2–2.75 in. (6–7 cm) long, adult males only about 1.5 in. (4 cm). Adult males possess a gonopodium, a long tube-like structure composed of fused rays of the anal fin and used to transfer sperm into the vent of females during mating. The gonopodium is usually held back along the belly, but during mating, it is turned down and forward.

Related or Similar Species. The two mosquitofish are easily confused with each other. In Arizona, the Sonoran topminnow (*Poeciliopsis occidentalis*) could be mistaken for a

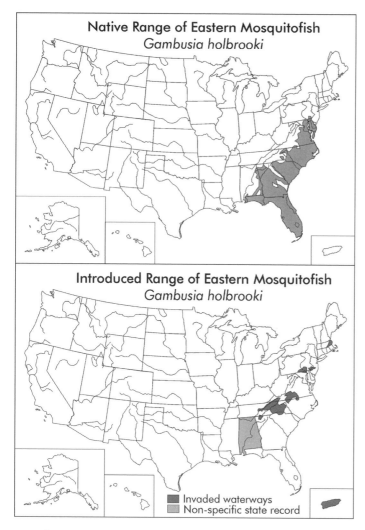

Native Range of Eastern Mosquitofish
Gambusia holbrooki

Introduced Range of Eastern Mosquitofish
Gambusia holbrooki

■ Invaded waterways
▨ Non-specific state record

Top: The eastern mosquitofish is native to drainages of the Atlantic Slope south of New Jersey and of the Gulf of Mexico. Bottom: *Gambusia holbrooki* has probably been transplanted only into bodies of water in the eastern United States. (Both maps adapted from Nico and Fuller 2009.)

mosquitofish. The gonopodia of the two are distinct, that of the topminnow is asymmetrical to the left, and that of the mosquitofish is symmetrical. In addition, the pelvic fins of male *Gambusia* are modified with a fleshy appendage. In Florida, the sailfin molly (*Poecilia latipinna*) looks superficially like a mosquitofish, but lacks the net-like scale pattern, and the position of the dorsal fin differs.

Introduction History. The first known introduction of mosquitofish in the United States occurred in the early 1890s. The western mosquitofish was taken from Texas to Hawai'i to test their effectiveness in consuming mosquito larvae in 1905. The same year, eastern mosquitofish from North Carolina were released into New Jersey waters to control mosquitoes. For several decades afterward, mosquitofish were routinely and widely introduced by government agencies; for example, the former U.S. Public Health Service introduced it as a cost-effective way to combat malaria. *Gambusia* continue to be promoted and intentionally stocked by local, state, and federal agencies for mosquito control and are viewed as attractive alternatives to the use of chemical insecticides. In some areas, range expansions have occurred once introduced populations became established.

Habitat. Mosquitofish can be found in slow-moving streams, ponds, and wetlands and in all sorts of artificial habitats. They thrive in shallow, often stagnant, ponds and at the edges of streams and lakes where vegetation is dense and the water shallow and warm. They will also live in slightly brackish backwaters. Mosquitofish tend to be much more tolerant of polluted water than many native fishes because they can survive at low oxygen levels. Generally subtropical to warm-temperate in distribution, they do not tolerate extremely cold temperatures, which may limit their success in northern states.

Diet. Best described as omnivores, mosquitofish prefer taking zooplankters and small aquatic invertebrates near the water surface. They also consume small fish, fish eggs, and the larvae of amphibians. When preferred items are in short supply, they consume diatoms and other algae. Although their name implies they are proficient feeders of mosquito larvae, they seem to be no more effective in mosquito control than some native fish. Indeed, in a laboratory experiment, mosquitofish fared poorly on a diet of mosquito larvae.

The two species of mosquitofish are very similar in appearance. Their small, protrusible mouths are upturned so they can feed at the surface of the water. (Gualtiero Boffi/Shutterstock.)

Life History. Mosquitofish are ovoviviparous, undergoing internal fertilization of the eggs, which hatch in the mother's body. Thus, they are livebearers and reproduce rapidly. Females can be sexually mature at six weeks of age, males at four weeks. When mating, the male swings its gonopodium forward and inserts it into the female's vent to transfer sperm. The female is able to store sperm for several months and fertilize several broods from one copulation. Gestation lasts 3–4 weeks. When the young are born, they are little more than 0.25 in. (6 mm) long. Females can produce 4–5 broods a year, with 1–300 young in each. At birth, the sex ratio is 1:1, but in the adult population, there are fewer males than females. The typical lifespan is 15 months.

Impacts. Outside their native range, mosquitofish are notoriously destructive, contributing to population declines among native fishes and amphibians. Adults are very aggressive and attack other fish, shredding their fins and sometimes killing them. They compete with or displace indigenous fishes. They also alter zooplankton communities, as well as insect and crustacean communities, by their selective predation on members of each. In some circumstances, they may actually benefit mosquitoes by eating predatory invertebrates.

Mosquitofish have had especially deleterious effects in western states, where they are implicated in the elimination or decline of endemic and federally endangered and threatened fish. They have displaced the plains topminnow (*Fundulus sciadicus*) from preferred habitat in shallow clear streams. They have caused the threatened Railroad Valley springfish (*Crenichthys baileyi*) to shift its habitat and decline in numbers in springs in Nevada. In Arizona, they are responsible for the local extirpation of the endangered Sonoran topminnow (*Poeciliopsis occidentalis*). Populations of the Gila topminnow (*P. o. occidentalis*) only survive where *Gambusia* is absent. Another subspecies, the Yaqui topminnow (*P. o. sonriensis*), is threatened with a similar fate. Mosquitofish are also responsible for the demise of the least chub (*Itichthys phlegethontis*) in parts of Utah. Desert pupfish (*Cyprinodon* spp.) populations are threatened by the introduction of mosquitofish in Nevada springs. Mosquitofish are at least partially responsible for population declines in the Chiricahuan leopard frog (*Rana chiricahuensis*) in southeastern Arizona, in the California newt (*Taricha torosa*) in California, and in native damselflies on O'ahu, Hawai'i.

Mosquitofish have posed problems for native species wherever in the world they have been introduced. This fish is on the IUCN's list of "100 of the World's Worst Invasive Alien Species."

Management. It may be possible to eradicate some populations of mosquitofish. More likely to be successful are efforts to prevent the further spread of the species. Application of rotenone to waters can be a means of eliminating mosquitofish from small areas of permanent water. The poison deprives the fish of oxygen, so they come to the surface for air and can be mechanically removed. Rotenone is a broad-spectrum piscicide, so nontarget species are also affected. Barrier construction is feasible in some areas to stop range expansion. Regulations to halt the transport and stocking of all nonnative fishes need to be drafted or enforced to prevent future introductions.

Selected References

Aarn, and Peter Unmack. "Gambusia control homepage." 1998. http://www.gambusia.net/.

IUCN/SSC Invasive Species Specialist Group (ISSG). "*Gambusia affinis* (Fish)." ISSG Global Invasive Species Database, 2006. http://www.issg.org/database/species/ecology.asp?fr=1&si=126.

Masterson, J. "*Gambusia affinis*. Mosquitofish." Smithsonian Marine Station at Fort Pierce, 2008. http://www.sms.si.edu/IRLSpec/Gambusia_affinis.htm.

Myers, G. S. "*Gambusia*, the fish destroyer," *Oecologia* 141: 713–21, 1965. Appendix A-73. Available online at http://wildlife.utah.gov/pdf/AIS_plans_2010/AIS_12oWesternMosquitofish-Jenny-final.pdf.

NatureServe. "*Gambusia affinis* (Baird and Girard, 1853): Western Mosquitofish." NatureServe Explorer: An Online Encyclopedia of Life, Version 7.1, 2009. NatureServe, Arlington, VA. http://natureserve.org/explorer.

Nico, Leo, and Pam Fuller. "*Gambusia holbrooki*." USGS Nonindigenous Aquatic Species Database, Gainesville, FL, 2009. http://nas.er.usgs.gov/queries/FactSheet.asp?speciesID=849.

Nico, Leo, Pam Fuller, and Greg Jacobs. "*Gambusia affinis*." USGS Nonindigenous Aquatic Species Database, Gainesville, FL, 2009. http://nas.er.usgs.gov/queries/FactSheet.asp?speciesID=846.

■ | Northern Snakehead

Scientific name: *Channa argus*
Synonym: *Ophicephaslus argus*
Family: Channidae (Snakeheads)

Native Range. China (Amur, Hwang, and Yangtze river systems), North Korea, and Russia (Amur River drainage system)

Distribution in the United States. Established in Maryland (Lower and Middle Potomac River and tributaries), New York (Meadow Lake, in Queens), Pennsylvania (Edgewood Lake, Philadelphia), and Virginia (Fairfax County in tributaries of the Middle Potomac River such as Dogue Creek, Kanes Creek, Little Hunting Creek, Massey Creek, Occoquan River; also Mulligan Pond and in Pohick Bay, Fort Belvoir; Dyke Marsh, Alexandria to Mason Neck National Wildlife Refuge; King George County in Upper Machodoc Creek). It has been collected in California, Florida, Massachusetts, and North Carolina, but is not established in these states. A 2008 capture in Arkansas suggests snakeheads may be established there.

Description. The northern snakehead has an elongated body with a long dorsal fin running almost the full length of the back and a long anal fin as well. The pelvic fins are close to the gills and set almost directly below the pectoral fins. The body is a golden tan with large dark irregular blotches on the flanks. The head is flattened and bears large scales resembling those of a snake. The eyes are set forward on the head, and nostrils are tubular. The mouth is large; the lower jaw is toothed and protrudes beyond the upper jaw; the roof of the mouth (i.e., palatine and prevomer bones) is also toothed. Some teeth look like the canines of mammals. Total length can reach about 40 in. (102 cm); adults may weight up to 15 lbs. (6.8 kg).

Related or Similar Species. Native bowfin (*Amia calva*) and burbot (*Lota lota*), both of which have long bodies and long dorsal fins, might be mistaken for snakeheads. Bowfin are distinguished by a short anal fin, pelvic fins set well behind the gills and pectoral fins, and a rounded tail. A black spot appears at the base of the tail in males and juveniles. Barbots display a split dorsal fin (a shorter segment lies in front of a longer one); and a single barbel or whisker hanging below the lower jaw. Other species superficially resembling snakeheads include the American eel (*Anguilla rostrata*) and the sea lamprey (*Petromyzon marinus*). Eels lack pelvic fins and their dorsal and anal fins converge with the caudal fin so that it looks like they have a single continuous fin. Sea lampreys have a round mouth and reddish eyes (see Fish, Sea Lamprey).

Introduction History. The snakehead is a popular fish in Asian food markets. Most introductions were probably deliberate efforts to raise the fish in local waters. The northern snakehead is not part of the aquarium trade, although other species in the family are. The first report of the species in the United States was in 1977. This represented a failed attempt to establish a population in Silverwood Lake, San Bernadino County, California. In 2000, three specimens were

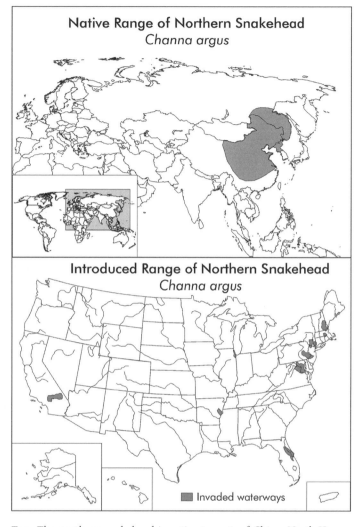

Top: The northern snakehead is native to parts of China, North Korea, and the Amur River system in Russia. Bottom: Northern snakehead populations are established in Maryland and Virginia. The fish has been collected in other states, but with the possible exception of Arkansas, does not seem to have established populations in those areas. (Adapted from Fuller and Benson 2009.)

taken from the St. Johns River in Florida. Subsquent collections were made in Burnham Harbor (Lake Michigan), Chicago, Illinois; two locations in Massachusetts (2001 and 2004); Lake Wylie and South Fork Catawba River, North Carolina; and FDR Pond, Philadelphia, Pennsylvania. In June 2008, a snakehead was taken from the Schuylkill River in Philadelphia. None of these captures have been interpreted as evidence of established populations.

Discovery of snakeheads in a pond in Crofton, Maryland, in June 2002 created intense media coverage and brought the problem of invasive fishes to the public's attention. This population was subsequently eradicated by state biologists, who applied the pesticide rotenone to the water.

Northern snakehead. The large scales on the head give this fish a snake-like appearance. (U.S. Geological Survey Archive, U.S. Geological Survey, Bugwood.org.)

In April 2004, several snakeheads were collected from the Potomac River in Maryland and Virginia. The source of these fish is unknown. Many others have since been collected in the Middle and Lower Potomac River basin around Dogue and Little Hunting Creeks in Virginia and from the Anacostia River in Maryland.

Habitat. The northern snakehead is a purely freshwater fish; it prefers shallow ponds and sluggish streams with well-vegetated or muddy bottoms. It has also been found in wetlands. Snakeheads inhabit waters with temperatures ranging from 32°F to 86°F (0–30°C). Snakeheads are air breathers and, if kept moist, can survive out of water for as long as four days. This enables them to wriggle across land and disperse to other bodies of water. It also allows them to live in oxygen-depleted water, since they can gulp air at the surface.

Diet. Snakeheads are top-level carnivores. Juveniles consume zooplankters, insect larvae, small crustaceans, and the fry of other fish species. Adults prey primarily on other fish, but will take crustaceans, frogs, small reptiles, and even small birds and mammals.

Life History. These fish reach sexual maturity at 2–3 years, when they are about 12–14 in. (30–36 cm) long. Spawning occurs 1–5 times a year. A breeding pair forms a nest by making a circular clearing in aquatic vegetation, sometimes weaving the plants they removed around the edge to protect a vertical column of water. The female releases 1,300 to 15,000 eggs near the center of the water column. The male wraps his body around hers as he releases milt and fertilizes the eggs when she releases them. Each egg contains a droplet of oil that lets it float. The snakehead pair, which remain together throughout the spawning season, guard the eggs and then the fry when they hatch 28–48 hours later. The fry remain in the nest until they are about 0.3 in. (8 mm) long, at which time the yolk sac is completely absorbed. Typically, the young stay together in a pack guarded by an adult until they reach early juvenile stage; they are then about 0.7 in. (18 mm) long and beginning to eat zooplankton.

Impacts. Unknown. As predators, they are expected to compete with native fishes for food and habitat.

Management. It is difficult to impossible to eradicate a population from a river system. Snakeheads can be removed from ponds by applying rotenone, but nontarget fish are killed too. Preventing new introductions is the main management strategy. Since 2002, federal law prohibits the introduction and interstate transport of live snakeheads or their eggs without a permit in the United States, the Commonwealth of Puerto Rico, or any territory or possession of the United States.

Selected References

"Do You Know the Difference?" Virginia Department of Game and Inland Fisheries, 2004. http://www.dgif.state.va.us/fishing/snakehead_comparisons_052004.pdf.

Fuller, P. F., and A. J. Benson. "*Channa argus* (Cantor 1842)." USGS Nonindigenous Aquatic Species Database, Gainesville, FL, 2009. http://nas.er.usgs.gov/queries/FactSheet.asp?speciesID=2265.

National Biological Information Infrastructure (NBII) and IUCN/SSC Invasive Species Specialist Group (ISSG). "*Channa argus* (fish)." ISSG Global Invasive Species Database, 2005. http://www.issg.org/database/species/ecology.asp?si=380&fr=1&sts.

"Recognizing Northern Snakehead." U.S. Fish and Wildlife
Service, 2003. http://www.ans
taskforce.gov/Education/Snake
head_Final%20from%20RO.pdf.
U.S. Fish and Wildlife Service.
"Injurious Wildlife Species:
Snakeheads (Family Chanidae):
Final Rule." *Federal Register* 67
(193), Rules and Regulations:
62193–62204, October 4, 2002.

■| Rainbow Trout

Also known as: Steelhead
(anadromous form)
Scientific name: *Oncorhynchus
mykiss*
Synonyms: *Salmo gairdnerii,
Fario gairdnereri*
Family: Salmonidae

Native Range. North America.
Pacific slope streams from the
Kuskokwim River, Alaska, south
to Rio Santa Domingo, Baja
California, Mexico. Also in the
upper Mackenzie River drainage
system, Alberta and British
Columbia, Canada; and in in-
ternal drainages of southern
Oregon.

*Distribution in the United
States.* Rainbow trout has been
stocked in all states.

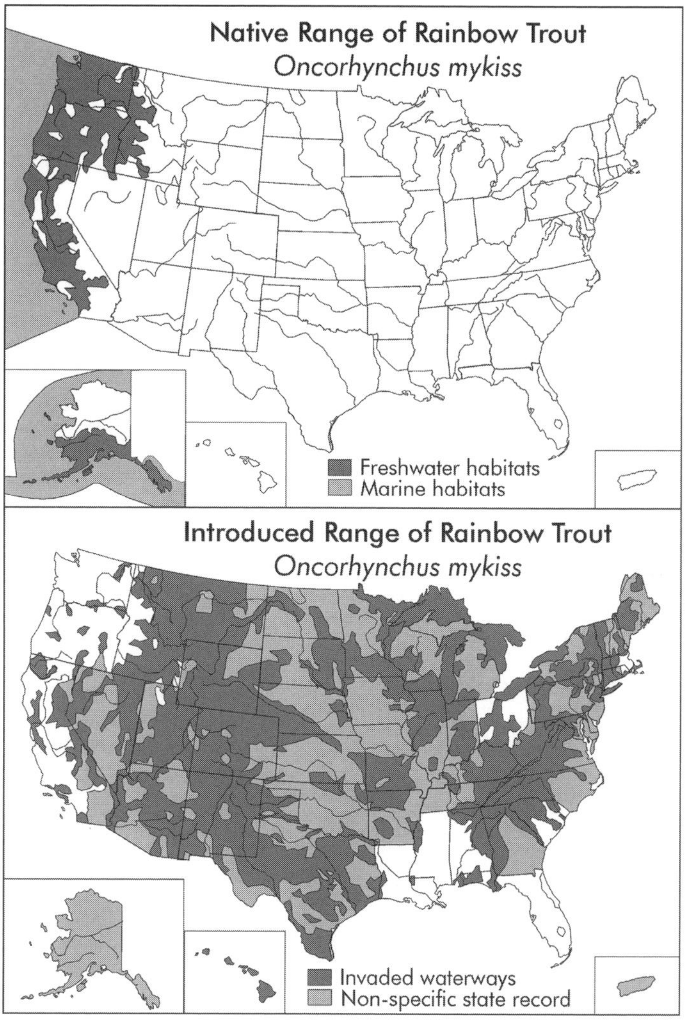

Description. Rainbow trout
are deep-bodied, compressed

Top: Rainbow trout are native to Pacific Slope streams in North America.
Bottom: Rainbow trout have been transplanted into streams in all 50
states and Puerto Rico. (Both maps adapted from Fuller 2009.)

fishes with large heads; mouths extend to the back of the eyes. Coloration is highly variable.
(There are hundreds of varieties and several subspecies of native rainbow trout.) The back is
bluish to dark olive-green. Black spots appear on the back and may extend down the sides as
far as the lateral line. Radiating rows of black spots appear on the unforked tail. The sides are
silver, and the belly and underside of head are white. Often a pink or reddish band runs
along the sides of the body and head. Colors become much more intense when rainbows
are on spawning runs. Resident riverine populations normally exhibit the most intense pink
stripe and heaviest spotting. Rainbows can be positively identified by the 8–12 rays in the
anal fins and lack of teeth at the base of the tongue. A typical four-year-old river rainbow
weighs about 1 lb. (0.5 kg) and is about 13 in. (33 cm) long. Lake-dwelling rainbow trout
will be larger.

Related or Similar Species. Brown trout (see Fish, Brown Trout).

Coloration on rainbow trout is highly variable. Usually a reddish band runs along the sides of the head and body. (Eric Engbretson.)

Introduction History. Stocking of this fish outside its native range began in the late 1800s. Introductions have been primarily into streams for sport-fishing. Around the world, rainbow trout are one of the most widely introduced fish species and important in aquaculture, as they are highly valued food fish.

Habitat. Rainbow trout is usually a freshwater species, although anadromous populations occur where rivers are open to the sea. Such fishes are called steelheads or coastal rainbows. Rainbows thrive in cool, clean, well-oxygenated streams and lakes where temperatures range from 50°F (10°C) to 75°F (24°C). A water temperature around 54°F (12°C) is preferred.

Diet. Rainbow trout consume zooplankters, crustaceans, larval and adult insects both aquatic and terrestrial, mollusks, fish eggs, and small fishes. Young trout feed on zooplankters.

Life History. Rainbow trout usually migrate into small tributary streams of lakes to spawn in the spring. The female uses her tail to excavate a redd or depression 4–12 in. deep and 10–15 in. (25–38 cm) in diameter in the streambed gravel into which she releases 700–4,000 orange-red eggs. The male then fertilizes the eggs. The female covers successive groups of fertilized eggs by digging at the upstream edge of the redd. When spawning is finished, the redd consists of layers of eggs and clean gravel, and the adults leave. Depending on water temperature, eggs incubate for 1–5 months. Several weeks after hatching, the fry wriggle through the gravel and swim free. They shelter and feed along the edges of streams or near lake shores for 2–3 years. In native populations, sexual maturity is usually reached at age 6–7.

Most introduced populations are bred in hatcheries and released into local streams just ahead of the fishing season. Fishing pressures typically deplete the population, so they must be restocked each year. They may also be restocked annually in areas where environmental conditions are not conducive to self-sustaining populations.

Impacts. A major problem with the introduction of rainbow trout outside their native range has been their ability to hybridize with other native salmonids. They thereby threaten the genetic integrity of several rare species and subspecies vulnerable to extinction. In California, they have crossed with Lahontan cutthroat trout (*O. clarki henshawi*), redband trout (*O. mykiss* subspp.), and California golden trout (*O. mykiss aguabonita*), the state fish. In the Lahontan drainage east of the Sierra Nevada and in several Rocky Mountain rivers, hybridization with rainbow trout has been a major cause of the decline of native cutthroat trout. In Nevada, it is implicated in the extinction of the Alvord cutthroat (*O. clarki alvordensis*), and in Arizona, it hybridizes with native Gila trout (*O. gilae*) and Arizona trout (*O. gilae apache*).

Rainbow trout negatively affect other species through predation and aggressive behavior. For example, rainbows push the Little Colorado spinedace (*Lepidomeda vittata*) from the undercut banks where it seeks shelter into open water, where it becomes much more vulnerable to predation. They are also known to drive suckers and squawfish off their feeding territories. In the Little Colorado River, rainbow trout prey upon endangered humpback chubs (*Gila cypha*). They are at least partly responsible for declines in the Chiricahuan leopard frog (*Rana chiricahuensis*) in southeastern Arizona and the Sierra Nevada yellow-legged frog (*R. sierrae*). They also compete with native brook trout (*Salvelinus fontinalis*) in many locations.

Through the stocking of hatchery-bred rainbows in rivers in more than 20 states, whirling disease has been introduced into open waters, where it threatens native species. Whirling disease is caused by a parasite (*Myxobolus cerebralis*) of salmonids. Introduced from Eurasia, the cnidarians-like organism infects juveniles and affects the nervous and skeletal systems, resulting in a curvature of the spine that affects the victim's ability to maintain its orientation in the stream. Infected fish swim in spirals, chasing their tails. Mortality rates are very high.

Rainbow trout has been nominated as one of "100 of the World's Worst Invasive Alien Species" by the IUCN/SSC Invasive Species Speciality group.

Management. Most management practices favor rainbow trout since they are an important element in sport fisheries across the country. Many continue to be raised in hatcheries and used in put-and-take operations.

Selected References

Fuller, Pam. "*Oncorhynchus mykiss*." USGS Nonindigenous Aquatic Species Database, Gainesville, FL, 2009. http://nas.er.usgs.gov/queries/factsheet.asp?SpeciesID=910.

National Biological Information Infrastructure (NBII) and IUCN/SSC Invasive Species Specialist Group (ISSC). "*Oncorhynchus mykiss* (fish)." ISSC Global Invasive Species Database, 2006. http://www.issg.org/database/species/ecology.asp?fr=1&si=103.

Root, Laurie. "Rainbow Trout (*Oncorhynchus mykiss*)." South Dakota Department of Game, Fish & Parks, 1994 http://www3.northern.edu/natsource/FISH/Rainbo1.htm.

"Steelhead /Rainbow Trout" Alaska Department of Fish and Game, 2011. http://www.adfg.alaska.gov/index.cfm?adfg=steelhead.main.

Talbot, Ret. "Native California Trout Species and Subspecies: At Risk Rainbow and Cutthroat Trout Indigenous to California." Suite 01.com. 2009. http://freshwater-fish.suite101.com/article.cfm/native_california_trout_species_and_subspecies.

■ |Round Goby

Scientific name: *Neogobius melanostomus*
Synonym: *Apollonia melanostomus*
Family: Gobiidae

Native Range. Eurasia. Native to brackish waters of the Sea of Azov, Black Sea, and Caspian Sea and their tributary streams, where it prefers brackish waters.

Distribution in the United States. Great Lakes and their tributary streams and connecting canals.

Description. The round goby is a small, soft-bodied fish with eyes raised above its head as in frogs. Its lips are thick. Body color is a mottled gray, brown, or greenish color, except in breeding males, which are black. The front dorsal fin usually has a conspicuous black spot. Pelvic fins are fused to form a suction disk. The front dorsal fin has 5–6 spines, and the posterior dorsal fin has one spine and 13–16 soft rays. The anal fin has one spine and 11–14 soft rays. Adult gobies are 4–10 in. (10–25 cm) long. In the Great Lakes, they rarely grow larger than 7 in. (18 cm).

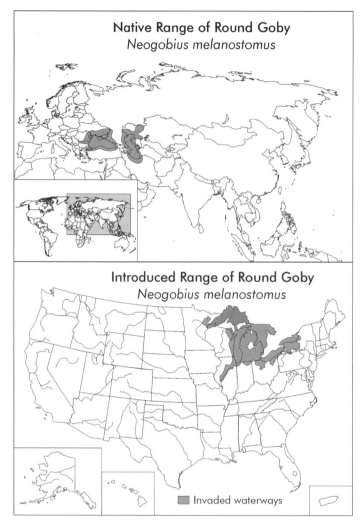

Native Range of Round Goby
Neogobius melanostomus

Introduced Range of Round Goby
Neogobius melanostomus

■ Invaded waterways

Top: The round goby is native to fresh and brackish waters of the Sea of Azov, Black Sea, and Caspian Sea, as well as their tributaries. (Adapted from Stepien, Carol, Joshua Brow, and Emily Sopkovich. "Round Goby Phylogeography," Lake Erie Center, University of Toledo, 2009. http://www.utoledo.edu/as/lec/goby/RoundPhy.html.) Bottom: Round gobies are invasive in all of the Great Lakes and many of their tributary streams. (Adapted from Fuller, Benson, and Maynard 2009.)

Related or Similar Species. Round gobies are similar in appearance to native sculpins (*Cottus* spp.), which have two separate pelvic fins instead of the fused one of gobies. They also look similar to tubenose gobies (*Proterorhinus marmoratus*), another exotic fish in the Great Lakes. The tubenose goby has nostrils that extend forward in tubes to or beyond the lower lip. There is no spot on the front dorsal fin, but oblique black lines. It usually is only about 4.5 in. (11 cm) long. Smaller and less aggressive than round gobies, the tubenose has only been found in Lake St. Clair, the St. Clair River, the Detroit River, and western Lake Ontario, although recently it appears to be spreading in Lake Erie.

Introduction History. The round goby was first collected in the United States in 1990 from Lake St. Clair. They most likely were introduced during ballast water exchange by trans-Atlantic ships coming from the Black Sea region. By 1993, it had entered Lake Erie. In 1994, it appeared in Lake Michigan near Chicago. It has since been collected near the terminus of the Chicago Sanitary and Shipping Canal, entryway to the Mississippi River drainage system. In 1995, gobies were collected from Wisconsin waters in Lake Superior and, in 1996, from Duluth Harbor, Minnesota. The spread to Lake Superior was probably aided by freighters operating on the lakes. By 1998, the round goby was reported from several Michigan sites on Lake Huron. Gobies were reported in the Erie Canal, Buffalo River, St. Lawrence River, Genessee River, Tonowanda Creek, and Lake Ontario in 2004 and 2005. Today there are large populations in Lakes Erie, Ontario, and St. Clair. Lake Michigan experienced a population explosion of gobies in 2008.

Habitat. Freshwater lakes, streams, and canals. These bottom-dwelling fish prefer hard substrates such as rock, gravel, or sand and areas shallow enough for aquatic macrophytes to grow. They hide in crevices. They tolerate a wide range of temperatures and are able to inhabit brackish water and degraded water quality.

Diet. Round gobies are carnivores and feed on zooplankters, mollusks, aquatic insects and benthic invertebrates, fish eggs, and small fish. They feed voraciously on zebra mussels (*Dreissena polymorpha*), another invasive species in the Great Lakes (see Mollusks, Zebra Mussel). Gobies are able to pluck mussels from their attachment sites on the lake floor and crush them with their

The round goby's pelvic fins are fused to form a suction disk. In their introduced range, they are rarely more than 7 in. long. (Eric Engbretson.)

pharyngeal teeth. They spit out the shells before swallowing the soft insides.

Gobies have well-developed lateral lines, which increase their ability to detect water movement and allow them to be nocturnal feeders.

Life History. Gobies spawn from April to September. Males migrate from wintering sites in deeper water to spawning areas ahead of the females. They establish territories and attract mates by releasing pheromones. Females attach 500–3,000 eggs to the underside of rocks, in pipes, or on other sheltered hard surfaces. They utilize several nests and spawn with several different males. Males aggressively guard the nests and newly hatched young. Gobies will spawn every 20 days during the long season. Females are sexually mature at 1–2 years; males at 3–4 years. Males die after spawning, so population sex ratios are usually skewed to favor females.

Impacts. Round gobies compete with native fishes for food and space and could potentially have serious impacts on aquatic ecosystems if they were to gain entry to the Mississippi drainage system. Their ability to feed at night and their ability to cling with their pelvic fins to rocks in fast currents give them a competitive advantage over native species, many of which have overlapping food preferences. They compete for small macroinvertebrates with rainbow darters (*Etheostoma caeruleum*), logperch (*Percina caprodes*), and northern madtoms (*Noturus stigmosus*). Since gobies were introduced to Lake St. Clair, populations of native sculpin and logperch have declined. They are known to eat the eggs and fry of lake trout and the eggs of lake sturgeon, and to prey on small darters in laboratory situations.

Since adult round gobies vigorously defend their spawning territories, they become dominant in prime spawning areas and displace natives. Mottled sculpin (*Cottus bairdi*) is one species that has been affected in that way by the establishment of round gobies. Large, mature gobies drive them from spawning sites; and small gobies, focusing on insects, compete with them for food.

Indirectly, gobies have had an impact on smallmouth bass (*Micropterus dolomieu*) in Lake Erie. Male bass guard nests and can successfully ward off goby attacks. However, if the males are removed through fishing, gobies immediately move in and gorge themselves on bass eggs. In order to preserve the smallmouth bass population, the State of Ohio has had to close the fishery in May and June, while bass are spawning. This has led to a significant economic loss as anglers traditionally took half the bass catch during those months. Gobies also affect recreational fisherman by snatching the bait from their hooks.

On the positive side, the diet of large adult round gobies overwhelmingly consists of zebra mussels. No other fish in the Great Lakes exploits this resource to such an extent. However, predation of zebra mussels by gobies is not sufficient to stem the spread of those aquatic invaders. And because zebra mussels are filter-feeders and accumulate contaminants in their body tissues, fish that eat gobies will biomagnify the concentrations, potentially rendering some sportfishes unfit for human consumption.

Some native fishes, including smallmouth bass and walleye, consume gobies. It is hoped that Atlantic salmon (*Salmo salar*), introduced into the Great Lakes in 1966 to feed on alewives (see Fish, Alewife), may switch to gobies as alewife populations decline in Lake Michigan, threatening the salmon fishery.

Management. Controlling or reducing the goby populations of the Great Lakes is impossible. Preventing the inland spread of the invader is the key goal of management. Electric barriers on tributary streams are under consideration. Good fishing and boating practices need to be applied and enforced. Dumping of ballast water in waterways needs to be regulated. Live bait should not be discarded into water. Boats, trailers, and fishing gear should be thoroughly inspected upon leaving any body of water and all visible organic debris removed. All water should be drained from the boat before leaving the shoreline. All boats should be washed and then sun-dried for five days before being put into another body of water.

Selected References

Crosier, Dani, and Dan Malloy. "Round goby (*Neogobius melanostomus*)," United States Federal Aquatic Nuisance Species Task Force, 2005. http://www.anstaskforce.gov/spoc/round_goby.php.

Fuller, P., A. Benson, and E. Maynard. "*Neogobius melanostomus*." USGS Nonindigenous Aquatic Species Database, Gainesville, FL, 2011. http://nas.er.usgs.gov/queries/FactSheet.asp?speciesID=713.

Hayes, R. "*Neogobius melanostomus*." Animal Diversity Web, University of Michigan Museum of Zoology, 2008. http://animaldiversity.ummz.umich.edu/site/accounts/information/Neogobius_melanostomus.

Morlock, Jerry W. "Invasive Species Round Goby Has Population Explosion in Lake Michigan." Michigan Live LLC, 2009. http://www.mlive.com/news/muskegon/index.ssf/2009/01/invasive_species_round_goby_ha.html.

National Biological Information Infrastructure (NBII) and IUCN/SSC Invasive Species Specialist Group (ISSG). "Neogobius melanostomus (Fish)." ISSG Global Invasive Species Database, 2006. http://www.issg.org/database/species/ecology.asp?si=657&fr=1&sts.

"Round Goby." Great Lakes Science Center, USGS, 2008. http://www.glsc.usgs.gov/main.php?content=research_invasive_goby&title=Invasive%20Fish0&menu=research_invasive_fish.

■ | Sea Lamprey

Also known as: Lamprey eel, lake lamprey
Scientific name: *Petromyzon marinus*
Family: Petromyzontidae

Native Range. The Atlantic coast of North America from Labrador to Florida and into the Gulf of Mexico. This is primarily a marine fish, but it ascends freshwater rivers to spawn. It is native to Atlantic Slope drainages, the St. Lawrence River, and probably Lake Ontario, and it is probably also native in the Finger Lakes of New York, and Lake Champlain, New York/Vermont. It also occurs naturally along the Atlantic coast of Europe and in the western Mediterranean Sea.

Distribution in the United States. Sea lampreys are nonnative transplants established and invasive in the four upper Great Lakes and in cold, clear streams in the region. Currently, populations are especially high in the St. Mary's River, which connects Lake Huron and Lake Michigan.

Description. The sea lamprey is a thin, eel-like fish without jaws. A member of a primitive group of agnathans, its skeleton is made of cartilage. The body is smooth and scaleless. It has two close dorsal fins situated toward the rear of the body, but no paired fins, no lateral line, and no swim bladder. There are seven gill slits. The back is gray-blue, with a metallic violet shimmer on the often mottled sides. The belly is yellow to silver-white. The seven gill openings are lined up behind the eyes, which are red. The mouth is a disk-like sucker containing whorls of 100 or more sharp, curved teeth made of keratin and a toothed "tongue." Total length of adults ranges between 12 and 20 in.

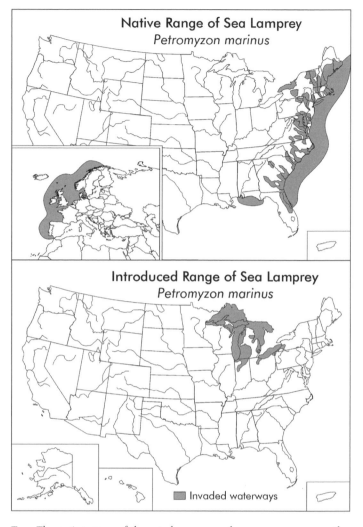

Top: The native range of the catadromous sea lamprey encompasses the Atlantic coasts of North America and Europe. It is probably also native to Lake Ontario, the Finger Lakes of New York, and Lake Champlain. Bottom: The sea lamprey is invasive in Lakes Superior, Huron, Michigan, and Erie and in their clear, cold tributary streams (Both maps adapted from Fuller, Nico, and Maynard, 2009.)

(30.5–50.8 cm); landlocked fish are usually smaller than native marine-dwelling individuals, which may attain lengths of 24–30 in. (61–76 cm). Adults weigh 8–13 oz. (0. 23 kg).

Related or Similar Species. There are four other native species of lamprey in the United States: the silver lamprey (*Ichthyomyzon unicuspis*), chestnut lamprey (*Ichthyomyzon castaneus*), northern brook lamprey (*Ichthyomyzon fossor*), and American brook lamprey (*Lampetra appendix*). The latter two, unlike the sea lamprey, are not parasitic. All are benign. Each has a characteristic structure to its mouth and arrangement of the teeth, which differ from that of the sea lamprey and are diagnostic.

American eels (*Anguilla rostrata*), true eels, might be mistaken for sea lampreys, but they have jaws and a sharp, pointed head.

A. The sea lamprey is a jawless fish that resembles an eel. (Andrei Nekrassov/Shutterstock.) B. The mouth of the sea lamprey is a sucker containing whorls of sharp teeth and a toothed "tongue." (U.S. Environmental Protection Agency, Great Lakes National Program office.)

Introduction History. Sea lampreys were first reported in Lake Ontario in 1835. It was not collected from Lake Erie until 1921. It was found in Lake Michigan in 1936, Lake Huron in 1937, and Lake Superior in 1946. Increasingly, genetic evidence points to the sea lamprey being native to Lake Ontario and, along with populations in Lake Champlain and the Finger Lakes, a relict of the last Pleistocene glaciation. Niagara Falls presented a barrier to dispersal from Lake Ontario into Lake Erie until the Welland Canal was opened in 1829 to bypass the falls. Still, it seems lampreys were not able to overcome the barrier until improvements were made on the canal and lock system in 1919. Within two years, it was in Lake Erie and within 25 years, it had spread to all the other Great Lakes.

Habitat. The larvae inhabit soft sediments at the bottoms of clear streams. Adults are parasitic and attach to larger fish in temperate cool-water lakes. They tolerate temperatures ranging from 40°F (5°C) to 68°F (20°C).

Diet. Larvae are filter-feeders, consuming invertebrates and detritus from the water. Juveniles are external parasites on healthy fish. A sea lamprey fastens onto its victim and uses its tongue to rasp a hole in the host fish's body. It then sucks out the blood, body fluids, and flesh, keeping the wound open for hours or even weeks by means of an anticoagulant in its saliva. One lamprey is estimated to eat 40 lbs. of fish in its lifetime. Mature adults do not feed.

Life History. In their native habitat, sea lampreys spawn in freshwater streams where the substrate is stony. They move upriver in late May and early June, by which time the adults are no longer feeding. The male and female excavate nests, or redds, about 6 in. (15 cm) deep and 2–3 ft. (60–90 cm) in diameter, removing stones with their sucker mouths, and piling them downstream. Both attach themselves to a large stone at the upstream edge of the nest. (This behavior is reflected in their Latin name: petromyzon means "stone sucker.") With the male wrapped around the female, the pair stirs up sand and deposits eggs and milt. The sand grains stick to the fertilized eggs, burying them; the larvae develop in the stream bottom. Adults die after spawning.

The larval (ammocoete) stage lasts 6–8 years. When 4–5 in. long, the larvae undergo an extreme metamorphosis and transform into adults. It is then that they migrate downstream into estuaries and the sea, where they remain as parasitic juveniles for 1–2.5 years. Populations introduced into the Great Lakes or that may be landlocked, ice age relicts in the Finger Lakes and Lake Champlain, of course, never go to sea; instead, juveniles migrate from the streams where they hatched and spend their long larval period in freshwater lakes, where they attach themselves to large lake fish, including lake trout (*Salvelinus namaycush*), yellow perch (*Perca flavescens*), burbot (*Lota lota*), walleye (*Stizostedion vitreum*), channel catfish (*Ictalurus punctatus*), and northern pike (*Esox lucius*). At the end of the juvenile period, they become sexually mature adults, cease feeding, and swim upriver to spawn.

Impacts. The introduced sea lamprey is considered to have been a major factor in the collapses of lake trout, whitefish (*Coregonus clupeaformis*), and chub (deepwater ciscos) fisheries in the Great Lakes (especially in Lakes Huron and Michigan and eastern Lake Superior) during the 1940s and 1950s. In its parasitic phase, its attacks and feeding on other fish result in high mortality among host species. It kills directly by sucking out fluids and tissues, and it kills indirectly by leaving open wounds vulnerable to infection. Circular scars on surviving host fish attest to many of them having been attacked more than once. The species' introduction to the Great Lakes coincided with major declines in several large native fish that had been important in commercial fisheries and sport fisheries. Combined with the effects of pollution and overfishing, the influx of sea lampreys led to the extinction of three endemic fish: the longjaw cisco (*Coregonus alpenae*), deepwater cisco (*C. johannae*), and blackfin cisco (*C. nigripinnis*). Since lamprey had so reduced populations of predatory fish, populations of the invasive alewife (*Alosa pseudoharengus*) exploded and affected the species composition of the native fish fauna (see also Fish, Alewife). Although sea lamprey populations have been reduced in most parts of the Great Lakes, they are still abundant enough to hamper restoration efforts directed at native game fish and the introduced Atlantic salmon.

Management. The Great Lakes Fishery Commission, U.S. Fish and Wildlife Service, and Fisheries and Oceans Canada are all involved in controlling sea lampreys in the Great Lakes. Mechanical weirs and electrical barrier were the first methods used to control the upstream migration of spawning sea lampreys. The release of sterile males also was implemented to reduce successful reproduction. In the late 1950s, an effective lampricide, TFM (trifluoro-methyl-4-nitrophenol), was developed. This chemical kills lamprey larvae in the stream bottom. It was successful in reducing populations by over 90 percent and allowing the recovery of some commercial fisheries. TFM must be used repeatedly to remain effective, and sometimes it is harmful to nontarget species such as walleye and native lampreys. Recently, a controlled-release version of the lampricide Bayluscide was developed at the Upper Midwest Environmental Sciences Center (USGS); it is supposed to kill sea lamprey larvae in bottom sediments without affecting other species.

Selected References

Fuller, Pam, Leo Nico, and Erynn Maynard. "*Petromyzon marinus*." USGS Nonindigenous Aquatic Species Database, Gainesville, FL, 2009. http://nas.er.usgs.gov/queries/FactSheet.asp?SpeciesID=836.

"Sea Lamprey." Upper Midwest Environmental Sciences Center, USGS, 2007. http://www.umesc.usgs.gov/invasive_species/sea_lamprey.html.

Summers, Adam P. "Sea Lamprey, *Petromyzon marinus*." Department of Biology, University of Massachuseatts, Amherst, 1997. http://www.bio.umass.edu/biology/conn.river/sealampr.html.

■|Silver Carp

Also known as: Flying carp
Scientific name: *Hypophthalmichthys molitrix*
Family: Cyprinidae

Native Range. Large Pacific drainages of eastern Asia. Silver carp are native to major lowland rivers from the Amur River of far-eastern Russia south to the Pearl River of China, and possibly into northern Vietnam.

Distribution in the United States. Established in the Mississippi River drainage system in Illinois and Louisiana. Silver carp have been reported in Alabama, Arizona, Arkansas, Colorado, Florida, Indiana, Kansas, Kentucky, Mississippi, Missouri, Nebraska, Ohio, South Dakota, and Tennessee, but these may not be reproducing populations.

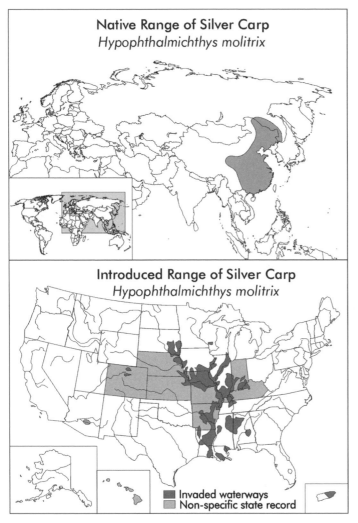

Description. The silver carp is a large, deep-bodied fish with a laterally compressed body. A long ventral keel extends forward from the vent almost to the junction of the gill membranes and is diagnostic. The lateral line is complete and bends down toward the belly. Large eyes are set low on the head. The mouth is large, toothless, and downturned. There are no barbels. Gill rakers are thin, branched, and fused into a sponge-like structure. The pharyngeal teeth are in one row, four on a side, at the back of the throat and have striated surfaces.

The short dorsal fin has a moderately stiff, non-serrated spine at its origin; it begins behind the origin of the pelvic fins. The anal fin is hooked and has a slightly stiffened spine at the origin. The caudal fin is forked. Pectoral fins have 15–18 rays and a stiff, hard spine with a finely serrated rear margin. Small specimens lack spines on their fins.

Scales are very small on the body, and the head and opercules are scaleless. Juveniles are silvery all over. The backs and

Top: Silver carp are native to the large lowland rivers of East Asia from the Amur River south to the Pearl River. Bottom: Silver carp are invasive in the Mississippi River system and have been reported elsewhere (Adapted from Nico 2009.)

upper sides of adults turn greenish, and gradually become silvery below the lateral line.

Silver carp grow to more than 3 ft. (1 m) long and may weigh 60 lbs. (27 kg) or more.

Related or Similar Species. Silver carp are most similar to another Asian carp, the bighead carp (*Hypophthalmichthys nobilis*).

In silver carp, a long keel on the belly is diagnostic, as is the long lateral line that bends toward the belly. (Michigan Sea Grant archives.)

The bighead has a shorter ventral keel and irregular dark blotches over its body (see Fish, Bighead Carp, for a more detailed description). The presence of a ventral keel separates both silver and bighead carp from all native cyprinids except the golden shiner (*Notemigonis crysoleucas*), which has larger scales and five pharyngeal teeth on each side of its jaw rather than the four found in Asian carps. Native shad could be mistaken for small juvenile silver carp, but shad have no lateral line and fewer than 14 rays in the anal fin.

Introduction History. Silver carp were first imported into the United States in 1973 by a private fish farmer in Arkansas for the purpose of controlling phytoplankton in his fish ponds. It was soon being raised at six state, federal, and private hatcheries and, by the 1970s, stocked in several municipal sewage lagoons. It was first discovered in natural waters in 1980, presumably the result of escapes from hatcheries and other facilities. Silver carp in the Ouachita River in Louisiana likely stem from escapes from an aquaculture site upstream in Arkansas. Florida received silver carp via a contaminated stocking of grass carp legally introduced to control aquatic plants. In Arizona, silver carp also came in a contaminated shipment of grass carp, but this was an illegal stocking. Ohio River fish may be products of the stocking of nearby ponds or from populations originating in Arkansas.

Habitat. Silver carp is a freshwater fish that inhabits large rivers, lakes, and ponds. It needs moving water for spawning and proper development of eggs. Flooded lowlands make good nursery areas for larvae and juveniles.

Diet. Adult silver carp filters phytoplankters, bacteria, and detritus from the water with its specialized gills. It reportedly also grazes aquatic plants. Juveniles feed on zooplankters. It feeds by rapidly gulping water as it swims, then closing its mouth and pumping the water out through its opercula. Those food particles taken in are ground by the pharyngeal teeth against a cartilaginous plate.

Life History. Silver carp are sexually mature at three years of age. In their native range, they migrate to communal spawning areas during the spring highwater period. They spawn in small groups of 15–25 fish at dusk and at dawn, when the water temperature is between 65° and 68°F (18–20°C). They require water moving enough to aerate the eggs and let them float downstream in the current.

Impacts. Silver carp is believed to have the potential to greatly alter native aquatic ecosystems because of its consumption of plankton, a food required by larval fish and native mussels. It could also compete with adults of some native fishes such as gizzard shad and bigmouth buffalo (*Ictiobus cyprinella*) and introduce diseases.

A greater impact occurs on fishermen and recreational boaters. Silver carp swim near the surface and are easily disturbed by boat motors. When startled, they leap high out of the water, the reason they are sometimes called flying carp. Being struck by a 40–60 lb. (18–27 kg) fish hurtling through the air can cause serious injury.

Management. The U.S. Fish and Wildlife Service lists silver carp as an injurious species; and under the Lacey Act, it is illegal to import it into the United States. The fear that this fish and its close relative the bighead carp could devastate Great Lakes fisheries has led the U.S. Army Corp of Engineers and the State of Illinois to construct electric barriers in the Chicago Sanitary and Ship Canal to halt its dispersal into the lakes from the Mississippi River (see also Fish, Bighead Carp).

Selected References

"Aquatic Invasive Species. Silver Carp." Indiana Department of Natural Resources, 2005. http://www.in.gov/dnr/files/SILVER_CARP.pdf.

Nico, Leo. "*Hypophthalmichthys molitrix*." USGS Nonindigenous Aquatic Species Database, Gainesville, FL, 2011. Revised January 11, 2011. http://nas.er.usgs.gov/queries/FactSheet.asp?speciesID=549.

"Silver carp. *Hypophthalmichthys molitrix*." Florida Integrated Science Center, USGS, 2005. http://fl.biology.usgs.gov/Carp_ID/html/hypophthalmichthys_molitrix.html.

■ | Spotted Tilapia

Also known as: Black mangrove cichlid
Scientific name: *Tilapia mariae*
Synonyms: *Tilapia dubia, T. meeki*
Family: Cichlidae

Native Range. West Africa, in coastal rainforest streams and brackish lagoons from the Tabou River in Côte d'Ivoire to the Pra River in Ghana and from southeast Benin to the Kribi River in Cameroon.

Distribution in the United States. Established in at least eight counties of southern Florida and in a few springs in Nevada. They are present in the Salton Sea, Colorado River, and Los Angeles area of California, and possibly established in Arizona.

Description. Spotted tilapia have compressed oval bodies. The two dorsal fins appear as a single fin; the forward one has 16 sharp spines and the rear one 12–13 soft rays. The tail is fan-shaped. The anal fin has three spines at the front and 10–11 rays that taper to a point at the rear. The mouth is terminal, and the large eye reddish.

Coloration is distinct. Juveniles bear no spots on the body but have a disruptive pattern of black bars on a yellow-green or gold background. There is a prominent black spot on the back of the dorsal fin, however. The bars fade with age, and 6–9 irregular square dark blotches appear along the midline of the flanks of adults. Some show pink or red coloration on chin or throat when spawning.

Some sexual dimorphism is present among spotted tilapia. Males have somewhat longer dorsal and caudal fins, both of which display shimmering white spots not found in females. Also, the foreheads of males have a steeper rise.

Spotted tilapia are smaller than most of the many other Africa cichlids established in Florida and other Gulf states, and it is the most abundant. Florida specimens typically are between 6 and 8 in. (15–20 cm), although they may grow to 12 in. (30 cm) or more and weigh up to 3 lbs. (1.4 kg). Most fish larger than 10 in. (25 cm) are males.

Related or Similar Species. Spotted tilapia resemble some native sunfishes in body and mouth shape. Occasionally, they are mistakenly called "oscars." They are very similar to the introduced redbelly tilapia (*T. zilli*), but that species has been eradicated in Florida.

Introduction History. The first known occurrence of spotted tilapia in the United States dates to April 1974, when it was discovered in Snapper Creek Canal in South Miami, Miami-Dade County. Florida. It quickly became established in canals throughout eastern Miami-Dade and southeastern Broward County. It was established in Everglades National Park and Big Cypress National Preserve by the late 1980s. These fish originated as escapes or intentional releases from aquarium fish producers in Dade County sometime between 1972 and 1974.

Similarly in Nevada, introduction was related to an aquarium release. Spotted tilapia have been abundant in Rogers Spring, a thermal spring in the Lake Mead National Recreation Area, Clark County, since 1980.

Habitat. Spotted tilapia prefer warm, slow-moving waters. They thrive in canals in South Florida and in warm springs in Nevada. Adults and juveniles stay close to sheltering structures such as vegetated shores or rock outcrops. They need hard, flat surfaces when spawning. Optimal water temperatures are 77–91°F (25–33°C); temperatures below 52°F (11°C) are not tolerated.

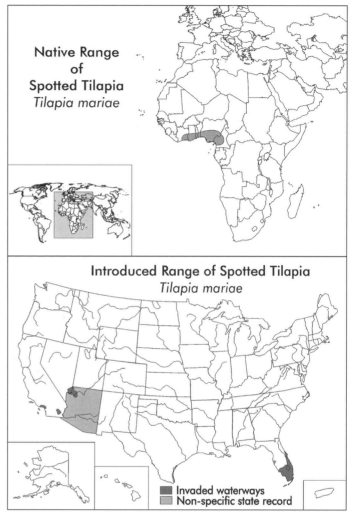

Top: Spotted tilapia are native to rainforest streams and brackish lagoons in West Africa. (Adapted from Robins, n.d.) Bottom: Spotted tilapia are invasive in Florida and also established in Arizona. They are present in the Colorado River and Salton Sea. (Adapted from Nico 2009.)

Diet. These fish feed primarily on "aufwuchs," the attached algae, detritus, and associated small invertebrates that grow on various plant and rock surfaces. They will also filter phytoplankton from eutrophic (plankton-rich) waters.

Life History. Spotted tilapia form pair bonds well before spawning begins and stay together to care for eggs and fry. Typically, breeding colonies assemble on the spawning grounds. In Florida, spawning generally occurs between November and March and may be synchronized with the lunar cycle, more batches of eggs being laid just before the full moon. These fish are substrate spawners and prefer gravel nests under flat rocks. Each female produces 200–2,000 turquoise eggs that stick to the gravel bottom. Two days after the eggs are laid, she removes all fertile eggs to a nearby pit and eats all infertile eggs. Early-stage larvae have "head glands" that produce a sticky strand by which the young attach themselves to

A. Adult spotted tilapia. (© Noel Burkhead.) B. Juvenile spotted tilapia. (© Noel Burkhead.)

the substrate so that they will not drift away. Both parents guard and feed the hatchlings until they are free-swimming some nine days later, when they are about 1 in. (2.5 cm) long and have developed the cryptic barring typical of juveniles. Spotted tilapia are sexually mature at 7 in. (18 cm).

Impacts. Spotted tilapia disperse rapidly and quickly become the dominant fish in both numbers and biomass in habitats to which they are introduced, thereby reducing biodiversity. By virtue of their numbers, they may compete with native fish for food. They may also compete with native fish such as sunfishes for spawning areas, since they aggressively defend their nests and broods.

Management. In the 1980s, Florida attempted to control spotted tilapia by introducing yet another exotic fish, the predatory South America peacock cichlid (*Cichla ocellaris*) to feed upon it. How successful this effort was is open to question; the two species seem to have reached an equilibrium in their predator-prey relationship.

Prevention of further spread of tilapia is the main goal of management. It is illegal to transport or possess live tilapia in Florida. This fish is edible, so fishermen wishing to eat tilapia should immediately kill them and put them on ice.

Selected References

National Biological Information Infrastructure (NBII) and IUCN/SSC Invasive Species Specialist Group (ISSG). "*Tilapia mariae* (Fish)." ISSG Global Invasive Species Database, 2009. http://www.issg.org/database/species/ecology.asp?si=1430&fr=1&sts=&lang=EN.

Nico, Leo. *Tilapia mariae*. USGS Nonindigenous Aquatic Species Database, Gainesville, FL, 2006. http://nas.er.usgs.gov/queries/factsheet.asp?SpeciesID=482.

Robins, Robert H. "Spotted Tilapia." Florida Museum of Natural History, n.d. http://www.flmnh.ufl.edu/fish/gallery/Descript/SpottedTilapia/SpottedTilapia.html.

■ | Walking Catfish

Also known as: Magur
Scientific name: *Clarias batrachus*
Family: Clariidae

Native Range. Southeast Asia. Native to Bangladesh, eastern India, Indonesia, Malaysia, Myanmar (Burma), Pakistan, Singapore, Sri Lanka, and Thailand. The fish introduced to Florida were from Thailand.

Distribution in the United States. Established throughout most of Florida, including Everglades National Park, Big Cypress National Preserve, Florida Panther National Wildlife Refuge, and Pelican Island National Wildlife Refuge. Reported from California, Connecticut, Georgia, Massachusetts, and Nevada.

Description. The walking catfish has a flat, broad head and an elongated body tapering toward the tail. It has a typical catfish appearance, with four pairs of barbels (whiskers) and wide, fleshy lips. The teeth are small and bristle-like. The eyes are small. The dorsal fin is long and continuous and extends two-thirds the length of the back. There is no adipose fin as in native catfishes. Anal fins are also long, but end in a lobe separate from the caudal fin (tail), which is rounded. Each pectoral fin has a strong, rigid spine at the front that helps the fish "walk." Somewhat ungainly terrestrial locomotion is achieved by using the pectoral spines to pull itself along while flexing the body back and forth.

The scaleless body is usually a drab gray or gray-brown with

Top: Walking catfish are native to parts of South and Southeast Asia. (Adapted from Robins n.d.) Bottom: Walking catfish are invasive in Florida. They have been collected in other states. (Adapted from Nico 2009.)

white flecks on the sides. Albino and calico colors are common in aquarium fishes, but wild populations revert to the natural gray. (All of the original walking catfish introduced into Florida were albinos.)

The gills of walking catfish are specially structured with tree-like organs to permit their breathing on land and in stagnant water.

Related or Similar Species. Native catfish, such as the marine hardhead catfish (*Ariopsis felis*) and gafftopsail catfish (*Bagre marinus*), as well as freshwater catfish such as the brown bullhead (*Ictalurus nebulosus*) and channel catfish (*I. punctatus*) could be mistaken for walking catfish. All native catfishes, however, possess a forked tail, an adipose fin just forward of the tail, and a dorsal spine.

Introduction History. Walking catfish were first imported to Florida in the early 1960s for the aquarium trade. The introduction into local waters occurred in the mid-1960s when adult fish escaped either from a fish farm in northeastern Broward County or from a transport truck carrying brood fish between Broward and Miami-Dade counties. Apparently purposeful releases were made by fish farmers elsewhere in Florida in 1967–1968 after the state prohibited the importation and possession of walking catfish.

By 1968, walking catfish were established in three counties; 10 years after the first escapes, they were in almost 20 counties in the southern half of Florida. Their spread was facilitated by the network of canals and drainage ditches in southeastern Florida and their ability to cross land, especially on rainy nights.

Aquarium releases are the likely source of introductions in other states.

Habitat. Walking catfish may inhabit a variety of habitats, but they are most commonly associated with freshwater habitats in which most native species do not thrive. They are bottom-dwellers that become the dominants in warm, muddy ponds, swamps, canals, and ditches and tolerate stagnant, low-oxygen conditions. Since they can "walk" from pond to pond and also burrow into pond and river banks for a period of dormancy, they can also inhabit intermittent streams and isolated temporary ponds. They are essentially tropical fish and vulnerable to cold. Winter kills have been reported in Florida when water temperatures drop below 50°F (10°C).

Diet. Walking catfish feed on large aquatic invertebrates, the eggs and larvae of amphibians and fish, and small fish. They are known to have invaded aquaculture farms to feed on fish.

Life History. Walking catfish become sexually mature at about one year of age. Mass migrations and spawning seem to be linked to the rainy season. Nests are hollowed out in detritus or submerged vegetation and guarded by the adults. The up to 1,000 eggs laid are adhesive and stick to the substrate. Once the eggs have been fertilized, the male guards the nest, and the female hovers nearby to drive off intruders. Embryos hatch in about 30 hours and the fry stay under the protection of the parents for about five days, at which time the yolk-sac is absorbed and the young begin to forage. Catfish grow rapidly and, in Florida, usually reach a length of about 12 in. (30 cm). The largest reported walking catfish in the United States was about 20 in. (50 cm) long and weighed about 3 lbs. (1.4 kg).

Impacts. When walking catfish were first released into Florida waters, they were forecast to become the most harmful introduction known so far. They were prolific, and because they could move across land, they dispersed rapidly. It was expected that they would outcompete native catfish species and cause the decline of some native amphibians. However, after an initial population explosion, numbers began to decline in the 1970s. Actual ecological impacts are unknown. They have not eliminated any native fish and indeed do not seem to have had any major negative effect on native fish or amphibian communities. It may be that their preference for ponds that partially dry up each year and stagnant bodies of water to which no natives are adapted has isolated them from native fish populations and reduced their effects. They do, however, have an economic impact on fish farmers, who must construct barriers to keep them out of their aquaculture ponds, and are still considered undesirable. The walking catfish has been nominated as among 100 of the "World's Worst" invaders by the Invasive Species Specialist Group (ISSG).

Management. All members of the walking catfish family are on the U.S. Fish and Wildlife Service's list of Injurious Wildlife Species. Under the Lacey Act, it is a violation of federal law to import them into the United States without a permit. Some states have laws that make it illegal to possess a live walking catfish. Any walking catfish caught by anglers must be quickly killed. Fish fences keep walking catfish out of aquaculture ponds.

Selected References

Masterson, J. "*Clarias batrachus* (Walking catfish)." Smithsonian Marine Station at Fort Pierce, 2007. http://www.sms.si.edu/IRLspec/Clarias_batrachus.htm.

Nico, Leo. "*Clarias batrachus*." USGS Nonindigenous Aquatic Species Database, Gainesville, FL, 2009. Revised April 11, 2006. http://nas.er.usgs.gov/queries/FactSheet.asp?speciesID=486.

Robins, Robert H. "Biological Profiles: Walking Catfish." Florida Museum of Natural History, n.d. http://www.flmnh.ufl.edu/fish/Gallery/Descript/WalkingCatfish/WalkingCatfish.html.

"Walking Catfish—*Clarias batrachus*." MyFWS.com, Florida Fish and Wildlife Conservation Commission, Tallahassee, n.d. http://myfwc.com/wildlifehabitats/profiles/fish/nonnative-fish/walking-catfish/.

■ Amphibians

■ | African Clawed Frog

Also known as: Upland frog, common platanna
Scientific name: *Xenopus laevis*
Family: Pipidae

Native Range. Highlands of sub-Saharan Africa in Angola, Botswana, Burundi, Cameroon, Central African Republic, Democratic Republic of the Congo, southwestern Kenya, Lestho, Malawi, Mozambique, Namibia, eastern Nigeria, Rwanda, South Africa, Swaziland, Uganda, Zambia, and Zimbabwe. Absent from Congo Basin and warmer lowlands of East Africa. Imports to the United States originated in South Africa.

Distribution in the United States. Disjunct populations are established in at least seven counties in southern California, and reported from others. A single population is established in artificial ponds at a golf course in Tucson, Arizona. African clawed frogs have been collected in Colorado, Florida, Massachusetts, North Carolina, Nevada, New Mexico, Texas, Utah, Virginia, Wisconsin, and Wyoming, but no established populations are known from these states.

Description. The African clawed frog has a flattened body and relatively small head. It lacks a visible external ear drum (tympanum) and has no tongue. Its eyes are small and lidless, located near the top of the head. The forefeet have four thin, unwebbed fingers that usually point forward. The large hind feet have five fully webbed toes, the inner three of which possess sharp, black claws. The skin is very smooth and slippery. The back is a mottled olive-brown or gray, and the belly is cream colored. The lateral lines seem stitched along the back. Snout to vent length is from 2.0 to 5.5 in. (5–14 cm.)

Tadpoles of this species are unique. They are transparent, so the internal organs are visible. Barbels occur at the mouth, making them resemble small catfish. The tail ends in a filament. They usually swim head down, vibrating the tail filament to stir up plankton. Tadpoles grow to about 1.5 in. (3.8 cm) long.

Related or Similar Species. None. No native American frog or toad has clawed hind limb toes. No native amphibians have transparent tadpoles or tadpoles with long barbells.

Introduction History. The African clawed frog was introduced to laboratories around the globe in the 1940s and 1950s after it was discovered it could be used as a reliable test for human pregnancy. An injection of urine from a pregnant woman stimulated a female frog

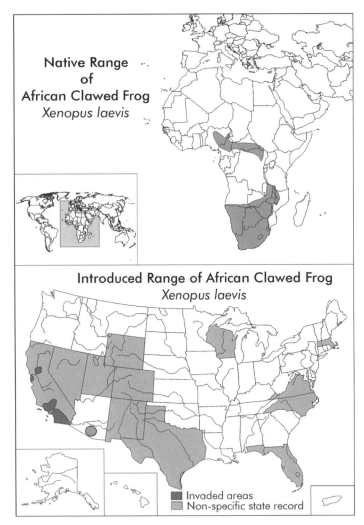

Native Range of African Clawed Frog
Xenopus laevis

Introduced Range of African Clawed Frog
Xenopus laevis

■ Invaded areas
■ Non-specific state record

Top: The African clawed frog is native to the uplands of sub-Saharan Africa. (Adapted from range map of *Xenopus laevis* at http://www.iucnredlist.org/apps/redlist/details/58174/0/rangemap.) Bottom: Self-sustaining populations of African clawed frog occur in California and Arizona. It has been collected in several other states, where, to date, no populations have become established. (Adapted from Somma 2005.)

to lay eggs. As the demand grew, the frog was bred in captivity, so that most free-living populations outside the natural range of the species are actually feral stock. The clawed frogs are easy to care for and disease resistant, and in the 1950s and 1960s, they gained a foothold in the pet trade. It is still an important research animal, used primarily for studies of developmental, cell, and molecular biology.

Most introductions to the wild were intentional releases from laboratories once other techniques to determine human pregnancy were devised. Pets, no longer wanted by their owners, were also released. The first feral African clawed frog was found in Orange County, California, in 1968. Independent introduction events occurred in five counties (Los Angeles, Orange, Riverside, San Diego, and Santa Barbara); in 25–30 years, the frogs spread throughout most of the drainage systems into which they were first released. San Bernadino and Ventura counties were invaded from neighboring counties. It is expected that they will continue to spread along the many irrigation canals and drainage ditches in southern California.

In Arizona, African clawed frogs were deliberately introduced to man-made ponds in Arthur Park Golf Course, in Tucson, Pima County, in the late 1960s, and they persist there. Spread to other bodies of water has been inhibited by the surrounding expanse of desert.

Other states where the frog was introduced but never became established, with the recorded dates of introduction, are: Colorado, 1990; Florida, 1964; Massachusetts, 1993; North Carolina, 1990s; Virginia, 1980s (eradicated 1987–1988); Wisconsin, 1972.

Habitat. The African clawed frog spends almost its entire life in aquatic habitats, leaving water only if forced to migrate when a water body dries up or is poisoned. They may be found in all types of aquatic situations except fast-moving rivers and water bodies with predatory fish. They tend to prefer eutrophic ponds, both natural and artificial, slow streams, and natural waterways that are frequently disturbed or experience high environmental variability. They are salt tolerant and can live in estuaries. Although they find optimal climatic conditions in Mediterranean climates and water temperatures between 60° and 80°F (15–27°C), adults can tolerate temperatures ranging from 32° to 86°F (0–30°C) and populations have persisted under winter ice. Tadpoles survive in water temperatures ranging from 50° to 86°F (10–30° C). Adults can aestivate for periods of up to eight months and go without food for a year. Eggs and tadpoles are viable in both acidic and alkaline water (pH 5–9).

African clawed frogs have flattened bodies and lack visible external ear drums. (Todd Battey.)

Diet. Adults feed on slow-moving aquatic invertebrates that they suck into their mouths. They will also scavenge dead and dying arthropods and take small fish and tadpoles, including those of their own species. They rely on the lateral line system to detect scents in the water and the movement of prey. Forelimbs gather food into the mouth, while the strong, clawed hind legs shred larger prey items into ingestible sizes.

Tadpoles are filter-feeders. They extract small phytoplankters from open water. Food includes one-celled algae such as diatoms, protozoans, and bacteria.

Life History. African clawed frogs in California breed from January through November. A peak occurs in April and May. Males vocalize underwater during the evening to call females. The mating call alternates long and short trills. The female either accepts him with a rapping sound or rejects him with a slow ticking sound. (Very seldom does this answering behavior happen with any other animal.) Mating usually takes place at night. Unlike most other frogs, the mating embrace or amplexus is pelvic. Over the course of the next 3–4 hours, the female releases, one at a time, hundreds of eggs that, fertilized by the male, adhere to plants or other underwater structures. The eggs grow into tadpoles and hatch in 3–4 days. The length of the larval stage is 5–12 weeks. Tadpoles are weak swimmers prone to predation by fish; they congregate in schools in deeper water to feed. As the tadpole metamorphoses, the tail is absorbed, providing nutrition for the developing froglet. Metamorphosis takes 4–5 days. Sexual maturity is reached 6–10 months after metamorphosis. African clawed frogs may live 9–15 years.

Impacts. The African clawed frog may be a threat to native amphibians and fish in California and the Southwest since it preys upon tadpoles and fish fry. Its range overlaps with such vulnerable species as western toad (*Bufo boreas*), red-spotted toads (*Bufo punctatus*), California red-legged frogs (*Rana draytoni*), Pacific treefrogs (*Pseudacris regilla*), and western spadefoot toads (*Spea hammondi*). It is known to eat larvae and recently metamorphosed western toads, but negative impacts on California red-legged frogs, though claimed, have not been verified. Observations of African clawed frogs preying on fish are rare, but do include consumption of the federally endangered tidewater goby (*Eucyclogobius newberryi*),

African Clawed Frog as Pharmacopoiea

The African clawed frog has long been used as a research animal because it is easy to care for and breeds rapidly, and the development of its transparent embryos is easy to observe. In 1987, Dr. Michael Zasloff discovered that the skin of the African clawed frog has glands that secrete a broad-spectrum antimicrobial chemical effective against viruses, bacteria, fungi, and protozoa. When first discovered, this defensive mechanism represented a totally new way that vertebrates were able to protect themselves against infection. Previously, it was believed only the immune system attacked invading pathogens. The frog skin chemicals work by creating a hole in the target's cell membrane, causing the cell contents to leak out.

Dr. Zasloff isolated two peptides from the skin of the African clawed frog and named them magainins, from the Hebrew word magain, meaning shield. Magainins were synthesized and manipulated in the laboratory to create thousands of related antibiotics. There was much hoopla in the 1990s about their potential use in new drugs to combat a variety of human infections. Experiments showed that magainins killed bacteria such as *E. coli*, staphylococci, streptococci, and enterobacteria. They gave promise in combating colon cancer. A topical cream proved effective as a treatment for diabetic foot ulcers. Magainins in toothpaste might fight dental plaque. They might control the tuberculosis bacillus or the plasmodium that causes malaria, or they could work as a spermicide that not only protects against pregnancy but also against sexually transmitted diseases. Although no magainin-containing drugs have completed the whole arduous course through animal trials and human safety tests to FDA approval and the drug store shelf, their discovery in the African clawed frog opened new paths of research in animal-derived antibiotics and pointed to future discoveries of antibiotics and other helpful compounds in what remains of Earth's biodiversity.

Since 1987 almost all frog species tested have been found to produce related chemicals in their skin and in their digestive tracts; and other animals have found to produce similar compounds in their saliva.

Source: Altman, Lawrence K. "Curiosity on Healing in Frogs Leads to a Gain in Antibiotics." *New York Times*, July 31, 1987. http://www.nytimes.com/1987/07/31/us/curiosity-on-healing-in-frogs-leads-to-a-gain-in-antibiotics.html?ref=lawrencekaltman

arroyo chub (*Gila orcutti*), and another federally endangered fish, the unarmored threespine stickleback (*Gastaerosteus aculeatus williamsoni*). Were African clawed frogs to get into the isolated pools containing desert pupfish (*Cyprinodon* spp.), those rare fish would likely be at severe risk.

Many scientists believe African clawed frogs carried the fungus *Batrachochytrium dedrobatidis* (see Fungi, Chytrid Frog Fungus) around the globe. This fungus is a major cause of the worldwide decline and extinction of frogs and toads. African clawed frogs are immune to the fungus, which is native to their African habitats. Museum specimens dating to 1938 show evidence of chytridiomycosis, the skin disease associated with it.

Management. Most efforts to eliminate clawed frogs in California have not been successful. Permanent eradication is known in only one instance, when an artificial pond on the University of California, Davis, campus was poisoned by the California Department of Fish and Game. Elsewhere, poisoning has not worked. At Vasquez Rock in the upper Santa Clara River drainage, the frogs simply left the water after rotenone was applied. Other methods of control include draining ponds, removing frogs by seine or traps, electroshocking, and introducing predatory fish; but none have proved effective in eliminating frogs or preventing their reintroduction.

Selected References

Crayon, John J. "*Xenopus laevis* (Daudin, 1802) African Clawed Frog." AmphibiaWeb: Information on amphibian biology and conservation. Berkeley, California, 2009. http://amphibiaweb.org/.

Garvey, Nathan. "*Xenopus laevis* African Clawed Frog." University of Michigan Museum of Zoology, 2000. http://animaldiversity.ummz.umich.edu/site/accounts/information/Xenopus_laevis.html.

Measey, John. "*Xenopus laevis* (Amphibian)." IUCN/SSC Invasive Species Specialist Group (ISSG) Global Invasive Species Database, 2006. http://www.issg.org/database/species/ecology.asp?si=150.

Rorabaugh, Jim. "African Clawed Frog *Xenopus laevis*." Online Field Guide to the Reptiles and Amphibians of Arizona, 2008. http://www.reptilesofaz.org/Turtle-Amphibs-Subpages/h-x-laevis.html.

Somma, Louis A. "*Xenopus laevis*." USGS Nonindigenous Aquatic Species Database, Gainesville, FL, 2009. Revised March 24, 2005. http://nas.er.usgs.gov/queries/FactSheet.asp?speciesID=67.

Willigan, Erin. "African Clawed Frog (*Xenopus laevis*)." Introduced Species Summary Project, Columbia University, 2001. http://www.columbia.edu/itc/cerc/danoff-burg/invasion_bio/inv_spp_summ/xenopus_laevis.htm.

■ American Bullfrog

Scientific name: *Lithobates catesbeianus*
Synonym: *Rana catesbeiana*
Family: Ranidae

Native Range. Central and eastern United States and southern parts of Ontario and Quebec provinces, Canada.

Distribution in the United States. The American bullfrog has become established outside its native range in Arizona, California, Colorado, Hawai'i, Nebraska, Nevada, Oregon, Utah, and Washington. It has also been introduced to locations where it was not historically found within its natural range in Massachusetts (on the islands of Nantucket and Martha's Vineyard, and in the Wellfleet Bay sanctuary on Cape Cod), Iowa (in the DeSoto National Wildlife Refuge on the Missouri River), and New Jersey (in the Cape May National Wildlife Refuge).

Description. Adult bullfrogs are green to brownish green, with dark blotches and bars on the back and legs. The upper jaw is often a light green. The belly is cream to yellow. During the breeding season, the throat of the male is yellow, while that of the female is white. The absence of dorsolateral folds is diagnostic. However, there is a short fold in the skin running from the eye, over the eardrum to the forelimb. The eardrum or tympanum is conspicuous and exhibits sexual dimorphism. In males, the tympanum is larger than the eye; in females, it is the same size as or smaller than the eye. The hindfoot is fully webbed with the exception of the last joint of the fourth toe.

The tadpole is large (up to 6 in. or 15 cm long). Its back is yellow green and has black spots; the belly is lighter. The dorsal fin is arched.

The call of the bullfrog is deep and very loud; it is usually described as "jug-o'rum" or "br-wum." They also sound a loud squeak of alarm before jumping into the water when

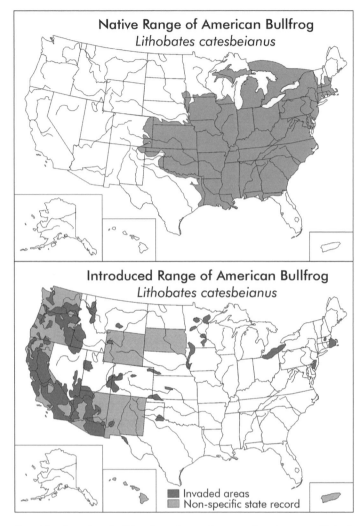

Top: The American bullfrog is native to the central and eastern United States and adjacent areas of Canada. Bottom: American bullfrogs have been transplanted beyond their native range into all western states (except Alaska), Hawai'i, and Puerto Rico. (Both maps adapted from McKercher 2009.)

surprised. Bullfrogs are active both day and night and are highly aquatic, rarely found out of the water.

The bullfrog is the largest frog native to North America. From snout to vent, the length of adults ranges from 3.5 to 8 in. (9–20 cm), the largest individuals being female. They may weigh more than 1 lb. (0.5 kg). They are celebrated jumpers, and their long legs are considered a delicacy; they are often hunted and farmed for meat.

Related or Similar Species. The green frog (*Lithobates* [=*Rana*] *clamitans*) is much smaller (adults only 2–4 in. or 5–10 cm, snout to vent) and has the prominent dorsolateral folds characteristic of most North American frogs. Its distribution is similar to that of the natural range of the American bullfrog.

Introduction History. Bullfrogs were first introduced to California in 1890s and Colorado in the early 1900s. By the end of the 1920s, they were common on the lower Colorado River near Yuma, Arizona. The pathways of introduction are largely unknown. In Colorado, bullfrog tadpoles at fish hatcheries may have contaminated fish stockings in trout streams and lakes. Other possible means of introduction include the aquarium trade, frog-farming operations, and the stocking of farm ponds to control agricultural pests. They may have been deliberately released to provide a new game animal or introduced as a living feature of ornamental ponds. Once in the wild, bullfrogs are able to disperse through entire drainage systems.

In some western states, the competitive advantage that the American bullfrog has over native frogs is enhanced by the introduction of nonnative fishes, such as bluegill sunfish (*Lepomis macrochirus*). In the Pacific Northwest, bluegills and other sport fish were stocked in lakes that had no native fishes. The nonnative fish eat the tadpoles of native frogs, but leave bullfrog tadpoles alone. Bullfrogs coevolved with these alien fishes and are unpalatable to them. Furthermore, the sunfish consume the nymphs of native dragonflies, species that do prey upon bullfrog tadpoles.

A. Adult American bullfrogs have no folds on the back or sides, but a short fold of skin does occur from the eye over the conspicuous external eardrum to the forelimb. (Ilias Strachnis/Shutterstock.) B. The tadpole is very large and has an arched dorsal fin. (Tommounsay/iStockPhoto.)

Bullfrogs have been introduced to Mexico, the Caribbean, South America, Europe, and Asia.

Habitat. Shallow, still waters are preferred, especially where abundant aquatic vegetation provides shelter. Bullfrogs are thus found in marshes and along the shores of lakes, ponds, slow-moving rivers, and reservoirs or other impoundments. They need warm water for breeding, and tadpoles prefer non-vegetated areas. Tadpoles are able to overwinter in water beneath a cover of ice. Adults hibernate in mud during cold weather.

Diet. Adult bullfrogs are carnivorous and tend to eat whatever they can catch with their large sticky tongues, including insects, crayfish, fish eggs, eggs of amphibians, tadpoles, and frogs, including other bullfrogs. They also take snakes, birds, bats, and mice.

Tadpoles are mostly herbivorous and feed on cyanobacteria, algae, other plant material, and small aquatic invertebrates.

Life History. American bullfrogs have a long breeding season, from May to July in northern states and from February to October in warmer areas. A female deposits up to 20,000 eggs in a broad foamy sheet in warm, quiet water. Fertilization is external. The raft of jelly-coated eggs may stretch 3 ft. (1 m) in diameter. It floats until just before hatching, when it sinks onto underwater vegetation. Tadpoles hatch out 3–5 days after fertilization. The gilled tadpoles develop slowly, and those in colder climes overwinter at least once. The transformation from tadpole to froglet, when the tail is slowly absorbed, may take 1–3 years. The new froglets measure about 2 in. (5 cm) long. Bullfrogs become sexually mature two years later, and live 7–9 years in the wild.

Impacts. The introduction of the American bullfrog to western states has likely exacerbated problems related to the demise of native frogs because bullfrogs feed on frogs and tadpoles. Tadpoles may significantly alter aquatic community structure because they feed heavily on algae. In Arizona, predation by bullfrogs has contributed to the extirpation of populations of native leopard frogs and garter snakes. Most vulnerable are the Mexican garter snake (*Thamnophis eques megalops*), and the federally threatened Chiricahua leopard frog (*Rana chiricahuensis*). Introduced bullfrogs may also spread chytridomycosis, a fungal disease of the skin implicated in the decline of amphibians worldwide (see Fungi, Chytrid Frog Fungus). ISSG has nominated them as one of 100 of the "World's Worst" invaders.

Management. Bullfrog populations are often managed as fisheries, and harvesting limits population growth. Adults can be taken with gigs, nets, traps, bow and arrow, and guns;

and they can be caught by hand. Spotlighting at night is a common practice since frogs are immobilized by bright lights. Larvae can be killed by the same poisons used against fish. Suctioning off the egg masses can be tricky but effective.

New introductions to areas where bullfrogs are not yet present should be prohibited.

Selected References

"Alien Species in Cahoots." U.S. Geological Survey News Release, March 13, 2003. http://fresc.usgs.gov/news/newsreleases.asp?NRID=3.

Bruening, S. "*Rana catesbeiana.*" Animal Diversity Web, University of Michigan Museum of Zoology, 2002. http://animaldiversity.ummz.umich.edu/site/accounts/information/Rana_catesbeiana.html.

Crayon, John J. "*Rana catesbeiana* (Amphibian)." Global Invasive Species Database, IUCN/SSC Invasive Species Specialist Group, 2005. http://www.issg.org/database/species/ecology.asp?si=80&fr =1&sts.

McKercher, Liz, and Denise R. Gregoire. *Lithobates [=Rana] catesbeianus.* USGS Nonindigenous Aquatic Species Database, Gainesville, FL, 2009. Revised March 7, 2011. http://nas.er.usgs.gov/queries/FactSheet.asp?speciesID=71.

Murphy, Martin. "North American Bullfrog (*Rana catesbeiana*)." Introduced Species Summary Project, Columbia University, 2003. http://www.columbia.edu/itc/cerc/danoff-burg/invasion_bio/inv_spp_summ/Rana_catesbeiana.htm.

Rorabaugh, Jim. "American Bullfrog *Rana catesbeiana.*" Online Field Guide to the Reptiles and Amphibians of Arizona, 2008. http://www.reptilesofaz.org/Turtle-Amphibs-Subpages/h-l-catesbeianus.html.

■ | Coqui

Also known as: common coqui, coquí común, Puerto Rican treefrog
Scientific name: *Eleutherodactylus coqui*
Family: Leptodactylidae

Native Range. Puerto Rico. Native to a variety of habitats and elevations.

Distribution in the United States. Coqui are established in Hawai'i on Hawai'i Island, Kaua'i, Maui, and O'ahu. They are established but noninvasive in Florida, where they tend to die off during winter freezes. Established populations also occur on St. Croix, St. John, and St. Thomas, U.S. Virgin Islands.

Description. A small but very loud frog. Coqui are quite variable in color. The back may range from light yellow to dark brown and may or may not be mottled with black spots. There may be a cream line running along the midline of the back from the snout to the hind legs or an indistinct dark "M" between the shoulders. The underside is light with brown stipples. Eyes are gold or golden brown. The toes have suction cup-like pads like those of true treefrogs (family Hylidae). Breeding males are slightly larger than 1 in. (34 mm) from snout to vent; females average about 1.5 in. (41 mm). It appears that the average size of coqui frogs is increasing in Hawai'i, since frogs continue to grow throughout their lives and may live up to seven years. Coqui have been on the Big Island for 10 years. They also tend to be larger at higher elevations. There is no tadpole stage.

The mating call of the male coqui is a rapid and loud "ko-KEE." Males begin to call at dusk and may continue all night, especially if it is raining. According to one report, the sound measures as much as 80–90 decibels, comparable to a lawn mower.

Related or Similar Species. In Hawai'i, the coqui most resembles another exotic amphibian, its close relative the greenhouse frog (*Eleutherodactylus planirostris*). Greenhouse frogs have narrower snouts, a narrower body shape, claw-like toes, red eyes, and a warty-textured skin. Adults are usually a mottled copper color and are less than 1 in. (2.5 cm) long. Most importantly, they have a soft bird- or cricket-like chirp. They live on the ground and are not considered invasive.

In Florida, coqui might be confused with native treefrogs.

Introduction History. The coqui was probably introduced to the island of Hawai'i accidentally around 1988 in a shipment of horticultural plants such as bromeliads from Puerto Rico. From a few infected nurseries, the coqui was likely spread in landscaping materials to other sites and other islands. Populations grew rapidly and expanded beyond the sites of initial entry due of lack of predators (e.g., owls, snakes, tarantulas) on the islands. By 2003, there were more than 200 infestation sites on the Big Island, 40 or more on Maui, 5 on O'ahu, and 1 site on Kaua'i. These sites include commercial plant nurseries, the grounds of resort hotels, ornamental plantings in parks, and in native forest.

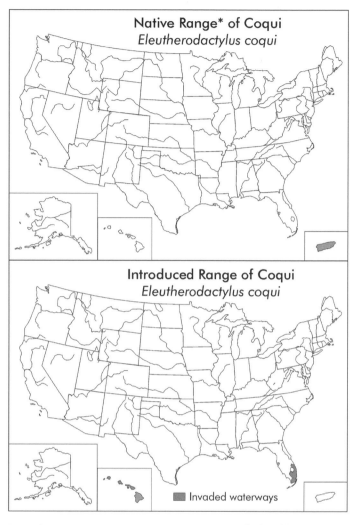

Top: The coqui is native to Puerto Rico. Bottom: The loud little arboreal frog is invasive in Hawai'i. The expansion of populations in Florida seems to be prevented by cold winter weather. (Adapted from Somma 2009.)

Habitat. Coqui frogs are found in Hawai'i in a variety of ecological zones at elevations ranging from sea level to 4,000 ft. (1,220 m). Most populations are in or near horticultural sites, although they spread from nurseries, resorts, and residential areas into natural areas on public lands. By day, they hide in shady, thick moist brush, leaf litter, dead leaves on banana plants, empty flower pots, or holes in porous rock. They nest in cavities, folded or rolled leaves, PVC pipe, or other sheltering structures. Coqui seem to especially like broad-leaf plants such as heliconias (*Heliconia* spp.), Koster's curse (*Clidemia hirta*), Loulu palms (*Pritchardia spp.*), silktree (*Albezia julibrissin*), split-leaf philodendrons (*Monstera* spp.), and ti (*Cordyline terminalis*). Adults prefer to be 3–9 ft. (1–3 m) above the ground; juveniles are usually found less than 4 ft. (1 m) above the ground. Since they have no tadpole stage, they do not require standing water for breeding.

Diet. Insects and other terrestrial invertebrates.

This coqui is resting in a tank bromeliad. (Rogeliao Doratt/USDA.)

Life History. Breeding occurs all year, but tends to concentrate during the rainy season. Coqui use internal fertilization. The female lays a clutch of about two dozen eggs in a protected, elevated cavity. Males guard the eggs and keep them from drying out. Four to six clutches may be produced in a year, usually one every two months. The fertilized eggs undergo direct development to tiny (approx. 0.5 in. or 1.25 cm long) froglets in 2–3 weeks. Coqui froglets are sexually mature in about eight months.

Impacts. Hawai'i has no native frogs, so any introduced frog poses the potential to disrupt native ecosystems. Experiments have shown that consumption of leaf-eating insects by coqui can increase foliage production and decomposition rates, thereby increasing nutrient cycling rates. It is possible that coqui will consume endemic insects and spiders, some of which are already on the verge of extinction and some of which are important pollinators of Hawai'i's endemic flora. It could compete with native Hawaiian birds, most of which are insectivorous. There is also concern that coqui might provide a food source for the nonindigenous brown tree snake (*Boiga irregularis*), should it invade the islands.

By far the greatest problem with the coqui is noise. Tourists and residents complain of sleepless nights. Property values have declined in infested areas. The tourist and real estate industries are threatened. Controls to prevent the further spread to the coqui could be deleterious to the floriculture and nursery industries, the main pathways of dispersal for this tiny invader. The combined value of the tourism, floriculture, and real estate industries which are at risk in Hawai'i is many millions of dollars.

For its potential ecological and economic impacts, the coqui has been nominated as among 100 of the "World's Worst" invaders by IUCN/Species Survival Commission's Invasive Species Specialist Group (ISSG).

Management. Small populations may be controlled by hand-capture and trapping. Larger infestations may be managed by spraying landscape plants with hot water (113°F) or dipping potted plants in hot water. A solution of citric acid will kill adults and eggs and is approved for use by the EPA. Prevention is key. Hawaiians need to eliminate the habitats that coqui prefer around their homes and vacation spots. They are encouraged to compost yard wastes, prune and thin shrubbery, remove dead leaves especially from large-leaved plants, and remove empty pots and other containers that collect rainwater during dry seasons.

The inter-island transport of coqui frogs is now prohibited in Hawai'i. If strict controls including certification of frog-free flowers and plants were to be placed on floriculture exports—a not unreasonable prospect, this would create a financial hardship on one of Hawai'i's major industries.

Selected References

Chun, S., A. H. Hara, and R. Y. Niino-DuPonte. "Greenhouse Frog or Coqui Frog?" College of Tropical Agriculture and Human Resources, University of Hawai'i at Manoa, 2003. http://www2.ctahr .hawaii.edu/oc/freepubs/pdf/coqui_id.pdf.

"Control of Coqui Frogs in Hawai'i." Department of Plant and Environmental Protection Sciences, College of Tropical Agriculture and Human Resources, University of Hawai'i at Manoa, 2008. http://www.ctahr.hawaii.edu/coqui/.

Kraus, Fred, Earl W. Campbell, Allen Allison, and Thane Pratt. "*Eleutherodactylus* Frog Introductions to Hawaii." *Herpetological Review* 30 (10): 21–25, 1999. Available online at http://www.hear.org/ articles/pdfs/herp_review_frogs_1999v30n1.pdf.

National Biological Information Infrastructure (NBII) and IUCN/SSC Invasive Species Specialist Group (ISSG). "*Eleutherodactylus coqui* (Amphibian)." ISSG Global Invasive Species Database, 2005. http:// www.issg.org/database/species/ecology.asp?fr=1&si=105.

Somma, Louis A. "*Eleutherodactylus coqui*." USGS Nonindigenous Aquatic Species Database, Gainesville, FL, 2009. Revised May 14, 2006. http://nas.er.usgs.gov/queries/FactSheet.asp ?speciesID=60.

■ | Cuban Treefrog

Also known as: Giant tree frog, rana platernera
Scientific name: *Osteopilus septentrionalis*
Synonyms: *Trachycephalus insulsus, Hyla septentrionalis*
Family: Hylidae

Native Range. Cuba, Isla de Pinos, the Bahamas, and the Cayman Islands.

Distribution in the United States. Established in Florida. Also established in Puerto Rico and St. Croix and St. Thomas, U.S. Virgin Islands; collected from Georgia, but apparently not established there. Single waifs have been reported from Colorado, Indiana, Maryland, and Virginia. A report of a population on O'ahu, Hawai'i, has not been verified.

Description. The Cuban treefrog is the largest treefrog in the United States. It has very large toe pads equal in size to the eardrum (tympanum). The very large eyes of the Cuban treefrog give them a "bug-eyed" look. The back is warty, like a toad, and the belly is granular. A skin fold extends from the eye rearward to the tympanum. Adults can be distinguished from all native treefrogs because the skin on the top of the head is fused to the skull. Color varies from gray to gray-green or tan-brown, or even cream, and individuals are able to change color rapidly. The body is usually heavily mottled and there are often stripes on the backside of the legs. A yellowish tinge occurs where the legs join the body. The rear toes are slightly webbed, and there is a distinct tarsal fold along the entire ankle (tarsus). Most adults in Florida are from 1 to 4 in. (2.5–10 cm) long, although adult females may grow to 6 in. (15 cm) or more.

Juveniles are difficult to identify because they lack warts and strong patterning. Sometimes they lack the lateral stripe that is found on many native treefrogs. Tadpoles have round bodies with dark or black backs. The intestinal coil is visible through the light-colored belly. The tail is pigmented with lighter areas near the point of attachment to the body. Scattered dark flecks appear on the wide fin.

The mating call of the male treefrog is a rasping snarl of varied pitch. It has been described as sounding like a squeaky door. The male's vocal sac looks like a double bubble. Males also will call during the day when it rains.

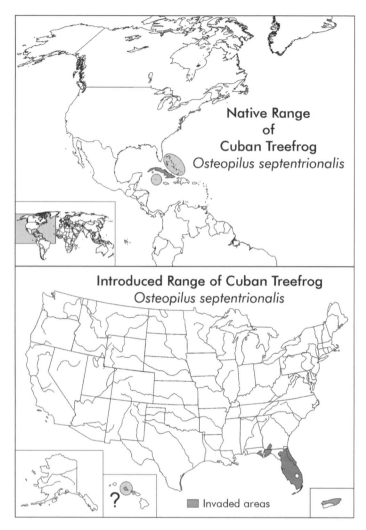

Cuban treefrogs secrete a sticky substance that can irritate the skin and mucus membranes of people. The burning and itching sensation that results can last for an hour or more. Wash your hands after handling one.

Related or Similar Species. In Florida, two native species could be mistaken for Cuban treefrogs. The green treefrog (*Hyla cinera*) has a smooth, bright-green skin. It has a pointed snout and a clear white line along each side of the body. The breeding call consists of a bell-like ringing, repeated many times. Maximum length is about 2.5 in. (6 cm). Squirrel treefrogs (*Hyla squirella*) are much smaller, adults being only about 1.5 in. (4 cm) long. They, too, can change their color, but are usually are muddy green on the back; belly and feet are light and spotted. Like all indigenous treefrogs, they have a single, rounded, balloon-like vocal sac. The breeding call sounds like a duck, while the rain call sounds like a chattering squirrel.

Introduction History. Cuban treefrogs were in Key West, Florida, sometime before 1928. Most likely they arrived as accidental hitchhikers in shipping crates from the Caribbean. By

Top: The native range of the Cuban treefrog consists of Cuba, Isla de Pinos, the Bahamas, and the Cayman Islands. Bottom: Cuban treefrogs are established in Florida and Puerto Rico and possibly on O'ahu in Hawai'i. Individual treefrogs have been collected in a number of other states. (Adapted from Somma 2009.)

the mid-1950s, they had spread through the keys to Key Largo. The first specimen from mainland Florida was collected in 1951. By the 1970s, they were being found throughout southern Florida, and in 2007, breeding populations occurred as far north as Cedar Key on the Gulf side of the peninsula and Jacksonville on the Atlantic side. Individuals have been reported in Georgia and South Carolina. Cuban treefrogs disperse as stowaways on horticultural plants (especially palm trees), building materials, cars, trucks, and boats. They respond to tropical storms and hurricanes by immediately breeding and dispersing. It is expected that they will continue to spread along the Gulf coast and then move southward into Mexico. Northward expansion into the United States seems to be checked by cooler climatic conditions.

Habitat. Cuban treefrogs thrive in both natural and man-made habitats. In Florida, they can be found in swamps and on hardwood hammocks as well as in pine forests. In

A. The skin on the head of the Cuban treefrog is fused to its skull. (Mike Pingleton, University of Illinois at Urbana-Champaign, Bugwood.org.) B. The very large toe pads are characteristic of this treefrog, which often invades homes. (Wayne Smaridge.)

residential areas, they are most frequently seen near ornamental ponds and near outside lighting, where they wait on walls, windows, porches, and potted plants to catch insects. When they get into homes, they are often encountered in the toilet. They also inhabit orange groves and plant nurseries. Active at night, by day they hide in tight, moist areas such as cellars, cisterns, drains, and plants.

Diet. These frogs are ambush predators that feed mainly on insects, spiders, and several of Florida's native treefrogs. They are also cannibalistic. Toads, lizards, and small snakes are additional items reportedly consumed by Cuban treefrogs.

Tadpoles are omnivores and may consume the eggs and tadpoles of native frogs and even those of their own kind.

Life History. The breeding season in Florida extends from May to October. Breeding is stimulated by warm summer rains. Breeding events usually last only one night. Eggs are deposited as a thin floating sheet in warm quiet waters of just about any size that lack predatory fish. Their gelatinous films of eggs can be found in ponds, ditches, abandoned swimming pools, and the stagnant waters of discarded containers. There is no parental care. Tadpoles hatch about two days later. The new hatchlings have a body length of about 0.05 in. (1.2 mm) and a tail about 0.15 in. (3.8 mm) long. The tadpoles metamorphose into small froglets in 3–8 weeks; the actual rate is determined by water temperature. At metamorphosis, they are about 1 in. (26–28 mm) long. Males mature when they are about 1.5 in. (40 mm) long; females mature more slowly and therefore are much larger than males at maturity. Mature males participate in all breeding events and typically do not live as long as females, which do not participate in every breeding event. Mature males may only live two months, whereas mature females may survive more than two years.

Impacts. Cuban treefrogs could negatively affect native treefrog species through predation and/or competition for food and space. Anecdotal evidence suggests they have replaced green frogs and squirrel frogs in residential areas. But mostly these frogs are a nuisance in and around homes and other buildings. Their calls, though not particularly loud, can be annoying. As they feed and defecate on windows and walls, they create unsightly messes. They enter homes on house plants, by jumping through an open window or door, or by making their way into bathroom vent pipes. They end up in the toilet, a startling surprise for whomever raises the lid. They are also known to have clogged sink drains. Perhaps the most serious feature of these invaders is the irritating secretions from their skin, which can cause strong reactions from people with asthma or allergies.

Management. Around the home, Cuban treefrogs can be caught by hand or attracted into traps of PVC pipe and then, when it is determined the culprit really is this exotic invader, humanely euthanized. Potential breeding sites can be eliminated by removing any containers that could collect water. Small-mesh aquarium nets can be used to scoop out eggs and dump them on the ground to dry out.

Other methods to prevent the spread or breeding of the Cuban treefrog include screening cisterns and fumigating imported plants. Monitoring its dispersal and taking quick action to eradicate new populations may slow its spread.

Selected References

Johnson, Steve A. "The Cuban Treefrog (*Osteopilus septentrionalis*) in Florida." University of Florida, IFAS Extension, 2007. http://edis.ifas.ufl.edu/uw259.

National Biological Information Infrastructure (NBII), Comité français de l'UICN (IUCN French Committee), and IUCN SSC Invasive Species Specialist Group (ISSG). "*Osteopilus septentrionalis* (Amphibian)." ISSG Global Invasive Species Database, 2008. http://www.issg.org/database/species/ecology.asp?si=1261&fr=1&sts=&lang=EN.

Somma, Louis A. "*Osteopilus septentrionalis*." USGS Nonindigenous Aquatic Species Database, Gainesville, FL, 2009. Revised February 11, 2009. http://nas.er.usgs.gov/queries/FactSheet.asp?speciesID=57.

■ Reptiles

■| Brown Anole

Also known as: Cuban brown anole, Bahamian brown anole
Scientific name: *Norops sagrei*
Synonym: *Anolis sagrei*
Family: Polychrotidae

Native Range. Cuba; the Bahamas; Cayman Brac; Little Cayman; Swan Island, Honduras.

Distribution in the United States. Established throughout peninsular Florida and in coastal counties of southernmost Georgia. Populations are reported from Hawai'i, Louisiana, North Carolina, South Carolina, and Houston, Texas; but establishment in those states is uncertain.

Description. This is a small anole lizard adapted for life on the trunks of trees and on the ground. It has long toes with relatively small toe pads that enable it to run swiftly and jump. The snout is short and rounded. It is distinguished as an anole by its dewlap, the red-orange or yellow throat fan. This is much larger in males than females and, when not extended, appears as a pale line on the throat. A crest along the neck and back can be erected. The somewhat laterally compressed tail sometimes bears a crest-like ridge. The stripe down the middle of the back is often boldly patterned with waves, zigzags, or diamonds in females, but it is indistinct in males. Body color varies from pale gray to dark brown and even black; mottling, spots, chevrons, or light-colored lines may be present. Brown anoles can change their skin color in a matter of minutes in response to environmental stimuli, aggression, or reproductive activity.

Adult males have a snout-vent length (SVL) of about 2.3 in. (6 cm) and weigh about 0.25 oz. (6–8 g). Females are usually less than 2 in. (5 cm) snout-vent, and weigh 0.1 oz. (3–4 g). The tail is normally longer than the body, giving a total length of 5–8 in. (13–21 cm); but it can be broken off and regrown, in which case it will be smaller.

Brown anoles are active during the day and commonly seen head-down on a low perch waiting to ambush prey or bobbing their heads and flashing their dewlaps in territorial and mating displays. When startled, these alert and quick lizards run away on the ground.

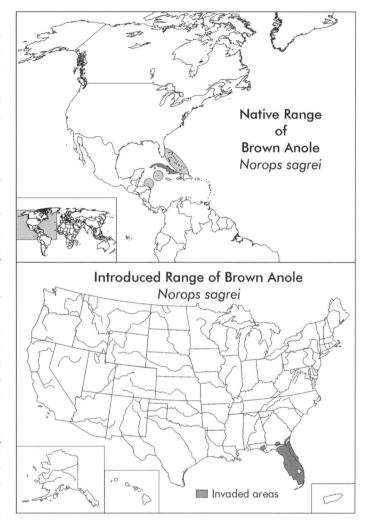

Top: The brown anole's native range consists of Cuba, Isla de Pinos, the Bahamas, Cayman Brac, Little Cayman, and Swan Island, Honduras. Bottom: Brown anoles have established populations throughout peninsular Florida and in some of the southernmost counties of Georgia. Although reported from a number of other states, they currently do not seem to have established populations elsewhere. (Adapted from Savannah River Ecology Laboratory, n.d.)

Related or Similar Species. The only anole native to the temperate southeastern United States is the green anole (*Anolis carolinensis*). It tends to be smaller and more delicately built than the brown anole. The head and snout are more slender. It is often bright green, but can change color to a brown that might lead one to mistake it for a brown anole.

In the Miami area, eight other introduced anoles might be confused with the brown anole. This is especially true of the bark anole (*A. distichus*), which is smaller and strictly arboreal.

Brown anoles are sometimes erroneously referred to as chameleons because of their ability to change color, but true chameleons are Old World reptiles with prehensile tails, mitten-like feet, and eyes mounted on turrets that move independently of each other.

A. The brown anole has a short, rounded snout. Its long toes with small toe pads enable it to run quickly and jump, adapting it for life in the trees and on the ground. (John Anderson/iStockPhoto.) B. The brown anole is typically seen in a head-down position on the lower trunks of trees. The large dewlap identifies it as an anole. (David Sischo/iStockPhoto.)

Introduction History. The brown anole was first reported from the Florida Keys in 1887. In the 1940s, it was introduced in at least six different ports in Florida. Most probably, it was a stowaway on ships. Both major subspecies, the Cuban brown anole (*N. sagrei sagrei*) and the Bahaman brown anole (*N. sagrei ordinatus*), arrived on the mainland, but through subsequent interbreeding, they have lost their genetic distinctiveness. The brown anole has continued to expand its range northward, hitchhiking on motor vehicles and in ornamental plants.

Habitat. The brown anole is a habitat generalist that prefers open vegetation in moist areas such as forest edge, disturbed sites, and the edge habitats created in urban and suburban settings. They spend much of their time on the ground or on trees just a few feet above the ground. This is a tropical to subtropical species; its northern limits seem to be controlled by cold winters.

Diet. Brown anoles are predators that feed on a wide range of animals, including annelid worms, amphipods, isopods, moths, crickets, beetles, flies, spiders, snails, and small vertebrates. They also eat the hatchlings of the green anole and probably of their own species.

Life History. Brown anoles begin to establish breeding territories in March and April and defend them for another 5–6 months. Within each male's territory may be two or more female territories. Males fight intensely to ward off other males, locking jaws and knocking one another off their perches. The dewlaps are displayed as part of the aggression ritual and as courtship behavior to attract females. Females lay a single round egg at time, once every week or two, throughout the breeding season. The egg is deposited under moist decaying leaf litter on the ground. It takes 2–3 months for them to hatch; the first usually

Anoles and Evolution

Anoles in general are excellent dispersers, and their spread and subsequent evolution around the Caribbean has made them the subject of much research. In many respects they are comparable to Darwin's finches on the Galapagos Islands and the honeycreepers of the Hawaiian Islands. They diversified into more than 300 species by adapting morphologically and behaviorally to a variety of niches. The brown anole itself has been used as a model for understanding ecology, animal behavior, and evolution. Recent work with brown anoles observed natural selection in action.

emerge in June. Recent hatchlings are 0.6–0.7 in. (15–18 mm) long, SVL. Both males and females mature the following summer, and most die during the subsequent winter when about 18 months old.

Impacts. The spread of the nonnative brown anole often appears to coincide with the decline of the native green anole, but much of the evidence is anecdotal. It has been demonstrated, however, that where the two species coexist, the green anole shifts its spatial niche higher into the canopy of trees and seldom utilizes ground perches. When the brown anole is absent, the green anole is a ground-trunk species. To what degree predation by brown anoles on green anole hatchlings affects populations of the latter is unknown. In places where the brown anole thrives, it can become the most numerous reptile in the area. Indeed, it is now one of the most abundant lizards, and possibly the most abundant vertebrate, throughout Florida.

Management. No attempts to control or eradicate brown anoles have been made. Any measures taken would likely prove fruitless since the lizard is so prolific and such a generalist.

Selected References

"Brown Anole." Lizards of Georgia and South Carolina, Savannah River Ecology Laboratory, Herpetology Program, University of Georgia, n.d. http://www.uga.edu/srelherp/lizards/anosag.htm.

Campbell, T. "The Brown Anole (*Anolis sagrei* Dumeril and Bibron 1837)." Institute for Biological Invasions: The Invader of the Month, February 2001. University of Tennessee, Knoxville, 2002. http://invasions.bio.utk.edu/invaders/sagrei.html.

Casanova, L. "*Norops sagrei*" Animal Diversity Web, University of Michigan Museum of Zoology, 2004. http://animaldiversity.ummz.umich.edu/site/accounts/information/Norops_sagrei.html.

National Biological Information Infrastructure (NBII) and IUCN/SSC Invasive Species Specialist Group (ISSG). "*Norops sagrei* (reptile)." ISSG Global Invasive Species Database, 2006. http://www.issg.org/database/species/ecology.asp?si=603&fr=1&sts=&lang=EN.

■| Burmese Python

Scientific name: *Python molurus bivittatus*
Family: Boidae

Native Range. Southern Asia. Found from northeastern India and Bangladesh, through Myanmar, Thailand, Laos, Cambodia, and Vietnam, into southern China. Also a number

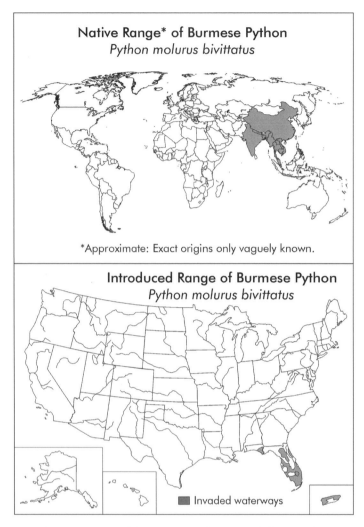

Top: The native range of the Burmese python extends from northeastern India through Southeast Asia into southern China. (Adapted from Barker and Barker 2008.) Bottom: Burmese pythons are invasive in southern Florida, including Everglades National Park and Big Cypress Preserve. They are also established on Puerto Rico. (Adapted from USGS 2007.)

of isolated populations in Nepal; Sichuan, China; and Java, Bali, and Sumbawa, Indonesia, that may represent relics of a formerly much larger distribution area or may represent introductions by humans.

Distribution in the United States. Established in southern Florida, including Everglades National Park and Big Cypress Preserve, and also in Puerto Rico. They can be found outside Everglades National Park along its eastern boundary and increasingly in more distant areas, such as Collier-Seminole State Park and Manatee County. Several individuals have been observed on Key Largo, but at the moment, they do not seem to be established in the Florida Keys.

Description. One of the world's largest snakes, the Burmese python may grow to lengths greater than 25 ft. (7.6 m) and weights close to 300 lbs. (137 kg). The largest found to date in the Everglades was 16 ft. (4.8 m) long and weighed 152 lbs. (69 kg). Females become longer and heavier than the males, which are distinguished by their larger cloacal spurs, the two projections on either side of the vent that may be vestigial hind limbs. The light-colored skin of this heavy-bodied snake is beautifully marked with a mosaic of large reddish-brown blotches that are outlined in black. The pattern begins on the large rectangular head as an arrow-shaped patch. The pet trade has bred a number of variations, including albino, that may be found in wild Florida populations, since they all stem from released or escaped pets. Burmese pythons are semiaquatic but also climb trees.

Related or Similar Species. Two large native pit vipers might be mistaken for young pythons. The eastern diamondback rattlesnake (*Crotalus adamanteus*) is also heavy-bodied and reaches lengths up to 7 ft. (2.1 m), but its head is triangular and the yellow-and-brown diamond pattern down its back is distinctive. Some Florida cottonmouths (*Agkistrodon piscivorus*) have a brown-and-black banded pattern and may reach lengths of

A. The Burmese python, one of the world's largest snakes, is beautifully marked with large reddish-brown blotches outlined in black. (Vassil.) B. The arrow-shaped patch on its squarish head is charactersitic. (Vassil.)

6 ft. (1.8 m). However, this snake has a triangular head and a dark cheek stripe that runs through the eye. Both of the pit vipers are venomous. Another brightly patterned snake that could be confused with a small python is a colubrid and a constrictor, the corn or red rat snake (*Elaphe guttata guttata*). It is common in residential areas and can reach lengths of 6 ft. (1.8 m). It has a spear- or arrow-shaped pattern on the head. Typically, the body is orange-brown with red or brown blotches outlined in black. The belly is marked with a black-and-white checkerboard, and two black stripes appear on the underside of the tail.

Introduction History. Burmese pythons were first collected from the wild in southern Florida in the 1980s. They were first deemed established in Everglades National Park in 2000 based on collections made along Main Park Road in the 1990s. Since then, the number of pythons in the park has increased dramatically and is now estimated to be 10,000 or more. The first python in the Florida Keys was documented in 2007 at Key Largo Hammock Botanical State Park. Pythons are well able to the swim the six miles from the Everglades to Key Largo.

Burmese pythons in Florida derive from the pet trade. They were either intentionally or accidentally released by pet owners. Burmese python hatchlings continue to be imported into the United States from Southeast Asia by the thousands, and domestic breeding of different color morphs also continues.

Habitat. Burmese pythons can live in a variety of open habitats such as swamps, marshes, grasslands, and woodlands, including the brackish glades and mangroves at the south end of the Everglades. In Florida, they generally inhabit the same places that alligators do. These excellent swimmers are usually close to water, which they must enter before shedding, and can stay underwater without breathing for as long as 30 minutes.

Diet. The Burmese python is a carnivore that prefers live prey but will consume carrion. Rodents and other small mammals are dietary staples, but it also takes amphibians, reptiles, and birds. Researchers in Florida have found alligators, limpkins, ibises, coots, House Wrens, rabbits, squirrels, mice and rats, muskrats, raccoons, Virginia opossums, bobcat, and white-tailed deer, as well as domestic cats and geese, in the digestive tracts of pythons captured in the wild.

Other Big Snakes in Florida

The Burmese python is perhaps the most publicized invasive snake established in Florida; however, it is not the only giant snake on the loose. A population of boa constrictors (*Boa constrictor*) inhabits a county park and possibly adjacent areas in southeastern Miami. (It is difficult to know if individuals found beyond the park are dispersers from the park population or recently released or escaped pets.) In 2010, researchers reported establishment of another large constrictor, the Northern African python (*Python sebae*), in a small area southeast the Tamiami Trail at State Route 997. Other constrictors with free-living individuals in Florida, but apparently not reproducing populations, include the green anaconda (*Eunectes murinus*), the yellow anaconda (*E. notaeus*), the reticulated python (*Broghammerus reticulates*) and the white-lipped python (*Leiopython albertisii*).

Sources: Reed, R. N., K. L. Krysko, R. W. Snow, and G. H. Rodda. "Is the Northern African Python (*Python sebae*) established in southern Florida?" *IRCF Reptiles & Amphibians* 17(1): 52–54, 2010.

"Northern African Python, African Rock Python [Non-Native]," Florida Museum of Natural History, http://www.flmnh.ufl.edu/herpetology/fl-guide/Pythonsebae.htm.

The python has poor eyesight and hunts by using its sense of smell. It also has heating sensing pits along the upper lip with which it detects the body heat of its prey. It may either stalk or ambush its victims. Prey items are killed by constriction and swallowed whole.

Life History. Burmese pythons are solitary except when they come together to mate between December and April. Then groups of one female and several males often form. Males locate the females chemically, detecting the pheromones secreted by the female. The male wraps himself around the female's body and internal fertilization takes place. Two to three months later, in May and June, the female lays a clutch of 12–36 eggs (up to 100), each weighing about 7 oz. (200 g). She incubates the eggs for two months by coiling on top of them. The mother rarely if ever leaves her eggs and will raise their temperature above ambient temperature by shivering.

The eggs hatch 2–3 months later, in July and August, and parental care ends. The hatchlings are 18–24 inches long and grow rapidly if food is abundant. Both males and females are sexually mature in 2–3 years and at lengths of about 8 ft. (2.6 m). Average lifespan is 15–25 years.

Impacts. Burmese pythons have been in Florida's Everglades too brief a time for definitive assessments of their impacts. The concern is that they will compete with native species and further endanger rare species upon which they prey. Among their known prey are two wading birds of special concern, the Limpkin (*Aramus guarauna*) and White Ibis (*Endocemus albus*). On Key Largo, prey included the endangered Key Largo woodrat (*Neotoma floridana smalli*). They may also may be preying upon native mangrove fox squirrels (*Sciurus niger avicennia*) and Wood Storks (*Mycteria americana*). Due to an overlap in diet, they could compete with the state and federally threatened eastern indigo snake (*Drymarchon couperi*) for food and space.

With the population explosion that pythons in the Everglades are currently experiencing, they could become a major ecological problem and hamper efforts to restore the greater

Everglades ecosystem. Their future establishment is a major concern in the biologically sensitive Florida Keys.

Pythons pose a threat to small children and pets in residential areas and to small livestock and poultry in agricultural areas.

Management. Once pythons are established, it becomes impossible to eradicate them. Efforts therefore focus on controlling numbers and preventing new infestations. The main strategy involves tracking, capturing, and euthanizing. In southern Florida, volunteer members of an Eyes & Ears Team spot pythons on roads and call in a person trained to capture and dispose of them. Among the people trained to find snakes are mail carriers and package delivery drivers who daily drive the region's roads. In March and April 2010, a special python hunting season was instituted in Florida. Permit holders could also take African pythons and Nile monitors in this effort to kill as many large exotic reptiles as possible. Research continues to learn the ways of the python and monitor its spread.

Recent legislation in Florida requires owners of pythons and some other large introduced reptiles to buy a permit and implant a microchip in the animal to identify its owner. A federal bill, the Non-native Wildlife Invasion Prevention Act (H.R. 669) that came before the Congress in 2009–2010, would prohibit the importation of species determined by the U.S. Fish and Wildlife Service to likely become invasive in the United States.

Selected References

Austin, Jill. "Stopping a Burmese Python Invasion." The Nature Conservancy, 2009. http://www.nature.org/wherewework/northamerica/states/florida/science/art24101.html.

Barker, David G., and Tracy M. Barker. "The Distribution of the Burmese Python, *Python molurus bivittatus.*" *Bulletin Chicago Herpetological Society* 43(3): 33–38, 2008. Available online at http://www.vpi.com/sites/vpi.com/files/Barkers.pdf.

Harvey, Rebecca G., Matthew L. Brien, Michael S. Cherkiss, Michael Dorcas, Mike Rochford, Ray W. Snow, and Frank J. Mazzotti. "Burmese Pythons in South Florida: Scientific Support for Invasive Species Management." Publication #WEC242, IFAS Extension, University of Florida, 2008. ttp://edis.ifas.ufl.edu/pdffiles/UW/UW28600.pdf.

National Biological Information Infrastructure (NBII), Puerto Rico Department of Natural and Environmental Resources, and IUCN SSC Invasive Species Specialist Group (ISSG). "*Python molurus bivittatus* (Reptile)." ISSG Global Invasive Species Database, 2010. http://www.issg.org/database/species/ecology.asp?si=1207&fr=1&sts.

Padgett, J. "*Python molurus.*" Animal Diversity Web, University of Michigan, 2003. http://animaldiversity.ummz.umich.edu/site/accounts/information/Python_molurus.html.

"*Python molurus bivittatus.*" USGS Nonindigenous Aquatic Species Database, Gainesville, FL, 2009. Revised October 24, 2007. http://nas.er.usgs.gov/queries/FactSheet.asp?speciesID=2552.

■| Green Iguana

Also known as: Common iguana, gallina de palo
Scientific name: *Iguana iguana*
Family: Iguanidae

Native Range. Mexico south through Central America to Ecuador on the Pacific side and southeastern Brazil on the Atlantic slope of South America. Also in the Lesser Antilles (Curacao, Grenada, St. Lucia, St. Vincent, Trinidad and Tobago, and Utila). Most iguanas imported as pets into the United States come from captive farming operations in Honduras, El Salvador, Colombia, and Panama.

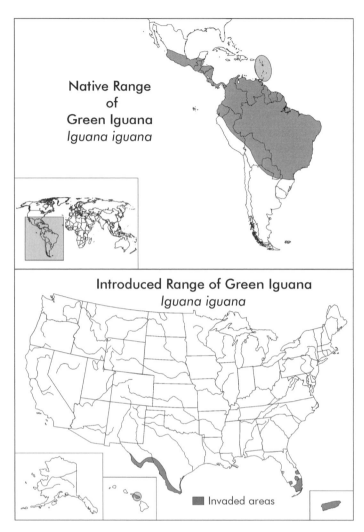

Distribution in the United States. Established in southern Florida. Feral populations are also established on Maui, Hawai'i; in the Rio Grande Valley, Texas; on Puerto Rico; and on the U.S. Virgin Islands.

Description. The green iguana is one of the largest lizards found in the United States. Only young animals are bright green; adults assume a uniform, grayish-green color, which they can alter somewhat in response to social and environmental cues. A month or so before courtship begins, the males acquire a bright orange wash on the neck and forelimbs that persists through the mating period. Color may also change diurnally, becoming darker when the body is cold so as to absorb more solar energy.

A crest of large spines along the back and tail is particularly well developed on males, as is the very large dewlap under the throat. The fleshy dewlap is used in threat and courtship displays, but also helps absorb and dissipate heat. The tail is long and tapering, and ringed with broad dark stripes. The tympanum is prominent and covered by a membrane. Large scales appear on both sides of the head; beneath the tympanum is a large rounded scale

Top: Green iguanas are native to southern Mexico, Central America, and tropical South America. They also are native to the Lesser Antilles. (Adapted from "Green Iguana," The Wild Ones Animal Index, http://www.thewildones.org/Animals/iguana.html.) Bottom: Released or escaped pets, green iguanas have established populations in southern Florida, on Mau'i in Hawai'i, in the Rio Grande Valley of Texas, and on Puerto Rico.

known as a subtympanic plate. The eyes are set on the sides of the head and covered by an immobile upper lid and a movable lower lid. On the midline of the top of the head is a white, light-sensing organ known as the parietal "eye." It helps coordinate photoperiod with reproductive status, but also detects the shadows of overhead predators. Sharp teeth are positioned along the inner sides of both jaws. Nasal glands help expel excess salts.

Adult males may achieve a total length of 6 ft. (1.8 m) and weight of as much as 17.5 lbs. (8 kg), although 9–12 lbs. (4–6 kg) is more common in Florida specimens.

Related or Similar Species. Some other large introduced lizards might be mistaken for young green iguanas. The black spinytail iguana (*Ctenosaura similis*) also has a crest of spines along the back, but has black banding on the body. Spiny scales circle around the tail. The young are bright green like young green iguanas, but have broken bands of black around the mid-body. The largest of these lizards may be 3–4 ft. (approximately 1 m) in total length. The brown basilisk (*Basiliscus vittatus*) has a large crest behind the head and a brown body; it only grows up to 2 ft. (60 cm) long. Knight anoles (*Anolis equestris*) and Jamaican giant anoles (*A. garmani*), are both green but only about 1 ft. (30 cm) long.

Introduction History. Baby green iguanas are imported into Florida by the tens of thousands each year in the pet trade. They were first reported outside captivity in the Miami area in 1966, but populations probably did not become established until the 1980s, when they became popular pets. Mature green iguanas become powerful animals that can bite, scratch, and whiplash owners with their tails. When they become too much to handle, they are often released into the wild. The earliest sightings of free-living iguanas were in Key Biscayne, Hialeah, and Coral Gables, and in the vicinity of Miami International Airport. Undoubtedly, these reptiles were intentional releases by pet owners or accidental escapes. Since that time, they have expanded northward on the Atlantic coast, establishing populations in Broward County by 2001 and in Palm Beach County by 2003. Along the Gulf coast, iguanas were reported as established in Monroe County (Coral Reef State Park) in 1995 and in Lee County (Fort Myers and Cape Coral) sometime in the 1990s. Increasing numbers are seen in Martin and St. Lucie counties, but the iguana does not yet seem to be established in those areas. Iguanas have also been reported in Pinellas County on the Gulf coast. Expansion much farther north seems to be halted by irregular freezes. Green iguanas are relatively common in the Everglades, and individuals have been reported as far south in the Keys as Stock Island.

Habitat. In Florida, the green iguana occurs in frost-free coastal areas. They thrive only where temperatures range from 79° to 95°F (24–32°C) and where sunlight is sufficient to let their bodies produce vitamin D. They are mostly arboreal and prefer large trees overhanging water, into which they dive to escape predators. Therefore, they are likely to be near canals, ditches, and ponds, and in mangroves lining the shores of shallow bays. Iguanas also inhabit urban and suburban yards and like to bask on tree branches, sidewalks, docks, and seawalls, or in open, mowed areas.

Diet. Green iguanas are herbivores and feed on leaves, flowers, and fruits. Juveniles consume insects and other animal matter in addition to plant foods. Digestion occurs in the hindgut, and juveniles eat the droppings of adults in order to acquire the microflora necessary to break down plant material. Among preferred food plants are hibiscus, orchids, roses, garden greens, and squashes and melons.

Life History. Green iguanas breed during the dry season so that the young hatch during the wet season when food will be most plentiful. Iguanas of both sexes come together in sandy areas to mate during a single nesting season each year. Dominant males scent-mark territories and females with pheromones secreted from femoral pores on the undersides of their thighs. Fertilization is internal. About 65 days after mating, each female excavates a burrow 1.5–3 ft. (0.5–1.0 m) deep and, over a period of three days, deposits 20–70 eggs in it. If nesting space is scarce, several females will use the same nest. No parental care is given to the nest or eggs. Eggs measure roughly 0.6 in. (15 mm) in diameter and 1.5 in. (35–40 mm) long. Hatchlings emerge 10–15 weeks later, usually during July and August, and look like miniature females without spiny crests. They are 6.5–10 in. (17–25 cm) in total length. The juveniles remain in family groups for a year; curiously, the young males protect the females from predators. Green iguanas become sexually mature at 3–4 years of age. The reproductive life of a female lasts several years. After mating, she is able to store sperm for several years and fertilize her eggs even if a male is not present.

A. The green iguana has a crest of long spines and a very large dewlap. (Susan Woodward.) B. This large lizard has a long tail ringed with dark bands and a prominent external ear drum. (Susan Woodward.)

Impacts. Threats to native lizards are probably slight, since adult green iguanas are herbivores and all native species are carnivores. Green iguanas sometimes use the burrows of the Florida Burrowing Owl (*Athene cunicularia floridana*), a federally endangered subspecies, and could prevent the birds from nesting or destroy their eggs and nestlings. Iguanas eat the native butterfly sage (*Cordia globosa*), a plant of special concern in Florida, and consume yellow nickerbean (*Caesalpinia bonduc*), a primary host for the larvae of the endemic and endangered Miami blue butterfly (*Hemiargus [=Cyclargus] thomasi bethunebakeri*). They are potential dispersers of any invasive plants upon which they feed.

In residential areas, iguanas can destroy vegetable gardens and ornamental plants and pose a health hazard by spreading salmonellosis when they defecate in pools or on docks and sidewalks. Their droppings are also unsightly and smelly.

In Puerto Rico, basking green iguanas are a runway hazard at the Luis Muñoz Marín International Airport and must be shooed away before planes can take off.

Management. Iguanas can be removed from private property without special permits. They can be caught legally in Florida by hand, noose pole, nets, or live traps. It is illegal, however, to release captured animals; so they must be kept as pets or captive breeding stock or be destroyed. Feral adults rarely tame down. The best control measures are to discourage iguanas by removing protective cover from the yard, sheathing trees with metal to prevent their climbing up to sun bathe, not planting their favorite food plants, and protecting valued plants in screened enclosures.

In open country, it is very difficult to capture iguanas because they are such strong swimmers and dive into water at the first sign of danger. Sometimes in cold weather, when they are sluggish, they can be picked off branches or off the ground. Local iguana populations along canals and in mangroves have been reduced using boats to collect those knocked off branches when temperatures are in the 40s.

In Florida, iguanas are protected only by anticruelty laws. In Hawai'i, strict laws regulate the importation and possession of green iguana, and violations can lead to fines of $200,000 and up to three years in jail.

Selected References

Gibbons, Whit, Judy Greene, and Tony Mills. *Lizards and Crocodilians of the Southeast.* Athens: University of Georgia Press, 2009.

Gingell, F., Biology of Amphibians and Reptiles, and J. Harding. "*Iguana iguana*" Animal Diversity Web, University of Michigan Museum of Zoology, 2005. http://animaldiversity.ummz.umich.edu/site/accounts/information/Iguana_iguana.html.

"Green Iguana." Wikipedia, 2009. http://en.wikipedia.org/wiki/Green_Iguana.

"Green Iguana—*Iguana iguana.*" Florida's Exotic Wildlife. Species detail. MyFWC.com, Florida Fish and Wildlife Conservation Commission, n.d. http://myfwc.com/wildlifehabitats/nonnatives/reptiles/green-iguana/.

Kern, W. H., Jr. "Dealing with Iguanas in the South Florida Landscape." Publication #ENY-714, IFAS Extension, Univeristy of Florida, 2009. http://edis.ifas.ufl.edu/in528.

Masterson, J. "*Iguana iguana*: Green iguana." Smithsonian Marine Station at Fort Pierce, 2007. http://www.sms.si.edu/irlspec/Iguana_iguana.htm.

National Biological Information Infrastructure (NBII) and IUCN/SSC Invasive species Specialist Group (ISSG). "*Iguana iguana* (Reptile)." ISSG Global Invasive Species Database, 2006. http://www.invasivespecies.net/database/species/ecology.asp?si=1022&fr=1&sts.

■ | Nile Monitor

Also known as: Money monitor
Scientific name: *Varanus niloticus*
Family: Varanidae

Native Range. Much of Africa between 15° N and 15° S. It occurs in forests and savannas near permanent water. It is absent from desert regions, but has been found at elevations greater than 6,500 ft. (2,000 m).

Distribution in the United States. Known to be established in at least two counties in southern Florida: Lee County (Cape Coral) and adjacent Charlotte County. Sightings of Nile monitors are reported from several other counties, including Broward, Collier, DeSoto, Miami-Dade, Orange, and Palm Beach.

Description. This is the largest lizard now occurring outside of captivity in the United States and the longest lizard in its native Africa. Nile monitors can reach SVLs averaging

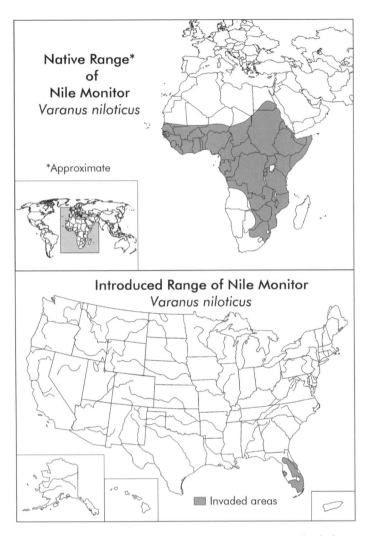

Top: The Nile monitor is native to forests and savannas in much of Africa. (Adapted from map at http://en.wikipedia.org/wiki/File:Nile_monitor_range.PNG.) Bottom: The Nile monitor is established in at least two counties in southern Florida and has been reported from several others. (Adapted from Somma 2007.)

23.5–31.5 in. (60–80 cm) and total lengths over 7.5 ft. (2.4 m). They may weigh more than 22 lbs. (10 kg). Like all monitors, they have long necks, long tails, and an obviously forked tongue. The head is narrow and wedge-like and shorter than the neck. The tongue is blue. The background color is gray-brown or olive green. The jaws and head bear cream-colored stripes that graduate into chevrons on the neck. Six to nine bands of yellow dots encircle the body. Said to look like coins, these dots are the origin of one of the African names for this species, the "money monitor." Juveniles are more brightly patterned than adults. Nile monitors have large, strong claws for digging and a muscular tail that is laterally compressed for swimming and usually 1.5 times as long as the body.

Related or Similar Species. The Asian water monitor (*Varanus salvator*) looks a bit like the Nile monitor and grows even longer, reaching lengths of 10 ft. (3 m). It lacks the bands of "coins" adorning the Nile monitor. As far as it is known, water monitors are not living free in Florida, but they are easily obtained through the pet trade, and it is entirely possible that they will become established in the state. Some reported observations of individual Nile monitors may actually have been escaped or released water monitors.

Introduction History. Free-living Nile monitors were first spotted at Cape Coral in 1990. Their origins are unknown but presumed to be related to the pet trade. One rumor has it that a pet store that went bankrupt in the 1980s released their Nile monitors into what was then a wilderness of saw palmettos, slash pine, and mangrove. Another story suggests that wholesalers of exotic animals purposefully released monitors to establish a breeding population from which they could capture young animals to sell. It is perhaps most likely that the animals came from pet owners who found that the cute little hatchlings they had purchased had turned into large, temperamental reptiles they no longer wanted or could manage, so they "humanely" released them. The animals are also quite able to escape from cages on their own. As the cape was developed, monitors came into more frequent contact with residents and the number of observations increased, especially along canals in the western part of the peninsula. By 2002, monitors had crossed Matlacha Pass onto Pine Island. Sanibel Island to the south lies within reach. Since some populations in Africa are adapted to temperate climate, monitors from those regions, if introduced to the United States, theoretically could spread throughout Florida and the Gulf states and possibly as far north as the Carolinas.

The Nile monitor has a narrow wedge-shaped head and a laterally compressed tail for swimming. (Dr. Gordon E. Robertson.)

Habitat. Nile monitors are well adapted to both terrestrial and aquatic habitats. They prefer cover near permanent bodies of water. In Florida, favorable conditions exist in mangrove swamps, along the edges of both freshwater and saltwater marshes, and on the banks of rivers and canals. In Lee County, they also live in pine flatwoods, on golf courses with ponds, and in suburban neighborhoods and urbanized areas. They shelter in burrows that they dig into canal banks or in those excavated by other animals. In Africa, they live in a variety of climates at different elevations and will hibernate during cold weather in more temperate regions.

Diet. Nile monitors are indiscriminate carnivores and scavengers. They will consume invertebrates such as cockroaches, mangrove tree crabs, snails, and clams; amphibians, including the introduced Cuban treefrog (see Amphibians, Cuban Treefrog); lizards, snakes, baby alligators, and turtles, as well as the eggs of reptiles; birds and bird eggs; and small mammals such as rodents and cats. They also eat carrion, garbage, and feces. Young monitors are more arboreal than mature adults and feed almost exclusively on fast-moving invertebrates and lizards. With increasing age, they consume less active prey that is more armored and must be crushed by strong jaws and teeth before being swallowed. These intelligent animals will hunt cooperatively.

Life History. What information is available on life history comes from studies done in the monitor's native habitats in Africa and may or may not be directly applicable to populations in Florida. In the Sahel region of Africa, females are sexually mature when they are two years old and have a SVL of about 14 in. (36 cm). They breed every other year, so during any given breeding season, only 50 percent of adult females lay eggs. In some areas, females feed and accumulate fat in spring and summer. Eggs are laid during the winter dry season, buried in the ground and in active termite mounds. Clutches average 35 eggs or more, with smaller females producing many fewer than larger ones. Each egg is about 3.5 × 1.6 in. (6 × 4 cm). Hatchlings emerge 6–10 months later, near the beginning of the wet season. They are 6–12 in. (15–32 cm) in total length and will more than double in size during the first year of life. Nile monitors can live more than 10 years in the wild.

Impacts. Nile monitors are potentially the most destructive lizard introduced to Florida. Nile monitors in Lee County, Florida, could have negative effects on native crocodilians, (the American alligator [*Alligator mississippiensis*] and, especially, the American crocodile [*Crocodylus acutus*]), for in Africa, Nile monitors not only have diets similar to those of crocodiles, but also are the major predators of crocodilian eggs and hatchlings. Their propensity to occupy existing burrows raises concern for two protected species on the Cape Coral peninsula, the Burrowing Owl (*Athene cunicularia*) and the gopher tortoise (*Gopherus polyphemus*), whose nesting burrows could be taken over and eggs and young consumed. If Nile monitors were to invade Sanibel Island, the large rookeries of pelicans and herons would be threatened. If they gain access to Sanibel's beaches, the nests of sea turtles would be vulnerable to their depredations. Should Nile monitors make their way into southeastern Florida and the Keys, a number of listed species could become prey, including the southeastern beach mouse (*Peromyscus polionotus niveiventris*), Key Largo cotton mouse (*Peromyscus gossypinus allapaticola*), Key Largo woodrat (*Neotoma floridana smalli*), and silver rice rat (*Oryzomys palustris argentatus*).

When cornered, Nile monitors become very defensive and attack with their teeth, sharp claws, and strong tails. Humans and pets are at risk of serious wounds that can become septic due to the bacteria in monitors' mouths. Small dogs and domestic cats are also at risk since they can be caught easily and eaten by monitors. Reports from Cape Coral suggest a decline in pet populations as the monitor population increased.

Management. Once established, Nile monitors are very difficult if not impossible to eradicate. Population reduction and prevention of further infestation are the main management strategies. Monitors may be captured with nooses or live traps or by digging them out of the ground. Arboreal hatchlings can be caught at night by hand. Regulations can keep monitors from being purchased in the first place by casual, untrained pet owners.

Selected References

Enge, K. M., K. L. Krysko, K. R. Hankins, T. S. Campbell, and F. W. King. "Status of the Nile Monitor (*Varanus niloticus*) in Southwestern Florida." *Southeastern Naturalist* 3(4): 571–82, 2004.

Gore, Jeff, Kevin Enge, and Paul Moler. "Nile Monitor (*Varanus niloticus*) Bioprofile." MyFWC.com, Florida Fish and Wildlife Conservation Commission, n.d. http://www.scribd.com/doc/26904049/NILE-MONITOR-Varanus-Niloticus-Bioprofile-Compiled-By.

Kruse, Michael. "Nile Monitor Lizards Invaded Florida and They're Winning the Battle." *St. Petersburg Times*, June 21, 2009. Available online at http://www.tampabay.com/news/environment/wildlife/article1011745.ece.

"Nile Monitor." Florida's Exotic Wildlife, MyFWC.om, Florida Fish and Wildlife Conservation Commission, Tallahassee, 2007. http://myfwc.com/wildlifehabitats/nonnatives/reptiles/nile-monitor/.

Somma, Louis A. "*Varanus niloticus*." USGS Nonindigenous Aquatic Species Database, Gainesville, FL, 2009. Revised August 28, 2007. http://nas.er.usgs.gov/queries/FactSheet.asp?speciesID=1085.

Youth, Howard. "Florida's Creeping Crawlers." *ZooGoer* 34(3), Smithsonian National Zoological Park, 2005. http://nationalzoo.si.edu/Publications/ZooGoer/2005/3/reptilefeature.cfm.

■ Birds

■ | Cattle Egret

Also known as: Buff-backed Heron

Bubulcus ibis

Family: Ardeidae

Native Range. Parts of Africa and Asia and southern Spain and Portugal.

Distribution in the United States. This bird has been reported in all states, but breeding takes place primarily in the southeastern United States as far north as Virginia in the east and Kansas in the center of the country. Resident populations can also be found in southernmost California, Hawai'i, and Puerto Rico.

Description. The Cattle Egret is a small white heron about 20 in. (46–56 cm) long with a wingspan of about 36 in. (88–96 cm). It is stocky and has a thick neck, shorter than its body length. When standing still, it typically assumes a hunched position. Most of the year its plumage is white, but for a short time during the mating season, orange-buff feathers occur on the crown, throat, and back. The bill is relatively short and yellow, and the legs yellow to gray-green during the non-breeding season, but both turn red during the breeding season.

In flight, the Cattle Egret holds its neck tight to the body. Cattle Egrets are gregarious and fly to and from feeding areas in flocks. When it walks, it tends to sway in an exaggerated strut and then suddenly dart forward to stab its prey. Most often, Cattle Egrets forage alongside grazing animals, waiting for the livestock to flush insects. They may also pick parasites off large herbivores and be seen standing on the animals' backs.

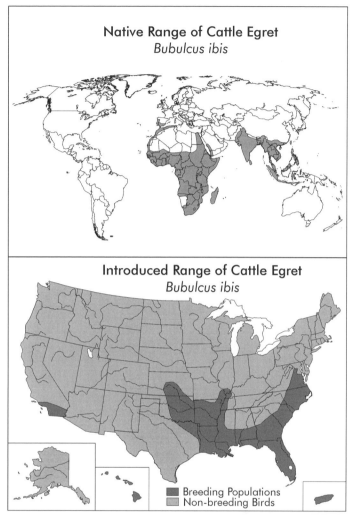

Top: The Cattle Egret's vast native range spans much of non-desert Africa, the southern Iberian peninsula of Europe, and South and Southeast Asia. (Adapted from map at http://wapedia.mobi/en/File:Ardea_ibis_map.svg.) Bottom: Cattle Egrets can be found in all states, but breeding populations occur primarily in the southeastern states, southern California, Hawai'i, and Puerto Rico. (Adapted from map by Cornell Lab of Ornithology. http://www.allaboutbirds.org/guide/Cattle_Egret/lifehistory.)

Cattle Egrets are usually silent. Their voice consists of a low, nasal "rik-rak."

Related or Similar Species. The Cattle Egret is smaller than any native herons or egrets that have white plumage. It might be confused with the somewhat taller Snowy Egrets (*Egretta thula*) and juvenile Little Blue Herons (*Egretta caerulea*). Snowy Egrets are distinguished by their black legs and yellow feet, as well as a black bill. They are adorned with long, lacy white plumes on the neck and back during the breeding season. Juvenile Little Blue Herons have dull, greenish legs and a pale gray or greenish bill. Both of these birds

A. The Cattle Egret is a small, stocky heron. (Robbie Taylor/Shutterstock.) B. Cattle Egrets commonly hunt insects flushed by cattle or other large mammals. (Donna Beeler/Shutterstock.)

are associated with freshwater, where they feed on fish and aquatic invertebrates; whereas the Cattle Egret is generally found in upland pastures and feeds mostly on insects.

Introduction History. The introduction of the Cattle Egret to the Americas appears to have been a natural event. It suddenly arrived in Suriname, South America, in the late 1870s and 1880s, presumably having simply flown across the Atlantic Ocean from the African continent. It spread throughout South America, filling a vacant niche as a unique terrestrial member of the heron family. By 1917, it was in Colombia, and in the 1940s, it showed up in Florida. In the United States, its long-distance wandering habits let it rapidly expand its range. By the 1960s, it was in California as well as Canada.

In contrast to the natural range expansion that carried the Cattle Egret to all of the continental United States, the bird was deliberately introduced to Hawai'i in 1959. Local ranchers funded efforts by the Hawaiian State Board of Agriculture and Forestry to establish Cattle Egrets as a biological control for flies and other insects that were cattle pests. Twenty-five egrets were released at Kipu Ranch on Kaua'i, and 105 others were distributed among ranches on Hawai'i, Maui, Moloka'i, and O'ahu. The Honolulu Zoo also received some birds. The egrets began nesting soon after release and established large populations on Kaua'i and O'ahu, where they came to attain pest status.

Habitat. Cattle Egrets commonly inhabit pastures, marshes, and ploughed fields. They thrive in altered habitats and tolerate busy roadsides and urbanized areas well. Most often, when feeding, they associate with cattle or other livestock. Cattle Egret roosts are usually in trees or shrubs near water. They are colonial nesters and frequently nest in dense rookeries with other herons and egrets.

Diet. Insects are the mainstay of the Cattle Egret diet. They consume live insects flushed by grazing mammals, but are also known to follow tractors, plows, lawn mowers, and even airplanes to snatch insects disturbed by the machines. Most of their food consists of grasshoppers, crickets, flies, and beetles, but when insects are not abundant, they will take spiders, certain moths, frogs, and crayfish as well as bird eggs and nestlings. They also scavenge in refuse for edible leftovers.

Life History. Males establish breeding territories within large colonial nesting areas from spring through early summer. Courtship displays attract females, and pair-bonds are established that last the season. Bulky nests of sticks are built by the female from materials carried to her by the male. Nest-building and mating commonly last only three days. The breeding colors are lost as soon as mating is over. When egg-laying begins, one pale blue egg is produced every other day. The eggs are not brooded until the last egg is laid. Clutches usually consist of 34 eggs. Both adults incubate the eggs for approximately 24 days. The eggs hatch in the order in which they were laid; usually only the first two survive to fledge. The siblings begin to compete with each other for food when less than a week old. Two to three weeks after hatching, they leave the nest to climb around the rookery but continue to beg for food from the parents. They begin to fly a week or so later and become independent about 2.5 months after hatching. Juveniles disperse hundreds of miles in apparently random directions. Even as adults, the egrets are highly migratory and wander widely. Cattle Egrets become members of the breeding population at 2–3 years of age.

Impacts. In most instances, Cattle Egrets have little or no impact on native heron species. They utilize different habitats for feeding, have a different diet, and breed after the nesting seasons of native birds. Some concern does remain that they could potentially displace native herons and egrets in rookeries, since they occur in such large numbers. Cattle Egrets are now more numerous in North America than all other herons and egrets combined.

In Hawai'i, the birds are not quite as benign as on the mainland. They are known to feed on the eggs and young of endangered wetland birds such as the Black-necked Stilt or A'eo (*Himantropus mexicanus*) and could compete with such insect-eating species as frogs, toads, and skinks. Furthermore, they have become a nuisance for aquaculture on Oahu because they feed on prawns and are a hazard at airports in Honolulu, Lihue, and Hilo.

Large rookeries anywhere can become nuisances due to noise, odor, and potential public health threats.

Management. Population reduction and control efforts in Hawai'i are likely to have only temporary effects, since the birds are so mobile and will return to areas from which they have been removed. Techniques that repel Cattle Egrets may have some success at airports and nuisance rookeries. Elsewhere, the bird is usually not viewed as a problem and may even be welcomed as a way of helping to control insect pests of cattle.

Selected References

Ivory, A. "*Bubulcus ibis.*" Animal Diversity Web, University of Michigan Museum of Zoology, 2000. http://animaldiversity.ummz.umich.edu/site/accounts/information/Bubulcus_ibis.html.

Masterson, J. "*Bubulcus ibis* (Cattle Egret)." Species Report, Smithsonian Marine Station at Fort Pierce, 2007. http://www.sms.si.edu/irlspec/bubulcus_ibis.htm.

National Biological Information Infrastructure (NBII) and IUCN/SSC Invasive Species Specialist Group (ISSG). "*Bubulcus ibis* (bird)." ISSG Global Invasive Species Database, 2006. http://www.issg.org/database/species/ecology.asp?fr=1&si=970&sts.

"The Cattle Egret." Hawaii Nature Focus, Booklet No. 10, Kilauea Point Natural History Association, n.d. http://www.kilaueapoint.org/education/naturefocus/hnf10/index.html.

■| Common Myna

Also known as: Indian Myna, House Myna
Scientific name: *Acridotheres tristis*
Family: Sturnidae

Native Range. Southern Asia from southeastern Iran and Afghanistan through Pakistan, India, Nepal, and Sri Lanka to Southeast Asia and southern China.

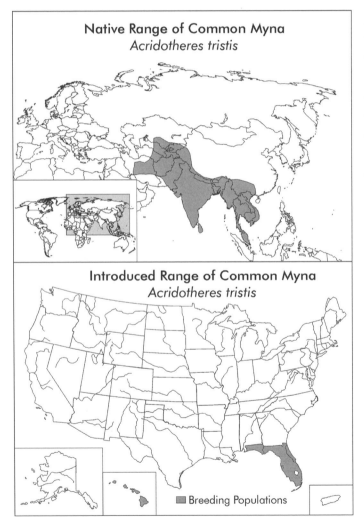

Top: The native range of the Common Myna stretches across southern Asia from southwest Iran to Vietnam. (Adapted from "Common Mynas aka Indian Mynas aka Talking Mynas." AvianWeb, 2006.) Bottom: Common Mynas are established in southern Florida and on all the major islands of Hawai'i. (Adapted from "Common Myna *Acridotheres tristis*" 2003.)

Distribution in the United States. Established on all the major islands of Hawai'i and in southern Florida.

Description. This member of the starling family is a stocky brown bird with a glossy black head and neck, yellow bill, and yellow legs. A patch of bare yellow skin occurs behind the eye. Adults are 9.0–9.8 in. (23–25 cm) long and have a wingspan of 18 in. (46 cm). In flight, the white tips of tail feathers and white wing patches are clearly visible. Immatures have duller colors and browner heads than adults. Common Mynas have a large repertoire of raucous calls, squeaks, clicks, and whistles. Both males and females sing and often fluff their feathers and bob their heads when vocalizing. At dawn and dusk they engage in loud choruses and at their communal roosts keep up a noisy chatter well after dark.

Common Mynas strut across the ground, rather than hopping as many birds do. They tend to travel in pairs.

Related or Similar Species. Common Mynas are related to the European Starling (see Birds, European Starling) and share some of their annoying behaviors. The Hill Myna (*Gracula religiosa*) is the more common pet talking myna. It is larger than the Common Myna and has a black body with fleshy yellow wattles on the head that extend back from below the eyes to the nape. The bill is red or orange. This mostly arboreal bird has escaped captivity and may be encountered in urban gardens and parks in Miami, Florida, and in several large cities in southern California and Hawai'i.

A bare patch of yellow skin behind the eye distinguishes the Common Myna. The white wing patch is more conspicuous in flight. (K. W. Bridges, "Common Campus Birds," University of Hawai'i at Manoa. http://www.botany.hawaii.edu/biology101/birds/common_myna.htm)

Introduction History. The Common Myna was brought to Hawai'i from India in 1865 by Dr. William Hildebrand in an effort to control army worms (*Spodoptera mauritia*) and army cutworms (*Euxoa auxiliaris*) that were devastating the islands' sugar cane and pastures. They did help control cutworms, but also adapted well to urban life and were abundant in Honolulu by 1879. Eventually, they inhabited all the main islands.

Common Mynas' ability to "talk," i.e., mimic human voices, and their general intelligence made them part of the pet trade. Mynas in Florida most likely derive from escapes or intentional releases of pet birds. The first sighting and confirmed breeding of the Common Myna in the state was in Miami in 1983. It has since been reported from 19 other counties within a 300-mi. (480 km) radius of that city and may have spread as far north as Sapelo Island, Georgia. Although the bird has dispersed rapidly, its numbers have not exploded, and small scattered populations seem to be the rule.

Habitat. Common Mynas prefer open country such as farmland and suburban and urban parks and yards. The small populations in Florida apparently prefer shopping mall parking lots. In their native range, they inhabit open lowland woodlands and the edges of settlements and roost in isolated stands of tall trees. They are cavity nesters, but are not restricted to holes in trees, building their bulky nests in any nook or cranny available.

Diet. Common Mynas are omnivorous. They take insects—especially grasshoppers, small vertebrates, and carrion, feeding mostly on the ground. Its generic name *Acridotheres* means "grasshopper hunter." Mynas also feed on grains and fruits and sometimes on the eggs and nestlings of other bird species. In Florida, they are known to beg for French fries at fast food establishments.

Life History. Common Mynas begin to build their nests of grass, leaves, and twigs in late February or early March. Both parents construct the nest, which may be placed in tree cavities, crevices in buildings, martin houses, and the tops of coconut and date palms, and aggressively defend their nesting territories. Between March and July, the female lays 2–5 blue eggs, and she and the male both incubate them. Chicks hatch in 13–18 days. They fledge when 3–4 weeks old. Parental care continues for another month while the young learn to forage for themselves. A female may produce 1–3 clutches a year. Both sexes reach maturity at one year of age.

Juveniles form small flocks once they become independent of their parents. Adults forage in loose groups of 5–6 birds. During the nonbreeding season, Common Mynas will roost together in huge flocks numbering more than 1,000 birds.

Impacts. The noise and droppings from large flocks of Common Myna can make them a nuisance in urban and suburban settings. However, due to the large numbers of insects they consume and their bold antics, many people like this bird. In Hawai'i, they serve as a reservoir for avian malaria (see Microorganisms, Avian Malaria) and are thus implicated in the decline of endemic island birds. They disperse the seeds of lantana (*Lantana camara;* see Volume 2, Shrubs, Lantana)—which is an invasive weed in both Hawai'i and Florida—and prey upon the eggs and nestlings of native songbirds and seabirds such as the Wedge-tailed Shearwater (*Puffinus pacificus*). In Florida, they compete with Purple Martins for nest sites, but do not congregate in the large flocks typical elsewhere, so are not as problematic as in Hawai'i.

The Common Myna has been introduced to many parts of the world, including many islands in the Pacific. It is such an agricultural pest in fruit- and wheat-growing areas (especially in Australia) that the IUCN has included the Common Myna on its nomination list for "100 of the World's Worst invasive alien species."

Management. Although the Common Myna has not become the pest it has in other parts of the world, its importation into the United States is now prohibited.

Selected References

"Common Myna *Acridotheres tristis*." Florida's Breeding Bird Atlas: A Collaborative Study of Florida's Birdlife, 2003. http://legacy.myfwc.com/bba/docs/bba_COMY.pdf.

"Common Myna—*Acridotheres tristis*." Florida's Nonnative Wildlife. Species detail. MyFWS.com, Florida Fish and Wildlife Conservation Commission, 2009. http://myfwc.com/wildlifehabitats/nonnatives/birds/common-myrna/.

"Common Mynas aka Indian Mynas aka Talking Mynas." AvianWeb, 2006. http://www.avianweb.com/commonmynas.html.

IUCN/SSC Invasive Species Specialist Group (ISSG). "Acridotheres tristis (Bird)." ISSG Global Invasive Species Database, 2006. http://www.issg.org/database/species/ecology.asp?si=108&fr=1&sts.

Lin, T., and T. Root. "*Acridotheres tristis*." Animal Diversity Web, University of Michigan Museum of Zoology, 2007. http://animaldiversity.ummz.umich.edu/site/accounts/information/Acridotheres_tristis.html.

Pranty, Bill. "Common Myna (*Acridotheres tristis*)." In Pranty, B., J. L. Dunn, S. C. Heinl, A. W. Kratter, P. E. Lehman, M. W. Lockwood, B. Mactavish, and K. J. Zimmer, "Annual Report of the ABA Checklist Committee: 2007–2008," 37. *Birding* 40: 32–38, 2008. Available online at http://www.aba.org/birding/v40n6p32.pdf.

"The Common Myna." Nature Focus Booklet #5. Kilauea Point Natural History Association, n.d. http://www.kilaueapoint.org/education/naturefocus/hnf5/index.html.

■ Eurasian Collared-Dove

Scientific name: *Streptopelia decaocto*
Family: Columbidae

Native Range. South Asia. Prior to the 1600s, the native range was apparently restricted to India, Sri Lanka, and Myanmar. Its range later expanded by natural means and through deliberate introductions to Turkey and southeastern Europe, and by the end of the twentieth century, it was found throughout Europe.

Distribution in the United States. Southeastern and central United States; Los Angeles region of California. The range is rapidly expanding, however, and it may soon be in all 49 continental states.

Description. Like several native doves, the Eurasian Collared-Dove is a medium-size bird with short legs and a small head. The body is pale gray with a slightly pinkish wash over the head and breast. The slender bill is black, the legs and feet are mauve, and the eyes are red. A black line forms a distinct "collar" on the back of the neck; above it is a white line. The square tail is white underneath. The sexes look alike. Adults are 12–13 in. (30–33 cm) long and have a wingspan of 18–22 in. (45–55 cm). Average weight is 7 oz. (200 g).

Juveniles are similar to adults except that their feathers have pale reddish edges and their eyes are brown. The legs are a darker brownish red. Until they are three months old, they display no clear collar.

Eurasian Collared-Doves have a rhythmic "*coo COOO cup*" resembling that of the native Mourning Dove (*Zenaida macroura*), but lower in pitch. During display flights, they produce a harsh, nasal "*krreew*" call. The wings do not whistle when the bird takes flight.

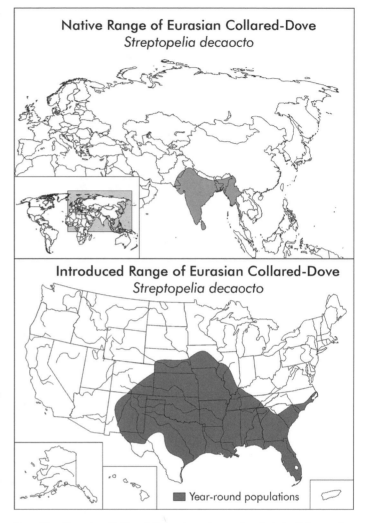

Top: The Eurasian Collared-Dove is native to South Asia. Prior to the 1600s, it seems to have been restricted to the area shown on the map, but since that time its range has expanded through Turkey to all of Europe. Bottom: The range of the Eurasian Collared-Dove is rapidly expanding in the United States. Currently found in southeastern and central states, it may eventually invade all of the continental United States. (Adapted from Sibley 2000.)

Related or Similar Species. The native Mourning Dove is smaller and lighter and has a long pointed tail. When it takes flight, the wings produce a whistling sound. The native White-winged Dove (*Zenaida asiatica*) is about the same size as the Eurasian Collared-Dove and has a square tail, but the tail is dark underneath with a white edge. Its white wing patch is distinct at rest and in flight. Neither native species has a black collar.

Closely related and very similar in appearance to the Eurasian Collared-Dove is the somewhat smaller, introduced Ringed Turtle-Dove (*Streptopelia risoria*) for which it was initially mistaken. The turtle-dove is a feral bird of extremely limited distribution in the United States. It is

The Eurasian Collared-Dove is pale gray with a distinct black "collar" on the back of the neck. (Gregg Williams/Shutterstock.)

distinguished from the Eurasian Collared-Dove by its very light color and white tail and undertail coverts as well as its call.

Introduction History. The Eurasian Collared-Dove came to the New World as a cage bird. It first became established in the wild in the Bahamas in 1974 when birds escaped from a breeder. They are excellent colonizers, as their history in Europe demonstrates; and by the 1980s, they had expanded their range into southern Florida. By 2009, they were well established in the southeastern United States, especially along the Atlantic and Gulf coasts; isolated populations were established in the southern Plains states and southern Rocky Mountain states as well as in southern California. It has been reported in Alaska and in the Great Lakes region, but these may represent local escapes of captive birds. In some or the disjunct areas, the dove has been deliberately released for sport hunting or represents an accidental introduction. At least some of these populations are expected to serve as nuclei from which further range expansion will occur.

Habitat. Eurasian Collared-Doves are most abundant in coastal, agricultural, and suburban habitats. They forage in open areas and tend to avoid forested areas and areas that are intensively cultivated. Nests are constructed in trees and shrubs and on man-made structures.

Diet. These birds are primarily seed and fruit eaters, but also consume small invertebrates. They generally forage on the ground but will eat at bird feeders.

Life History. Eurasian Collared-Doves are monogamous and have long breeding seasons; they may breed year-round in warm climates. In Florida, a pair may raise three or more broods each year. Each clutch consists of two eggs, the first laid being significantly larger than the second. Both parents incubate the eggs until they hatch about 15 days after laying. Hatching is asynchronous, with the second egg hatching 12–40 hours after the first. The young are fed regurgitated crop milk; as they get older, seeds become part of the nestlings' diet. Juveniles fledge at about 18 days of age and are independent of the parents when 30–40 days old. The young are sexually mature by their first spring.

In the nonbreeding season, the birds congregate at communal roosts. Eurasian Collared-Doves live a relatively long time; the oldest bird on record was more than 13 years old.

Impacts. When they occur in large numbers, Eurasian Collared-Doves can deter other birds from using bird feeders and may even aggressively defend the food source. Large flocks also become a noise problem and produce large amounts of unsightly droppings. They can transmit the parasite *Trichomonas gallinae* to native doves using the same feeders or birdbaths and to hawks that prey upon them. Potentially, they could damage crops, but such impacts have not been reported.

Management. Eradication is no longer a possibility. Numbers may be controlled by hunting. As an introduced species, it is not protected by law; but state and local hunting regulations still apply.

Selected References

Johnson, Steve A., and Gay Donaldson-Fortier. "Florida's Introduced Birds: Eurasian Collared-Dove (*Streptopelia decaocto*)." Document WEC256, Department of Wildlife Ecology and Conservation, University of Florida/ Institute of Food and Agricultural Services, 2009. http://edis.ifas.ufl.edu/uw301.

National Biological Information Infrastrucutre, IUCN French Committee, and IUCN SSC Invasive Species Specialist Group. "*Streptopelia decaocto* (bird)." ISSG Global Invasive Species Database, 2008. http://www.issg.org/database/species/ecology.asp?si=1269&fr=1&sts=&lang=EN.

Sibley, David Allen. "Eurasian Collared-Dove." In *National Audubon Society, The Sibley Guide to Birds*, 256. New York: Alfred A. Knopf, Inc., 2000.

Sibley, David Allen. "Ringed Turtle-Dove." In *National Audubon Society, The Sibley Guide to Birds*, 256. New York: Alfred A. Knopf, Inc., 2000.

■ | European Starling

Also known as: Common Starling
Scientific name: *Sturnus vulgaris*
Family: Sturnidae

Native Range. Europe. In lowlands from the United Kingdom westward across Ukraine and Central Asia to western China. The Starling is a summer visitor in the northern part of its range where the climate is continental. Migratory birds winter in the Mediterranean region in Spain, Portugal, North Africa, and across the Middle East into Pakistan.

Distribution in the United States. The European Starling is common in all of the contiguous 48 states and is also found in Hawai'i. In Alaska, resident populations occur around Anchorage and the Kenai Peninsula; breeding birds summer in the Yukon River valley.

Description. European Starlings are robin-size birds that at a distance appear to be black. They have short, squared tails, pointed wings, and long, thin, pointed bills. The feathers of nonbreeding males and females are glossy black with white tips, giving the bird a spotted appearance. Bills are grayish-black, and the eye is dark-colored. Legs are a dull red. During the breeding season, the feathers become iridescent purple and green, and most of the spots vanish. The male sports long feathers on the breast. The bill of both sexes becomes yellow when they are in breeding condition, but males have a blue spot at the base of the beak whereas females have a reddish pink spot. Juvenile birds have uniformly gray-brown plumage and brownish-black bills.

Full grown Starlings are 8–9 in. (20–23 cm) long, have a wingspan of 12–16 in. (31–40 cm), and weigh 2–3 oz. (60–90 g). Their calls tend to be harsh squeaks, whistles, and gurgles, but they are good mimics of the calls of other birds and can imitate other sounds in the environment as well.

Starlings walk; they do not hop. In flight, they resemble tiny fighter planes, with their triangular wings and short tails. Nonbreeding birds gather in huge flocks and perform spectacular mass aerial displays, especially at dusk as they ready themselves to settle at a roosting site.

Related or Similar Species. European Starlings often occur in mixed flocks with other "blackbirds" with which they could be confused. The Common Grackle (*Quiscalus quiscala*) is larger, has a more elongated body and a proportionately much longer tail. Its iridescent feathers are not speckled, and its bill is never yellow. The eye, however, is bright yellow. The male Brown-headed Cowbird (*Molothrus ater*) is roughly the same size as a Starling,

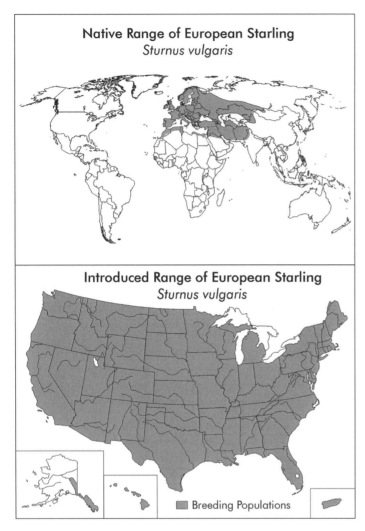

Native Range of European Starling
Sturnus vulgaris

Introduced Range of European Starling
Sturnus vulgaris

■ Breeding Populations

Top: the European Starling is native to Europe and western Asia. (Adapted from map at http://en.wikipedia.org/wiki/File:Sturnus_vulgaris_map.png.) Bottom: The European Starling is found in all 50 states and Puerto Rico. (Adapted from "European Starling" [map]. Cornell Lab of Ornithology. http://allaboutbirds.org/guide/european_starling/id.)

but its head is distinctly brown, the feathers are not spotted, and the bill is conical in shape and always dark brown in color. Male Brewer's Blackbirds (*Euphagus cyanocephalus*) also have glossy black plumage with a green or dark-blue sheen in the breeding season. They are similar in size to Starlings and have straight bills, but have bright yellow eyes.

Introduction History. The arrival of the European Starling in North America traces to late nineteenth-century attempts by acclimatization societies to create a familiar landscape for European immigrants. The American Acclimatization Society, under the leadership of Eugene Scheiffelin, released 80–100 Starlings into New York's Central Park in 1890–1891. Scheiffelin's goal was to have all the birds mentioned in the works of William Shakespeare brought to the United States. (He also brought over the House Sparrow from England. See Birds, House Sparrow.) In *Henry IV*, Hotspur says, "I'll have a Starling . . . to speak nothing but 'Mortimer.' "

Several attempts to introduce Starlings were made from 1850 to 1900, but the first success came with the release of 60 birds in Central Park in 1890. Fifteen pairs survived. The following year, another 40 were set free. During the first 10 years, the population was contained in the greater New York City area. Thereafter, European Starlings rapidly increased their numbers and expanded their range north, south, and west. They reached Alaska by 1970. Within 75 years of introduction, Starlings had dispersed across the entire continent.

Habitat. Starlings are commonly seen in urban areas and other disturbed sites. They feed on the ground in open areas with short grass, including suburban lawns and city parks. They also infest feed lots and agricultural areas, where they feed on grains and other crops. They may fly great distances from roosting sites in tree tops, under bridges, or in other open structures to their feeding grounds. They are secondary cavity-nesters: they do not excavate their own nest sites, but occupy those made by other birds, or mammals or humans. They

A. European Starlings are about the size of a robin but walk, rather than hop. (Steve Byland/Shutterstock.)
B. Nonbreeding birds assemble in huge flocks and perform what appear to be well-coordinated mass aerial displays. (Vasily A. Ilyinsky/Shutterstock.)

are not fussy about location and will build nests in holes in trees, crevices in buildings, rain gutters, under the eaves of roofs, in roof vents and attics, and in birdhouses.

Diet. European Starlings are omnivores. They consume a variety of invertebrates, including earthworms, beetles, grasshoppers, spiders, and snails, and also feed on seeds, berries, and fruits such as apples, pears, and cherries. By inserting their bills into the ground or into food items and then opening their beaks, they can pry fruits open to extract any seeds or insects inside. Favorite foods reportedly include the berries of poison ivy and Virginia creeper, blackberries, mulberries, and elderberries.

Life History. Breeding pairs form and begin to select nesting sites in late winter or early spring. The nests are constructed of dried grasses and other materials and typically fill the nest cavity. The female lays a clutch of 4–6 blue eggs. She may produce two or three clutches during the breeding season, which extends into July. The eggs incubate for 15 days, and hatchlings remain in the nest another 21–23 days. Both parents incubate eggs and feed the young birds. Fledglings follow their parents and beg for food for several days after leaving the nest. When they become independent, the juveniles congregate in flocks with other young Starlings. At the end of the breeding season, the parents again become gregarious and spend the nonbreeding season feeding and roosting as members of large flocks.

Impacts. Much of the trouble associated with Starlings comes from their being nuisances due to the fact that they congregate in such large, noisy flocks. In cities, their acidic guano can coat buildings and sidewalks and corrode statues and the paint on cars. Diseases such as salmonella and histoplasmosis can proliferate at established roosts. Droppings have the potential to contaminate animal feed and water sources. Large flocks are able to inflict serious damage to crops such as grain, grapes, olives, and cherries. Starlings massing near airport runways pose a real danger to arriving and departing airplanes because they can clog engines, damage planes, and, when an entire flock collides with a plane, can cause planes to crash.

The ecological impacts of the European Starling in the United States are related to their aggressive takeovers of nest cavities sought or occupied by other bird species. Declines of woodpeckers, martins, tree swallows, and bluebirds have been blamed on the presence of Starlings. Starlings are also implicated in the spread of the seeds of exotic weeds.

The European Starling has also been introduced to South Africa, Australia, and New Zealand. It has been nominated by the Global Invasive Species Programme as one of the 100 "World's Worst" invaders.

Management. Direct population reductions may be accomplished through poisoning with Starlicide Complete, trapping, or shooting, but most efforts try to repel or exclude Starlings. At airports, livestock facilities, and some urban roosting sites, frightening the birds with noises, including recordings of their own distress calls, has had at least temporary success. Barricading entrances to cavities with screens or other covers will eliminate nesting sites. Habitat modification involving the removal of water sources and foraging sites has had some success at airports.

Selected References

Adeney, Jennifer Marion. "European Starling (*Sturnus vulgaris*)." Introduced Species Summary Project, Columbia University, 2001. http://www.columbia.edu/itc/cerc/danoff-burg/invasion_bio/inv_spp_summ/Sturnus_vulgaris.html.

Johnson, Steve A., and Walter Givens. "Florida's Introduced Birds: European Starling (*Sturnus vulgaris*)." Document WEC255, Department of Wildlife Ecology and Conservation, University of Florida/IFAS, 2009. http://edis.ifas.ufl.edu/UW300.

Withers, David Ian. "Origins of the European Starling in the United States." Tennessee Department of Environment and Conservation, 2000. http://www.state.tn.us/environment/tn_consv/archive/starlings.htm.

■ | House Finch

Also known as: Linnet
Scientific name: *Carpodacus mexicanus*
Family: Fringillidae

Native Range. Western North America. Prior to the late nineteenth century, House Finches were found in the southwestern states. They expanded their range northward and were first reported in the Columbian Basin of Oregon in 1885. Dam construction and irrigation projects apparently facilitated their spread into eastern Washington State in the early twentieth century; they were found in the western part of the state by the 1950s. Today, naturally occurring populations inhabit the drier habitats west of the Rocky Mountains from southern British Columbia to southern Mexico. Originally occupying undisturbed desert scrub, desert grasslands, chaparral, oak savannas, and low elevation open coniferous forests, they adapted to human-dominated environments and moved into suburban and agricultural areas. They also crossed the Rockies onto the High Plains.

Distribution in the United States. House Finches are now common birds throughout the eastern United States and have also been introduced into Hawai'i.

Description. House Finches are small, sparrow-like songbirds. Adults are 5–6 in. (13–14 cm) long and have wingspans of 8–10 in. (20–25 cm). The male usually has a rosy-red forehead, stripe over the eye, breast, and rump. The brown back has dark brown streaks, and light-brown streaks appear on the flanks and belly. Some males are orange or even yellow; plumage color depends on diet. Females are brown with a plain brown head, finely streaked underparts, and two pale wingbars. They have no eye stripe. Both sexes have squarish, slightly notched brown tails and blunt, rounded beaks. Juveniles look like females; males acquire their color in their second spring.

House Finches have a melodic warbling song with a few harsh notes and a downward trend.

Related or Similar Species. The most similar bird in the eastern United States is a congener, the Purple Finch (*Carpodacus purpureus*). Purple Finches are somewhat more robust. Males have more red on their heads and breast than House Finches and lack brown streaks on their flanks and bellies. Females have broad white stripes over and under the eye, a larger beak, and bolder striping on breast and belly than female House Finches. Female House Sparrows (see Birds, House Sparrow) could also be mistaken for female House Finches. The former have light brown stripes on the back and unstreaked undersides. Pine Siskins (*Carduelis pinus*) are considerably smaller and more heavily streaked than female House Finches; they have yellow patches on the wings (not always obvious) and slender, pointed bills.

Introduction History. House Finches were introduced into Hawai'i in the 1880s. Finches in the eastern United States trace their origins to the 1930s, when House Finches were sold

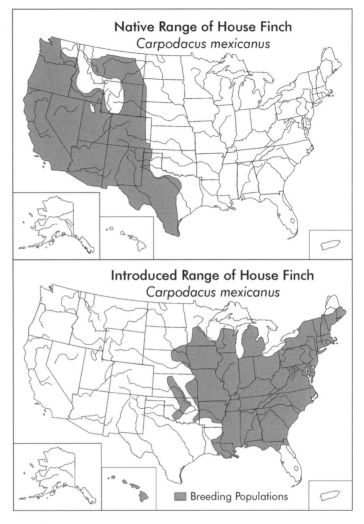

Top: The House Finch is native to the western United States, where its range has been expanding since the 1960s. Bottom: The House Finch is a native transplant and now a common songbird in the eastern half of the United States. It has also been introduced into Hawai'i. (Both maps adapted from All About Birds, 2009.)

in pet stores in New York City as "Hollywood Finches." The selling of wild birds became illegal in 1940, and some dealers apparently released their captive finches on Long Island to avoid legal action. For the first few years, the population's survival was tenuous, but by the 1950s, House Finches were established in New York City. The population then irrupted, and finches spread rapidly up and down the East Coast, probably helped by the increasing presence of bird feeders in suburban and urban areas of the United States. By the 1980s House Finches had reached the Mississippi River, and today the eastern and western distribution areas seem to have merged; House Finches are now found across the continent.

The population explosion ceased in the early 1990s with appearance of mycoplasmal conjunctivitis. This infection causes the eyelids to become red, swollen, and encrusted and can lead to blindness and starvation. The disease first appeared in the Washington, D.C.,

A. The male House Finch has a rosy-colored forehead and light brown streaks on the flanks and belly. (Stubblefield Photography/Shutterstock.) B. The female House Finch has a plain brown head and finely streaked underparts. There are two pale wing bars. (Chris Hill/Shutterstock.)

area during the winter of 1993–1994. Two years later, it had spread across the Appalachians, and by the following winter, it was in the Midwest and beginning to show up on the Great Plains. The disease is most prevalent in winter when the birds flock at feeders. By 1999, scientists were reporting that House Finch numbers were roughly 40 percent less than expected before the epidemic began. The rapid decline in House Finch populations coincided with an increase in the number of House Sparrows in some affected areas. The virulence of the disease may be abating. Current estimates are that only 5–10 percent of the population is infected, whereas at the peak of the disease, 50 percent of the eastern finch population may have been died from it. (In 2004, the disease appeared in the Pacific Northwest and has reached epidemic proportions there.)

Habitat. Where they have been introduced, House Finches inhabit backyards, urban parks, farmland, and forest edge. They congregate at bird feeders and like to perch in high trees close by. Nesting sites include ledges on buildings, shrubs, debris piles, hollows in trees, and birdhouses.

Diet. House Finches consume almost any type of seed and also eat ripening fruits, flowers, and buds. In addition to bird seed at feeders, where sunflower seeds are preferred, the finches also extract seeds from thistle, dandelion, and mistletoe and forage on the ground. The males acquire their rosy color after the postnuptial molt by eating carotenoid-rich red berries and fruits. Cherries and mulberries are among those preferred.

Life History. Pair formation begins in late winter. Females tend to choose the males with the brightest coloration. The breeding season starts in early spring when the female constructs a small cuplike nest of fine grasses, hair, or other fiber. She lays 3–6 small, pale blue eggs that are flecked with black on the larger end. The female incubates the eggs for 12–14 days, while the male feeds and guards her. The nestlings fledge when 11–19 days old. The male will feed the fledglings for several days until the female has built a new nest. Most House Finches produce at least two clutches during the breeding season. The young of the year disperse widely and congregate at food sources. Nonbreeding adults and juveniles are especially gregarious in fall and winter and may form large, mobile foraging flocks. Some in the northeastern United States migrate in winter. House Finches are mature at one year of age and may live more than 10 years.

Impacts. Most people welcome and encourage these colorful little songsters at their backyard feeders. However, as the House Finch spread through the eastern United States, dramatic declines in Purple Finches and House Sparrows took place in newly invaded areas. The range of the Purple Finch has shifted northward, confining it more and more to the conifer forests avoided by the House Finch. Part of this range shift may be due to climatic change and not solely the influence of increasing House Finch populations.

House Sparrow numbers declined wherever House Finches invaded, suggesting the latter were outcompeting the former. (Later, when finch numbers decreased, the sparrows rebounded, reinforcing this interpretation.) Recent research has also found evidence of an evolutionary response among House Sparrow populations, a phenomenon known as character displacement. At bird feeders, the sparrows prefer smaller seeds such as millets, whereas the finches prefer sunflower seeds. Where the two species both occur, sample measurements indicate that House Sparrow beak depths are smaller than before the finch invasion and they are becoming better adapted to specialize on the small seeds neglected by finches.

Since 1993, a bacterial eye infection (mycoplasmal conjunctivitis) has swept through eastern House Finch populations. There is a possibility that this disease could be spread at feeders to other wild birds, such as Blue Jays (*Cyanocitta cristata*) and American Goldfinches (*Carduelis tristis*), as well as to domestic poultry.

Management. Since most people like having these birds around, management is largely directed at reducing the infection rate of conjunctivitis among House Finches. This can be accomplished by homeowners cleaning bird feeders and disposing of old seed and accumulated bird droppings. For those who find House Finches to be a nuisance, fruits can be protected from finch depredation with plastic netting, and potential nesting sites can be eliminated.

Selected References

Cooper, Caren, Wesley Hochachka, and André Dhondt. "Why Did House Sparrow Numbers Rise, Then Fall?" *Birdscope* 21(2): 2007. http://www.birds.cornell.edu/Publications/Birdscope/Spring2007/sparrow_numbers.html.

Dewey, T., K. Kirschbaum, and J. Pappas. "*Carpodacus mexicanus*," Animal Diversity Web, University of Michigan Museum of Zoology, 2002. http://animaldiversity.ummz.umich.edu/site/accounts/information/Carpodacus_mexicanus.html.

"House Finch." All About Birds, Cornell Lab of Ornithology, 2009. http://www.allaboutbirds.org/guide/house_finch/id.

Johnson, Steve A., and Jill Sox. 2009. "Florida's Introduced Birds: House Finch (*Carpodacus mexicanus*)." Document WEC 253, Department of Wildlife Ecology and Conservation, University of Florida/IFAS. http://edis.ifas.ufl.edu/uw298.

Kammermeier, L. "Population Dynamics of the House Finch." *Birdscope*, 13(2): 15, 1999.

National Biological Information Infrastructure (NBII) and IUCN/SSC Invasive Species Specialist Group (ISSG). "*Carpodacus mexicanus* (bird)." ISSG Global Invasive Species Database, 2005. http://www.invasivespecies.net/database/species/ecology.asp?si=485&fr=1&sts.

Wootton, J. Timothy. "Ecology and Evolution of Invading and Restored Species." Department of Ecology and Evolution, University of Chicago, n.d. http://woottonlab.uchicago.edu/index/invasive-species/ecology-and-evolution-of-invasive-and-restored-species.

■ | House Sparrow

Also known as: English Sparrow
Scientific name: *Passer domesticus*
Family: Passeridae

Native Range. Eurasia and North Africa. Occurs from the United Kingdom eastward through Siberia and southeastward to the Arabian Peninsula and Indian subcontinent, but

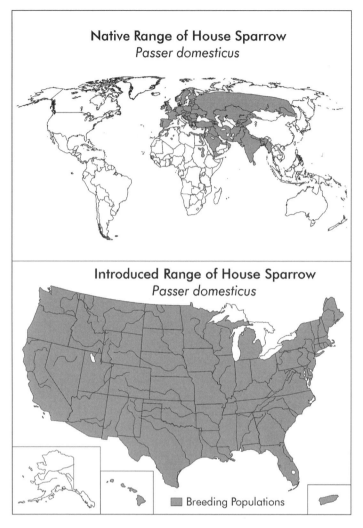

Native Range of House Sparrow
Passer domesticus

Introduced Range of House Sparrow
Passer domesticus

■ Breeding Populations

Top: The native range of the House Sparrow includes all of Europe, much of Asia, and parts of North Africa. (Adapted from map at http://en.wikipedia.org/wiki/File:PasserDomesticusDistribution.png.) Bottom: The House Sparrow is a common bird of the built environment in all 48 contiguous states, Hawai'i, and Puerto Rico.

is not native to Italy. In North Africa, it is found in coastal areas and at isolated oases in the Sahara. They have been associated with human settlements for centuries.

Distribution in the United States. Established in all 48 of the contiguous states and in Hawai'i and Puerto Rico.

Description. This Old World sparrow is a small brown bird common in towns and cities. It has a relatively large head and short wings. The legs are pink. The male has a black bib and mask, gray cap and rump, and chestnut-colored nape. The side of the head and underparts are pale gray, and the brown back is streaked with black. The conical bill is black in summer and yellowish in winter. Non-breeding and immature males have less black on the throat and breast. The female is brown all over, with tawny streaks on her back and a strong buff-colored eye line edged below by a dark brown stripe. Her throat and breast are light and unstreaked. The bill is yellow. Young birds resemble females.

House Sparrows are 5.25–6.25 in. (133–159 mm) long. In the relatively short time they have been naturalized in the United States, they have evolved in accord with regional climatic differences across North America. Northern birds are larger than southern ones; desert populations have much lighter coloration than populations in the humid east or Pacific Northwest.

The song of the House Sparrow is a series of chirps.

Related or Similar Species. Only one other Old World sparrow has become established in the United States: the European Tree Sparrow (*Passer montanus*), restricted to St. Louis, Missouri, and nearby Illinois. It can be distinguished by its rufous crown, white collar, and prominent black patch in the white cheek.

Introduction History. Some early introductions of the House Sparrow (then called the English Sparrow) were attempts to control insect pests, especially the elm spanworm (*Ennomos subsignaria*), a defoliator of shade trees. Many introductions were the results of

A. The male House Sparrow has a black bib and mask. (Francis Boose/Shutterstock.) B. The female House Sparrow is brown with tawny streaks on the back and a strong buff-colored eye line. (Stubblefield Photography/Shutterstock.)

activities of acclimatization societies; and at first, citizens tenderly cared for the birds. The first introduction into the United States consisted of eight pairs from England released in Brooklyn, New York, in 1851. More birds were imported the following year, and 50 of them were released. Neither group thrived. In 1853, birds held and nurtured in a tower in the Greenwood Cemetery were released and protected. These sparrows did survive; they multiplied, and people carried them to other towns and cities in the region. Other deliberate introductions, commonly by acclimatization societies, include those in Portland, Maine (1854), Peace Dale, Rhode Island (1858), Boston, Massachusetts (1858), and New Haven, Connecticut (1867). House Sparrows were brought to San Francisco, California, in 1875. The birds prospered in part by eating seeds in the dung of urban horses. They rapidly dispersed along carriage roads into agricultural areas, where they became pests. House Sparrows are now found in almost all populated parts of the country. Numbers are believed to have peaked in the early twentieth century before motorized vehicles replaced horses. Another period of decline followed the conversion to modern industrial agriculture in the 1960s.

Habitat. These birds are creatures of human-modified environments. They are found in cities, suburbs, and agricultural settings, but not in undisturbed forests, grasslands, or deserts. By preference cavity nesters, they build their nests in crevices in and on buildings, in the hollow posts of traffic signs, in the rafters of porches, in birdhouses, in holes in trees, or in any other sheltering location.

Diet. House Sparrows primarily forage on the ground for seeds, but also feed on fruits, insects, and garbage. It is not unusual for them to extract dead insects from the grills of automobiles and to hop about the parking lots of fast food restaurants looking for discarded food.

Life History. House Sparrows begin to nest as early as February, when monogamous pairs form. The nests, usually in small colonies, consist of dried plant material, feathers, string, and shreds of paper. The female usually lays 3–6 speckled pale blue eggs per clutch and may produce up to four clutches a year. Incubation starts after the last egg in a clutch has been laid and continues for 10–12 days. The young fledge about 14 days after hatching. They may live more than 12 years.

Impacts. Today, the House Sparrow is considered mainly a nuisance species in cities and towns when large groups create noise and messy, corrosive droppings. In the early years of their invasion of the United States, they were an agricultural pest, eating ripening grains, destroying fruits, and consuming or fouling seeds and other feed. To some extent, the negative impacts were balanced by the birds' consumption of insect pests, including cabbage worms and cotton caterpillars. House Sparrows will evict native birds from their nest cavities and were implicated in declines of the Eastern Bluebird (*Sialis sialis*), Purple Martin (*Progne subis*), and some woodpeckers during the twentieth century. Sparrow populations have declined since the advent of large monoculture farms in the 1960s. Declining numbers of sparrows in the central and eastern United States and recovery programs targeting bluebirds and martins have mitigated their effects on native birds.

Management. Habitat modifications that eliminate roosting and nesting sites can reduce local House Sparrow populations. So, too, can reduction of food sources by cleaning up garbage dumps and protecting berries and other small crops with bird netting. The birds can be repelled with noise and scarecrows. Since these nonnatives are not protected by the Migratory Bird Act, they may be trapped or shot.

Selected References

"House Sparrow." All about Birds, Cornell Lab of Ornithology, 2009. http://www.allaboutbirds.org/guide/House_sparrow/id.

Laycock, George. *The Alien Animals. The Story of Imported Wildlife.* New York: Ballantine Books, 1966.

National Biological Information Infrastructure (NBII) and IUCN/SSC Invasive Species Specialist Group (ISSG). "*Passer domesticus* (bird)." ISSG Global Invasive Species Database, 2006. http://www.issg.org/database/species/ecology.asp?fr=1&si=420.

Roof, J. "*Passer domesticus*." Animal Diversity Web, University of Michigan Museum of Zoology, 2001. http://animaldiversity.ummz.umich.edu/site/accounts/information/Passer_domesticus.html.

■ | Japanese White-Eye

Also known as: Me-jiro, Dark-green White-eye
Scientific name: *Zosterops japonicus*
Family: Zosteropidae

Native Range. Eastern Asia. Native to Japan, eastern and southern China, Vietnam to Burma in Southeast Asia, Taiwan, Hainan Island, Ryukyu Islands, and the northern Philippines.

Distribution in the United States. Established on all the major islands in Hawai'i, where it is now the state's most abundant bird.

Description. This extremely active and acrobatic little bird has an olive-green head, neck, and back. The white eye-ring appears almost to be embroidery. The throat and underside of the tail are yellow, the breast gray, and the belly dull white. Flanks are brownish. The wings and tail are dark brown; upper tail feathers and wing feathers are outlined in green. Legs and feet are black. They measure 4–4.5 in. (10–12 cm) long. Juveniles lack the white eye-ring. White-eyes forage in trees in small flocks and often hang upside to down to reach food.

Introduction History. The Japanese White-eye was first introduced to O'ahu in 1929 by the (Hawaiian) Territorial Board of Agriculture and Forestry to control insects. It was taken to the island of Hawai'i in 1937 and now can be found on all the major islands.

Habitat. In Hawai'i, the Japanese White-eye lives in trees and shrubs in rainforest, open deciduous forests, agricultural areas, towns, and city parks. It occurs from sea level to treeline in both arid and humid parts of the islands.

Diet. Japanese White-eyes probe foliage at all levels for insects, including beetles, fly larvae, and spiders. They also eat seeds, nectar, and fruit. They switch from one food source to another depending on availability.

Life History. The breeding season extends from February to December, but there is a distinct peak in activity in July and August. Neat cuplike nests are built at various levels in trees. They are constructed of grass, string, cobwebs, leaves, and mosses and attached to the fork of a branch with spider webs. If human habitation is nearby, the nest is often lined with human hair. The female lays 3–4 white eggs that hatch after an 11-day incubation period. Newborn chicks are altricial; their eyes open about five days after hatching. The young fledge 10–12 days after hatching. Often the young cannot fly when they first leave the nest, but they acquire this

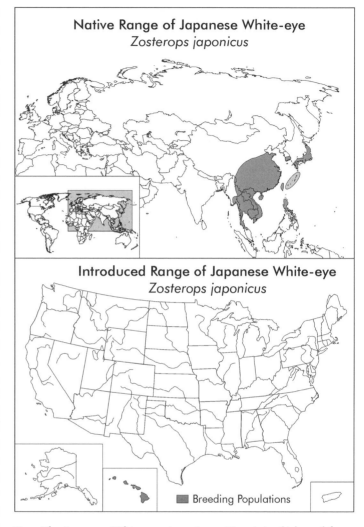

Top: The Japanese White-eye is native to East Asia. (Adapted from MacKinnon, J., and K. Phillips, *A Field Guide to the Birds of China.* Oxford: Oxford University Press, 2000.) Bottom: The Japanese White-eye is invasive only in Hawai'i, where it has become the state's most abundant bird.

ability in 1–6 days. Fledglings remain with their parents for 15–20 days, at which point the adults begin to build a new nest and force the juveniles out of the breeding territory. The white eye-ring is apparent at 23 days of age; and by 30 days, the young look like adults. Young birds gather in flocks until the next year, when they will begin to form pairs and breed.

Impacts. The Japanese White-eye, unlike many exotic birds in Hawai'i, has invaded native montane forest and is currently the most abundant land bird on the main islands. It appears to compete directly with some endemic Hawaiian Honeycreepers for food. As a generalist, the White-eye can switch its feeding specialty to those items most prevalent in a variety of forest types or to less desirable insects, fruits, and flowers when preferred food sources become depleted. In wet forests, it competes with the Elepaio (*Chasiempis sandwichensis*) for foliage insects on low branches and the Apanane (*Himatione sanguinea*) for nectar. In high-

The Japanese White-eye is an active and acrobatic little bird that sports a distinct white eye-ring. (K. W. Bridges, "Common Campus Birds," University of Hawai'i at Manoa. http://www.botany.hawaii.edu/biology101/birds/japanese_white_eye.htm.)

altitude forests, it competes with the Iiwi (*Vestiaria coccinea*) for nectar. Evidence suggests that the presence of White-eyes leads to undernourishment of the young of some native birds such as the endangered Akepa (*Loxops coccineus*), resulting in stunted adults more vulnerable to diseases such as avian malaria, higher infestations of chewing lice, higher mortality rates, and declining populations. It may have lesser impact on the Common Amakihi (*Hemignathus virens*), which has a similarly broad niche.

The Japanese White-eye has the potential to spread invasive plants into undisturbed native Hawaiian forests. It is known to disperse, for example, the velvet tree (*Miconia calvescens;* see Volume 2, Trees, Velvet Tree), lantana shrubs (*Lantana camara;* see Volume 2, Shrubs, Lantana), and the fire tree (*Myrica faya;* see Volume 2, Trees, Fire Tree). On a positive note, the Japanese White-eye may be replacing extinct species of honeycreeper that dispersed native shrubs.

Management. There appears to be no effort to control the spread of these birds, and indeed, it is unlikely any measure would be effective.

Selected References

Foster, Jeffrey T., and Scott K. Robinson. "Introduced Birds and the Fate of Hawaiian Rainforests." *Conservation Biology* 21(5): 1248–57, 2007.

"Introduced Japanese White-Eyes Pose Major Threat to Hawaii's Native and Endangered Birds." *ScienceDaily,* 2009. http://www.sciencedaily.com/releases/2009/09/090917131540.htm.

National Biological Information Infrastructure (NBII) and IUCN/SSC Invasive Species Specialist Group (ISSG). "*Zosterops japonicus* (Bird)." ISSG Global Invasive Species Database, 2006. http://www.issg.org/database/species/ecology.asp?si=954&fr=1&sts.

"*Zosterops japonicus.*" Senior Seminar: Introduced Species in Hawaii. Department of Biology, Earlham College, 2002.

■ | Monk Parakeet

Also known as: Quaker parrot, Quaker conure
Scientific name: *Myiopsitta monachus*
Family: Psittidae

Native Range. South America. Four subspecies are recognized, three in the lowlands east of the Andes: *M. m. monachus* from southeastern Brazil (Rio Grande do Sul) into Uruguay and northeastern Argentina; *M. m. cotorra* from eastern Bolivia south through Paraguay;

and *M. m. calita* in the Patagonian region of Argentina. The disjunct upland population in Bolivia, currently recognized as the subspecies *M.m. luchsi*, is probably a separate species as evidenced by its different habitat, appearance, and nesting behavior. Genetic evidence traces the origins of the Monk Parakeets in the United States to the region from Entre Rios, Argentina, to Rio Grande do Sul, Brazil, near the Uruguayan border. Other records confirm that most animals trapped for the pet trade were exported from eastern Argentina and Uruguay.

Distribution in the United States. Populations come and go, but the bird appears to be established in Alabama, Connecticut, Delaware, Illinois, Florida, Louisiana, New Jersey, New York, Oregon, Rhode Island, Texas, and Virginia; and perhaps also in Colorado, Missouri, Ohio, and South Carolina. The largest populations are in Florida and Connecticut, and in both states, they are increasing in numbers and expanding their ranges. Monk Parakeets were eradicated from California in the 1970s. Monk Parakeets have been observed

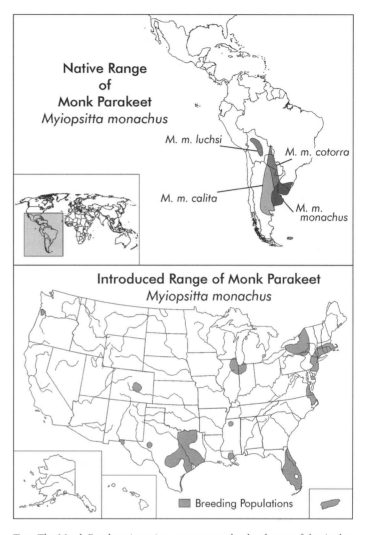

Top: The Monk Parakeet is native to temperate lowlands east of the Andes Mountains in South America. (Adapted from Russello et al. 2008.) Bottom: Monk Parakeets are established in several urbanized locations across the United States. (Adapted from Johnson and Logue 2009.)

in Hawai'i, but have not become established. They are well established in Puerto Rico.

Description. The Monk Parakeet is a small, stocky, mostly green parrot about 11.5 in. (30 cm) from head to tail. The wingspan is about 19 in. (53 cm), and the weight is 2–3 oz. (90–120 g). The forehead, throat, and breast are gray with white barring. The lower abdomen and vent areas are yellow, and the flight feathers are dark blue. The eyes are brown, the bill pale yellow or orange, and the legs are gray. Immatures have green foreheads.

Monks are highly social and noisy; they chatter continually at the nest. They possess a variety of calls and squawks and are especially loud when flying.

Monk Parakeets weave large communal nests of sticks, usually 30 ft. (10 m) or more above the ground. They utilize trees, power poles and towers, and the nests of other birds as supports for these constructions, which may become 3 ft. (1 m) in diameter and weigh up to 440 lbs. (200 kg). Each pair has its own nest cavity within the structure, the entrance

The Monk Parakeet is a stocky, mostly green parrot about a foot long. They are highly social. (Steve Baldwin, BrooklynParrots.com.)

pointing downward. The nest is used for breeding and for shelter year round, and is probably a major reason that the Monk Parakeet has been able to survive in northern states. It is the only parrot that builds a stick nest; others species are cavity-nesters.

Related or Similar Species. Three other parakeets introduced to the United States are about the same size and might be confused for Monk Parakeets. The White-winged Parakeet (*Brotogeris versicolurus*) and the Yellow-chevroned Parakeet (*B. chiriri*) are both smaller (8.75 in. or 222.5 mm total length) than monks and have green heads and bodies with yellow patches showing on their folded wings. In flight, both display yellow bands on the underwing (i.e., the greater coverts are yellow), which are lacking in the Monk Parakeet. The Nanday Conure or Black-hooded Parakeet (*Nandayus nenday*) has a black head and bluish breast. Budgerigars (*Melopsittacus undulatus*), a fourth exotic parrot, are much smaller (7 in. or 178 mm total length) than Monk Parakeets and lack the gray foreheads and throats. Their long pointed tails and yellow wing stripe distinguish them in flight.

Introduction History. From the late 1960s into the early 1970s, some 64,000 Monk Parakeets were imported into the United States. This intelligent species ranks as one of the 10 best talkers among parrots, and its cost is reasonable. Birds were exported from Argentina, perhaps because they were considered agricultural pests in that country. Their importation into the United States was not banned until the Wild Bird Conservation Act of 1992, but these popular cage birds continue to be bred by aviculturalists here. Releases, intentional or not, continue to replenish feral populations and start new ones.

Feral populations had become established in at least 21 urban locations in seven states by the early 1970s. All stemmed from accidental or intentional releases by zoos, the pet industry, or pet owners. The concern over the species becoming an agricultural pest led the U.S. Fish and Wildlife Service between 1970 and 1975 to capture or shoot free-roaming parrots and try to eradicate feral populations. They eliminated monks from California and reduced the overall number of populations to seven in five states. Nonetheless, the number of and size of populations grew after 1975. By 1995, the bird could be found at 76 sites in 15 states.

Monk Parakeets were first reported from Florida in 1972. They were widespread by 1992 and today may number 50,000 to 150,000 birds in 16 counties. The urbanization of eastern Florida with the planting of thousands of ornamental plants that together produced fruit and nectar year round provided an optimal habitat for the bird. Birdseed at feeders supplemented the diet, particularly in winter.

Monk Parakeets in New Jersey are believed to be descendents of birds released in New York in the 1960s. In 1968, parakeets escaped from a damaged crate at John F. Kennedy

Carolina Parakeet

The eastern United States was once home to North America's only native parakeet, the Carolina Parakeet (*Conuropsis carolinensis*), and some have suggested that the Monk Parakeet might come to fill its empty niche—and, like it, become an agricultural pest. Well into the nineteenth century, the colorful bird was an abundant resident of sycamore and bald cypress bottomland forests from the Gulf coast northward almost to the Great Lakes. Like most parrots, it was highly social, noisy, and fed on a variety of plant foods, including cockleburs, sandspurs, pine seeds, and bald cypress seeds. It also foraged on grain and orchard crops.

The Carolina Parakeet was up to 13 in. (33 cm) long from the head to the tip of the tail. Its body was bright green, and its head yellow. The forehead and lores were bright orange. Unlike Monks, Carolina Parakeets nested in tree cavities. It became popular in the early pet trade because of its colorful plumage, but was a poor talker. Its bright feathers were also in demand for a time for women's hats. The birds had the unfortunate habit of flocking to trapped or injured parakeets in response to their distress calls, and this made it relatively easy for farmers or hunters to shoot to trap large numbers at a time.

John James Audubon noted a rapid decline in the number of parakeets in 1831. The last confirmed sightings in the wild were made in 1904 at Lake Okeechobee in Florida by the respected ornithologist Frank Chapman. The last known captive bird died in 1918 at the Cincinnati Zoo, and the Carolina Parakeet was officially declared extinct in 1939.

The demise of the Carolina Parakeet was likely due to a combination of factors. Slaughter by farmers trying to protect crops usually is identified as the cause of extinction, but the birds were also trapped for food, feathers, and pets. Destruction of their bottomland habitat was undoubtedly a contributing factor. Other possibilities include introduced diseases—especially those originating in poultry—and competition for nesting and roosting cavities by feral honey bees, a species introduced to North America during the colonial period.

Airport. The first parakeet was recorded from Oregon in 1977. In 1980–1981 a nest was built near the Portland International Airport, and a small population has persisted since that time.

Habitat. In their native lands, Monk Parakeets prefer open country with tall, isolated trees for nesting. They are found in savannas and woodlands, and in farmland, open eucalyptus forest, and palm groves. The widespread planting of eucalyptus trees as windbreaks in the pampas allowed them to expand their range onto the grasslands. In the United States, all populations are urban.

Diet. Monk Parakeets are largely granivorous and consume primarily the seeds of wild grasses, sedges, and sunflowers. They also feed on berries, fruits, and occasionally insects. In agricultural settings, they forage on domesticated grasses such as maize, wheat, rice, and sorghum and are considered major agricultural pests in South America. In the cities of North America, they feed on the buds, seeds, and fruits of ornamental plants and, in winter, almost exclusively on birdseed put out for them on feeders.

Life History. Breeding begins in North America in spring as the photoperiod increases. Clutches contain 4–7 eggs. Eggs hatch asynchronously, beginning about 20 days after the first one was laid. The nestlings are initially covered with yellow down and remain in the nest for about 40 days, fed by the parents. Fledglings rarely move far from their birthplace, and often add their own chambers to the communal nest. Immature birds become sexually mature at age 2, but may not breed. Instead, they stay with their parents and help maintain the nest and care for the next generation of their siblings.

Impacts. In the United States, the Monk Parakeet is not the agricultural pest it was predicted to become in the 1970s. All populations are, to date, urban. There is little or no evidence suggesting they affect native wildlife. Instead they seem to have brightened the lives of many urban dwellers who delight at seeing these colorful birds at their winter bird feeders. The large nests do pose a problem on transmission line towers and distribution poles, which can be damaged. When they are wet, the nests can cause short circuits and disruption of electric power service. It is costly and can be dangerous to remove nests and repair poles and towers.

Monk Parakeets can also be a nuisance because of the noise they make.

Management. Control of monk parakeets is difficult. The U.S. Fish and Wildlife Service's efforts, which came early in the history of introductions of the birds to the United States, only resulted in eliminating the monks in California. Destruction of nests just causes the birds to build new ones, often nearby. Controlling population sizes may be achieved by limiting accessible food resources, for example, by removing bird feeders or not growing fruit-bearing ornamental plants. Trapping and selling the parakeets might be an option, since there remains a strong market for the birds. The issue of management is complicated by the fondness many people have for these birds and the political strength of animal rights organizations. Since any eradication program would be labor intensive, costly, and controversial, and since the birds are not currently an ecological problem, many in wildlife management think the effort to reduce populations simply is not worth the trouble.

Selected References

Campbell, T. S. "The Monk Parakeet, *Myiopsitta monachus*." Institute for Biological Invasions, University of Tennessee, Knoxville. Invader of the Month, 2000.

Gluzberg, Yekaterina. "Monk Parakeet (*Myiopsitta monachus*)." Introduced Species Summary Project, Columbia University, 2001. http://www.columbiauniversity.org/itc/cerc/danoff-burg/invasion_bio/inv_spp_summ/Myiopsitta_monachus2.html.

Johnson, Steve A., and Sam Logue. "Florida's Introduced Birds: Monk Parakeet (*Myiopsitta monachus*)." Document #WEC257, Department of Wildlife Ecology and Conservation, University of Florida/IFAS, 2009. http://edis.ifas.ufl.edu/UW302.

Russello, Michael A., Michael L. Avery, and Timothy F. Wright. "Genetic Evidence Links Invasive Monk Parakeet Populations in the United States to the International Pet Trade." *BMC Evolutionary Biology* 8: 217, 2008.

Stafford, Terri. "Pest Risk Assessment for the Monk Parakeet in Oregon." Oregon Department of Agriculture, 2003. http://www.oregon.gov/OISC/docs/pdf/monkpara.pdf.

■ | Mute Swan

Scientific name: *Cygnus olor*
Family: Anatidae

Native Range. Eurasia. The Mute Swan breeds in temperate areas from Northwest Europe to Russia, Ukraine, and Kazakhstan. It occurs south of the taiga in southern Siberia and northern China. Wintering grounds are primarily in subtropical regions that lie south of nesting areas. These areas include the eastern Mediterranean, the Black and Caspian seas region, the Persian Gulf region, Central Asia, and the Yellow Sea coast.

Wild populations in Western Europe were virtually eliminated by overhunting from the thirteenth through the nineteenth centuries. Large landowners preserved groups of semidomesticated birds on their estates; so most populations today descend from feral swans. Conservation efforts in the late nineteenth century allowed Mute Swans to regain much of the former territory of their wild ancestors. Birds in the United Kingdom and Western Europe are semimigratory, in winter moving only short distances, often to the coast, where they congregate in flocks. During severe winters, when lakes and rivers freeze over, they will fly much longer distances to open water.

Distribution in the United States. Mute Swans occur in the eastern United States from southern Maine to South Carolina and from the Atlantic coast inland to the Mississippi River drainage. Small numbers are found in western states. Mute swans are considered established in southern New England, Long Island Sound, the Chesapeake Bay, and the Great Lakes region. The largest populations are in Rhode Island, Connecticut, New York, New Jersey, and Maryland. They are not truly migratory in any part of their adopted range.

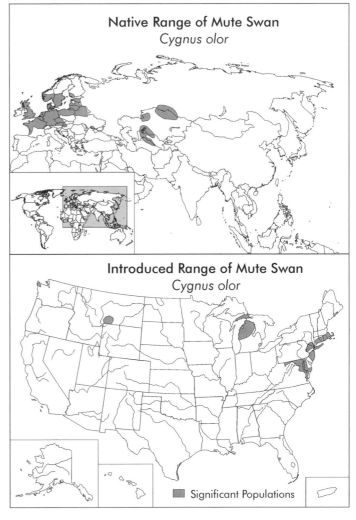

Top: The Mute Swan was once widely distributed in the temperate regions of Eurasia, but populations were largely extirpated by overhunting from the thirteenth through the nineteenth centuries. Many populations existing today derive from semidomesticated stock. (Adapted from Delany 2006.) Bottom: Mute Swans have established populations in southern New England, Long Island Sound, Chesapeake Bay, and Great Lakes areas. Small numbers occur in some western states. (Adapted from "Winter Christmas Bird Count" [map]. http://www.discoverlife.org/mp/20q?search=Cygnus+olor.)

Description. Mute Swans are very large white waterfowl with long, gracefully curved necks. When swimming, the bill points downward. The Mute Swan has an orange bill with a diagnostic black knob at the base and a black nail at the tip. The lores are also black. Adults may have a total length greater than 5 ft. (1.3–1.6 m) and weigh 20–25 lbs. (8–11 kg). The wingspan can be 7 ft. (1.8–2.5 m). Males ("cobs") are larger than females ("pens").

Mute swans have long, gracefully curved necks. When swimming, they point their bills downward. (Ozerov Alexander/Shutterstock.)

Immature swans (cygnets) have two morphs, greyish brown or white. The gray cygnets have slate-gray bills, legs, and feet. The white morph has a tan bill and pinkish tan feet. Both lack the basal knob on the bill that characterizes adults.

Related or Similar Species. Native swans have long necks that are held straight when swimming, and black bills. The Trumpeter Swan (*Cygnus buccinator*) is equal in size to the Mute Swan. Its range is generally north of that of the Mute Swan. The Tundra or Whistling Swan (*Cygnus columbianus*) is considerably smaller and has yellow lores. Its distribution overlaps with that of the Mute Swan on its wintering grounds and during migration. The white Snow Goose (*Chen caerulescens*) might also be mistaken for a swan on the wintering grounds, but it is a much stockier bird with a shorter neck and is about half the size of a swan. Its black wing tips are diagnostic in flight.

Introduction History. By most accounts, Mute Swans were introduced into the United States from Europe in the late 1800s as ornamental birds on private estates, in urban parks, and in zoos. Aviculturalists bred the birds, esteemed for their beauty and grace. Originally, swans had their flight feathers cut to keep them from flying off. However, some owners neglected to pinion their birds, and escapes established feral populations. The earliest known free-ranging populations lived along the Hudson River (1910) and on Long Island, New York (1912). These populations expanded their numbers and distribution within the states of the Atlantic Flyway. New Jersey had feral Mute Swans by 1919, Rhode Island by 1923, and Maryland by 1954. Mute Swans began to colonize the lower Great Lakes in the mid-1960s and were first sighted in South Carolina in 1993. Many populations in the eastern United States grew especially rapidly in the 1990s, but have shown signs of decline since that time. Mute Swans in the California and the Pacific Northwest are the products of local releases and escapes.

There is no credible evidence to the support the hypothesis that Mute Swans were native to North America and suffered continent-wide extinction during the Pleistocene Epoch. Nor is there scientific support for the belief that Mute Swans regularly migrated from Siberia to North America. Furthermore, none of the great early American naturalists and wildlife artists, such as Mark Catesby, John James Audubon, Spencer Baird, or Elliot Coues, encountered these large birds.

Habitat. Lakes, ponds, rivers, bays, estuaries, fresh and saltwater marshes. They prefer waters up to 4 ft. (1.2 m) deep where they can reach submerged vegetation, but will move to deeper water when the shallow water freezes in winter.

Diet. Mute Swans consume submerged aquatic vegetation (SAV). Studies in the Chesapeake Bay revealed a preference for widgeon grass (*Ruppia maritima*). They also consume eel grass (*Zostera marina*), wild celery (*Vallisneria americana*), and several pondweeds (*Potamogeton* spp., and others). They will pull up whole plants to eat or to feed to cygnets.

Life History. Defense of large (6–13 ac. or 2.4–5.3 ha) nesting territories begins in February, and nesting starts in March or early April. The mating ritual involves what is known as "busking," when males raise their wings and fluff their feathers while twirling in place. Pair formation is not for life as legend has it, but swans are monogamous for the season and pairs may be stable for years at a time.

The female performs most of the nest building, using rushes, cattails, reeds, and stems of other marsh plants to construct a large nest 4–6 ft. (1–2 m) in diameter, often on a mound safely above water level. Both males and females are highly aggressive toward intruders in their territories. Clutch size varies from 4 to 10 eggs. The young cygnets hatch about 35 days after the last egg is laid. In the Mid-Atlantic region, hatching usually takes place in early June. Mute swans generally nest only once a year, but if the eggs are destroyed early in the season, they will nest a second time.

Cygnets are precocial and begin swimming within a day or two of hatching. Families typically remain on the breeding territory away from other family groups or nonbreeding adults during the first two months or so. Cygnets grow rapidly and can fly at 4–5 months of age; they are fully grown at 6 months. Juveniles are forced out of the nesting territory before the next breeding season. Juveniles and subadults often gather in large flocks near open salt water and molt, becoming flightless for a short time. Mute Swans become sexually mature at age 2, but most will not nest until they are 3–5 years old. Swans can live in the wild up to 19 years, but the average lifespan is about 7 years.

Impacts. In the Chesapeake Bay, the Mute Swan's consumption of SAV is of major concern. SAV provides essential cover and nursery areas for a variety of fish and shellfish and feeding grounds for a large number of native waterfowl, and it has already been seriously depleted in estuaries of the bay due to development and pollution. Whereas native ducks, geese, and swans use the area as wintering grounds and consume SAV for only a season, Mute Swans are year-round residents and consumers. Plants such as wild celery reproduce and recover from wintering grazing pressures when the native waterfowl are absent. Mute Swans could impede the seasonal and long-term recovery of SAV and reduce winter forage for native birds. Loss of SAV is implicated in population declines of Canvasbacks (*Aythya valisineria*) and Redheads (*A. americana*).

The larger Mute Swan may also compete with native Tundra Swans, recently brought back from near extinction, for food and shelter in winter. Aggression on nesting territories has been directed against breeding Canada Geese (*Branta canadensis*), Mallards (*Anas platyrhynchos*), and American Black Ducks (*Anas ubripes*).

In Maryland, molting mute swans occupied a beach that was the last nesting area in the state for two state-threatened birds, the Black Skimmer (*Rynchops niger*) and the Least Tern (*Sterna antilarium*). The loafing swans trampled eggs and nestlings on the beach, leading to the abandonment of the area by both skimmers and terns for three seasons.

Swans are very aggressive toward people, pets, and other waterfowl. They can inflict serious damage with their wings and can prevent recreational use of some shorelines. Disturbance of SAV by swans can limit crabbing and fishing activities. Together with other waterfowl that defecate in water, flocks of Mute Swans contribute to degraded water quality and an increased coliform bacteria count. Cranberry farmers in New Jersey and Massachusetts complain that swans get into their bogs and uproot cranberry plants as they browse other submerged aquatic plants.

Management. In 2001, Mute Swans became a protected species under the Migratory Bird Treaty Act. The U.S. Fish and Wildlife Service (USFWS) received the authority for managing the birds. In 2002, the USFWS issued depredation permits to states for Mute Swan population control; and in 2003, it established the Atlantic Flyway Mute Swan Management Plan, which encourages states to develop and implement location-specific management plans. The chief means of controlling Mute Swans is by the labor-intensive practice of addling eggs. Eggs are coated with oil to suffocate the embryos with the adults unaware of any disturbance to their clutch. Trapping and relocating or killing adults are other population control measures. A hunting season could be a viable option; swans were originally domesticated in Western Europe for food. Several states prohibit the establishment of new populations by outlawing the sale or importation of birds and requiring that those already owned be pinioned. Every three years, extant populations in the Atlantic Flyway are monitored by air in midsummer, when native swans and Snow Geese are in the tundra.

Selected References

Delany, Simon. "The Mute Swan in Europe—a Preliminary Assessment of Numbers, Distribution and Potential Risks in Dissemination of HPAI–H5N1." Wetlands International, Feb. 14, 2006. http://global.wetlands.org/LinkClick.aspx?fileticket=0hONijRYbSs%3D&tabid=56.

Ivory, A. "*Cygnus olor*." Animal Diversity Web, University of Michigan Museum of Zoology, 2002. http://animaldiversity.ummz.umich.edu/site/accounts/information/Cygnus_olor.html.

"Mute Swan." New York State Department of Environmental Conservation, Fish, Wildlife and Marine Resources, n.d. http://www.dec.ny.gov/animals/7076.html.

"Mute Swan *Cygnus olor*." Invasive Species in the Chesapeake Watershed. Summary, Chesapeake Bay Program, 2002. http://www.mdsg.umd.edu/issues/restoration/non-natives/workshop/mute_swan.html.

"Mute Swan Management Plan." State of Rhode Island and Providence Plantations, Department of Environmental Management, Division of Fish and Wildlife, 2006. http://www.dem.ri.gov/programs/bnatres/fishwild/pdf/muswan07.pdf.

■ Rock Pigeon

Also known as: Common Pigeon, Feral Pigeon, Rock Dove
Scientific name: *Columba livia*
Family: Columbidae

Native Range. Europe, North Africa, southwestern Asia. At their maximum pre-domestication range, wild Rock Pigeons could be found from the Faeroe Islands and southwest Norway south through the United Kingdom and coastal France, to Spain and Portugal and eastward along the shores of the Mediterranean in southern Europe and North Africa. Its range also extended eastward across inland Europe into Russia, the Middle East, and India and Nepal. Its natural habitat was cliffs, usually along coasts, where it nested in caves and on ledges. Rock Pigeons were domesticated more than 5,000 years ago in the eastern Mediterranean, and since then, feral pigeons have lived close to people in urban and agricultural settlements. Most northern European populations of wild pigeons were driven to extinction by overhunting or by genetic dilution from feral birds.

Distribution in the United States. Rock Pigeons are common birds, especially in cities, throughout the United States and in Puerto Rico.

Description. Rock Pigeons are medium-sized birds with small round heads and large chests. They are 12–13 in. (30–35 cm) long and have a wingspan of about 25 in. (62–68 cm). When they walk, their heads bob forward and backward. The plumage is extremely varied, a product of their domestication. Among the two dozen well-recognized colors and patterns are those named blue-bar or "wild type," blue checker, blue-T, dark (melanistic) checker, spread, white, pied, and Ash red variants. The wild type is a bluish-gray body with a dark head. Iridescent feathers of blue, green, and purple circle the neck. The rump is white. Two black wing bars mark the folded wing, and a black band occurs on the end of the tail. The eyes are reddish, and the feet pink. A white cere or fleshy swelling appears at the top of the short beak. Females tend to select mates with a pattern different from their own and so perpetuate the variety of phenotypes in a population. There also tend to be geographic trends in color: city populations tend to have more checker and spread birds than rural populations (except in Honolulu, where 80% of the pigeons are white); and pigeons with red plumage are more frequently encountered in the American Southwest than in northern and eastern states.

Pigeons are strong, agile flyers with pointed wings similar to falcons. When they glide, the wings are held in a V. They appear tame and often forage and roost in flocks. Rock Pigeons produce a cooing

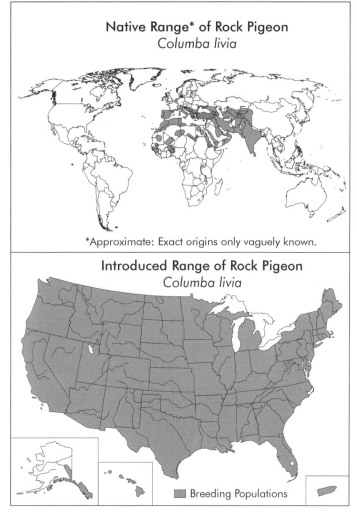

Top: The wild Rock Pigeon was a native species from Iberia to India and in mountainous areas of North Africa, where it inhabited cliffs. Domesticated for more than 5,000 years and probably mixed with feral stock for the same amount of time, the true native range is uncertain. (Adapted from Kravtchenko, Viktor. "Columba livia distribution map." http://en.wikipedia.org/wiki/File:Columba_livia_distribution_map.png.) Bottom: Rock Pigeons are common birds in cities and other human-modified habitats in all 50 states and Puerto Rico.

sound, but this call is mainly heard when they are on their nests or during courtship.

Related or Similar Species. The native Band-tailed Pigeon (*Patagioenas fasciata*) inhabits oak and pine woodlands in California and parts of Arizona and New Mexico. It will frequent waterholes, but is not a city bird. Another native pigeon, the White-crowned Pigeon (*Patagioenas leucocephala*), breeds in mangroves and forages in inland hardwood forests. In the United States, it is restricted to the southern tip of Florida and the Florida Keys and is also found in Puerto Rico.

Introduction History. Domestic pigeons were once important sources of meat and eggs, and people carried them around the world. The first captive birds came to North America

Rock pigeons are familiar denizens of urban landscapes. The plumage of these feral birds is extemely varied, a product of domestication. (Jaimaa/ Shutterstock.)

in 1606 with the settlement of Port Royal in Nova Scotia. The first pigeons to arrive in what is now the United States came with the settlers of Jamestown, Virginia, in 1607. All Rock Pigeons living free in the United States derive from escapes of domestic birds.

Habitat. Rock Pigeons are feral birds found in proximity to humans in farms, towns, and cities. They require structures with ledges for nesting, loafing, and roosting; and most buildings serve well as surrogate cliffs and caves. Nests will be built wherever there is a dark opening, on a ledge on a skyscraper, in the rafters of an abandoned building, or under a bridge.

Diet. Rock Pigeons forage on the ground, mainly for medium-sized seeds. Preferences include dried peas, wheat, oats, barley, millet, and maize. In urban settings, they also consume grass seed, berries, bread crumbs, popcorn, and other particles of discarded food.

Life History. Pigeons in cities may breed all year. Most nesting, however, occurs between March and October. Males will fluff their feathers and bow-coo in courtship of females at any time. The pair will mate for life. The male selects the nesting site, and the pair constructs a crude nest. The male brings nesting materials to the female, and she tucks the straw, twigs, and feathers he brings around her body. The female typically lays two white eggs 40 hours apart. The eggs hatch asynchronously 17–20 days later. The first egg hatches about a day before the second. Male offspring are larger than females and often the first to hatch. The larger nestling has a distinct advantage over its smaller nest mate, which will die in years of food shortages. The younger sibling can be viewed as insurance for the parents if something should happen to the first one, and in good years, both will survive and fledge. For the first 4–5 days, nestlings are fed pigeon "milk," a cheesy secretion of the crops of both parents that is rich in protein and fats. The crop milk is regurgitated to the young birds and ceases to be produced after 10 days. As the pigeon milk supply declines, more and more seeds are given to the squabs (birds 1–30 days old).

Squabs leave the nest in 4–5 weeks and are fully independent when about 7 weeks old.

A pair of pigeons may breed several times a year, sometimes constructing new nests on top of old ones encrusted with droppings from previous broods. A typical pair in Kansas produces 10 squabs a year from an average of 6.5 nests. Young pigeons are able to breed at six months of age. The large reproductive potential of feral pigeons is a legacy of their history as a domesticated species.

Impacts. Feral pigeons are no longer considered an agricultural pest in the United States. In cities, they can be a nuisance, littering building lofts, façades, and sidewalks with their droppings and nesting materials. Their dung can harbor the fungus *Histoplasma capsulatum*; cleanup programs aimed at getting rid of pigeon and starling droppings have caused histoplasmosis infections in the lungs of humans in Arkansas, Ohio, Missouri, and Wisconsin.

Nonetheless, many people derive pleasure from feeding pigeons in urban parks and at feeders, and eradication programs are few.

Management. The most effective way to control feral pigeon populations is to reduce or eliminate food sources and barricade nesting and roosting sites. Cleaning up vacant lots and reducing litter on urban streets and in parks will help with the former, as will using feeders that minimize the amount of seed that scatters onto the ground. Netting, metal spikes on ledges or other perches, and electric exclusion fences can help with the latter.

Selected References

Hetmanski, Tomasz, and Anna Jarosiewicz. "Plumage Polymorphism and Breeding Parameters of Various Feral Pigeon (*Columba livia* Gm.) Morphs in Urban Area." Gandsk, north Poland. *Polish Journal of Ecology* 56(4): 683–91, 2008. Available online at http://www.pol.j.ecol.cbe-pan.pl/article/ab56_4_12.pdf.

Johnston, Richard F., and Marian Janiga. *Feral Pigeons*. New York: Oxford University Press, 1995.

"Rock Pigeons." AvianWeb and Wikipedia, 2006. http://www.avianweb.com/rockpigeons.html.

Roof, J. "*Columba livia*." Animal Diversity Web, University of Michigan Museum of Zoology, 2001. http://animaldiversity.ummz.umich.edu/site/accounts/information/Columba_livia.html.

Youth, Howard. "Pigeons: Masters of Pomp and Circumstance." *Zoogoer*, 276. Friends of the National Zoo, 1998. http://nationalzoo.si.edu/Publications/ZooGoer/1998/6/pigeons.cfm.

■ Mammals

■ | Black Rat

Also known as: House rat, roof rat, ship rat
Scientific name: *Rattus rattus*
Family: Muridae

Native Range. Most authorities agree that black rats originated on the Indian subcontinent and possibly in neighboring areas of Southeast Asia. They have been so long associated with humans, migrating with people around the world, that it is difficult to know their origins with any certainty.

Distribution in the United States. Currently, the black rat is most common in the southern United States, although before the introduction of the Norway rat in the eighteenth century, it was the common rat in towns and on farms in the Northeast. In the lower 48 states, it is largely found in seaports, but small, isolated populations do occur inland. Larger populations are known from California's Central Valley and along the Pacific coast north into the Puget Sound region of Washington State. In the eastern United States, they are found from Norfolk, Virginia, south along the Atlantic coast and throughout the Gulf states. Black rats also occur in Hawai'i, where they live in moist natural forests.

Description. The black rat is slender and has a tail longer than its head-body length. The tail is used for balance, making this rat a most agile climber on overhead wires and tree limbs. Its ears are relatively large and hairless. The body may be black all over, or the back may be brown to gray and the underside a lighter tone. The tail is hairless, scaly, and uniformly colored. Head-body length ranges from 6 to 8 in. (160–220 mm) and tail length from 8.5 to 10 in. (190–240 mm). Black rats weigh 5–10 oz. (140–280 g). Males are larger than females.

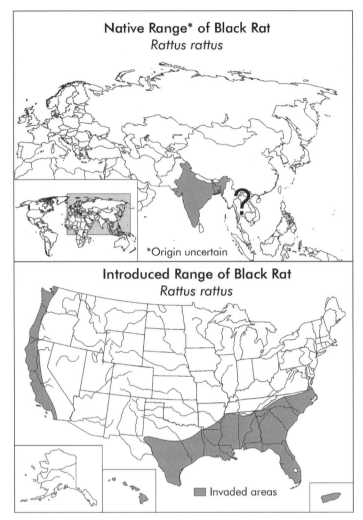

Native Range* of Black Rat
Rattus rattus

*Origin uncertain

Introduced Range of Black Rat
Rattus rattus

■ Invaded areas

Top: The Indian subcontinent is the most probable home of the black rat, but it may also have occurred originally in Southeast Asia. The original source area is obscured by the animal's long dispersal history in the company of humans. Bottom: Since the introduction of the Norway rat to the continental United States, the black rat has become almost exclusively a resident of southern states, where it is found primarily in seaports. In Hawai'i, however, it is found in moist forests. (Adapted from Marsh 2005.)

Related or Similar Species. The closely related Norway rat (see Mammals, Norway Rat) is larger and its tail is shorter than its head-body length. The ears are relatively small. Woodrats (*Neotoma* spp.) have white undersides. The rice rat (*Oryzomys palustris*) has small hairy ears; its gray back is streaked with black, and its belly and feet are whitish and its long tail is pale below.

Introduction History. The black rat spread along the trade routes of Asia and Europe, entering the eastern Mediterranean in Roman times and Europe by early medieval times. It is the infamous carrier of the Black Death (bubonic plague) that ravaged Europe several times during the Middle Ages. In the 1500s, it arrived with Europeans in Central and South America. Black rats were aboard the earliest ships arriving at Jamestown, Virginia; Captain John Smith complained of thousands of rats devastating grain stores in 1612. They subsequently moved from port to port and then dispersed inland wherever humans lived and farmed. The invasion of the Norway rat about the time of the American Revolution led to a range reduction among black rats in North America, as the larger newcomer replaced black rats in regions with more prolonged winters. Coastal and insular populations may be constantly replenished today as black rats continue to arrive on incoming vessels.

Habitat. Black rats prefer warm climates and areas inhabited by people. However, their rarity in the more temperate regions of the United States may be due more to their displacement by the later-arriving Norway rat than to environmental conditions, as they were formerly common in towns and farms in New England and elsewhere in the northern United States. These rodents are agile climbers and often live in treetops in natural settings or the upper floors and attics and rafters of buildings everywhere from the inner city to small towns. The latter is particularly true where their range overlaps with Norway rats, which occupy cellars and lower floors.

Black rats also commonly occupy ships, gaining access by climbing up ropes or stowing away in the cargo. While they may live in riparian vegetation, black rats rarely swim.

Diet. Omnivores, these rodents prefer grains, nuts, and fruits, but also prey on invertebrates and the eggs and chicks of birds. They also like poultry and livestock feed and cat and dog food. Black rats will cache solid food to consume later and must drink water daily.

Life History. These are social animals living in mixed groups of adult males and females. Nonetheless, one male is dominant, and usually two or three females are dominant to all others in the group except the dominant male. Males breed with more than one female and defend their territories and mates. In warm climates, breeding occurs all year; in more seasonal climates, breeding may peak in summer and fall. Nests are constructed in tree cavities, among palm leaves, in hedgerows, in protected niches in

The agile black or roof rat has a tail longer than its head-body length. (Falcon Scallagrim/iStockPhoto.)

buildings, or in other dense cover. After a gestation period of 20–22 days, females give birth to 5–8 altricial pups. Their eyes open and pellage begins to show in about two weeks, at which time they begin to move around. They will be weaned in at 3–4 weeks, but stay at the nest until they reach adult size. They will be reproductively mature at 3–4 months of age. The young of the season leave the area when mature; it is unknown how far they disperse. Under prime conditions, a female will produce five litters a year. Average lifespan for a wild rat is about one year.

Impacts. Black rats have been implicated, directly or indirectly, in the extinction of native species, particularly on islands. In Hawai'i they contributed to the extinction of some of the islands' endemic honeycreepers (Drepaniidae) in the 1800s and now threaten the survival of seabirds such as Bonin Petrels (*Pterodroma hypleuca*) through predation on the birds' eggs. Similar depredations on islands around the world led to their nomination as one of the 100 "World's Worst" invaders by the IUCN/SSC Invasive Species Specialist Group.

Apparent proof of the rat's impact can be seen on Anacapa Island in the Channel Islands off California, where the rare Xantus's Murrelet (*Synthliboramphus hypoleucus*) is increasing in numbers since the eradication of the black rat in 2000–2001.

As a vector for dangerous human diseases, the black rat is infamous. In addition to bubonic plague, caused by the bacterium *Yersinia pestis* that is transmitted by rat fleas, black rats also carry leptospirosis, toxoplasmosis, trichinosis, and typhus.

Damage to tree crops such as citrus, avocado, and some nuts is a problem in some places. Sugarcane is damaged where rats feed because they open stalks to insects and pathogens.

Like their cousins the Norway rats, black rats are major pests in grain and other feed storage facilities, contaminating food with their urine and feces. In buildings, they will gnaw electrical wiring and damage insulation. In backyards, they will consume ornamental plants, fruits, and vegetables.

Management. The best practice is to prevent the establishment of a rat colony in the first place. Rat-proofing roofs; clearing dense, overgrown shrubs and vines; eliminating water sources and removing pet food left outside; and securing food and feed in rat-proof containers are first lines of defense. Trapping and poisoning black rats can be problematic as they distrust new objects in their environment and will avoid them. Successful eradication programs on islands have relied on anticoagulant baits.

Black rats are not a protected species and may be killed or captured at any time using mechanical or chemical means. However, the chemicals used must be registered for rat control by federal and state agencies and used in accordance with label directions.

Selected References

"Black Rat, *Rattus rattus*." eNature.com, 2007. http://www.enature.com/fieldguides/detail.asp?recnum=MA0096.

Gillespie, H., and P. Myers. "*Rattus rattus*." Animal Diversity Web, University of Michigan Museum of Zoology, 2004. http://animaldiversity.ummz.umich.edu/site/accounts/information/Rattus_rattus.html.

IUCN/SSC Invasive Species Specialist Group (ISSG). "*Rattus rattus* (Mammal)." ISSG Global Invasive Species Database, 2006. http://www.issg.org/database/species/ecology.asp?si=19.

Marsh, Rex E. "Roof Rats." Internet Center for Wildlife Damage Management, 2005. http://icwdm.org/handbook/rodents/RoofRats.asp.

■| Feral Burro

Also known as: Wild donkey
Scientific name: *Equus asinus*
Family: Equidae

Native Range. The ancestor of the burro, the African wild ass (*Equus africanus*), evolved in the deserts of eastern Egypt, Sudan, and the Horn of Africa. The domestic breeds, which were the founding stock for feral burros, were Spanish in origin and most probably came into the United States from Mexico.

Distribution in the United States. Most feral burros are on public lands managed by the Bureau of Land Management (BLM) in Arizona, California, and Nevada. Almost half are located in the Lower Colorado River valley. A few small populations can be found in Oregon, Utah, and in Custer State Park, South Dakota.

Description. The typical burro is gray with a white muzzle, white eye-rings, black shoulder cross and black leg barring. Many, however, are black or spotted (paint), and white individuals are not uncommon. Average shoulder height is 47 in. (120 cm). Their long ears are distinctive.

Feral burros congregate in small groups with transient members. Groups generally consist of bachelor males or females and their immature offspring. Only mature jacks (males) are solitary.

Introduction History. Most feral burros stem from domestic animals used by prospectors and miners during the western gold rushes of the nineteenth century. By the end of that century, the development of roads and railroads marked the obsolescence of pack animals, and the boom time of small-scale gold, silver, copper, and lead mining was largely over. Many burros were abandoned to fend for themselves at that time.

Habitat. Feral burros thrive in warm desert habitats as long as permanent water sources are available. In spring and summer, burros tend to congregate in riparian habitats where shade and food are available; in winter, they frequent interfluves where annual forbs may be abundant.

Diet. Burros are primarily browsers. In a study area along

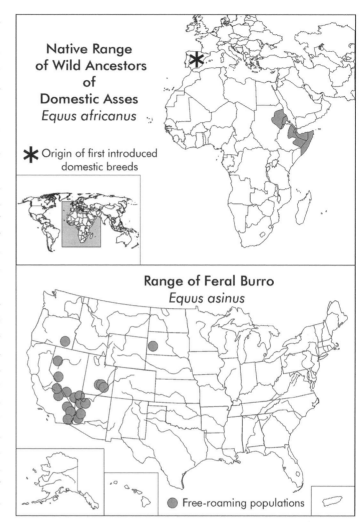

Top: Two subspecies of wild African asses occurred historically in the desert east of the Nile and in the Horn of Africa. They were likely first domesticated in Upper Egypt some time prior to 5,000 years ago. In the colonial period, Spain was known for its fine breeds of domestic ass, which were transported to its colonies in the New World. (Adapted from Woodward, S. L. "Feral burros of the Chemehuevi Mountains, California: The biogeography of a feral exotic." PhD diss., University of California, Los Angeles, 1976.) Bottom: The current range of free-roaming feral burros in the United States consists mostly of desert areas managed by the U.S. Bureau of Land Management. (Based on U.S. Bureau of Land Management's herd management areas maps and data.)

the Lower Colorado River in California, more than 60 percent of the diet of feral burros was comprised of desert shrubs such as palo verde (*Cercidium floridanum*), mesquite (*Prosopis* spp.), and arrowweed (*Pulchea sericea*). Thirty percent was composed of forbs, particularly winter annuals such as woolly plantain (*Plantago insularis*). Grasses were also consumed.

Life History. Feral burros breed year round in warm deserts. The jenny (female) gives birth to a single foal, who accompanies her for a year or longer. Juvenile males typically leave

A. The typical feral burro is gray with a white muzzle, white eye-rings, and black shoulder cross. However, black, white, and spotted individuals are not uncommon. (Chris Curtis/Shutterstock.) B. Most burros occur in the deserts bordering the lower Colorado River. (Susan Woodward.)

the mother when about one year of age and assemble in bachelor groups of varying size and membership. Occasionally, a juvenile female will tag along with a small bachelor group or single jack for a time. Mature jacks may or may not establish temporary territories near sources of water and attempt to breed with jennies passing through it. However, many breeding attempts also occur beyond the limits of the territory by non-territorial males.

Females are sexually mature by 10 months, males by 12 months; but they usually do not breed successfully until older. Full physical growth is not achieved until about two years of age. Females come into estrous immediately after giving birth, but usually do not conceive at that time. Gestation lasts almost 12 months, and jennies may nurse their foal for up to a year. It is common for females to foal only once every two years even in optimal habitat. Lifespan may be up to 15 years.

Impacts. Burros establish well-developed trails along the contours of desert slopes, which, along with their dust wallows, could accelerate erosion. Trampling can lead to compaction of desert soils and destruction of archeological sites. Feral burros could compete with desert bighorn sheep (*Ovis canadensis nelsoni*) for forage and water and could also compete with the federally threatened desert tortoise (*Gopherus agassizii*) for forage and crush its burrows. They are accused of fouling precious desert water holes with their feces; but domestic cattle are more likely perpetrators, since burros are loathe to get their feet wet.

Management. The 1971 federal Wild Horse and Burro Act mandated that the BLM protect and manage feral burros on the public lands under their jurisdiction as living symbols of our national heritage. At that time, an estimate 14,400 burros roamed the deserts of the Southwest. Herd Management Areas were established, and appropriate population levels were determined for each managed population. Excess animals were removed and put up for adoption. Today an estimated 4,700 feral burros remain in a free-roaming state. Nearly half of these are in Arizona. The other half is nearly evenly divided between California and Nevada.

To mitigate the potential impact of burros on bighorn sheep, water tanks have been established in some areas and fenced so as to permit sheep to enter but not burros.

Feral burros are also managed on special sanctuaries or preserves. Many more are held in short-term corrals and long-term pastures awaiting adoption. Most feral burros have been removed from those national parks in which they once occurred under the National Park Service's mandate to eliminate nonnative species.

In 2010, the secretary of the interior announced plans to more aggressively use fertility control measures and manage sex ratios of free-roaming populations in the management of feral horse—and presumably feral burro—herds.

Selected References

MacDonald, C. R. "Wild Burros of the American West: 2006 National Burro Status, A Critical Analysis of the Current Status of Wild Burros on Public Lands." 2007. http://www.wildhorsepreservation .com/pdf/BurroAnalysis-2006-Public.pdf.

"National Wild Horse and Burro Program." Bureau of Land Management, 2010. http://www.blm.gov/ wo/st/en/prog/wild_horse_and_burro.html.

Woodward, S. L. "The Social System of Feral Asses (*Equus asinus*)." *Zeitshrift fur Tierpsychologie* 49: 304–16, 1979.

Woodward, S. L., and R. D. Ohmart. "Habitat Use and Fecal Analysis of Feral Burros (*Equus asinus*), Chemehuevi Mountains, California," *Journal of Range Management* 29: 482–85, 1976.

■|Feral Cat

Also known as: Alley cat, house cat
Scientific name: *Felis silvestris catus*
Family: Felidae

Native Range. The domestic cats from which feral cats derive were primarily European animals transported with colonists to North America and Hawai'i. Recent genetic studies point to the Near East as the site of the earliest domestication of the cat and its ancestor as the Near Eastern wildcat, *F. s. lybica*. The propensity of free-roaming cats to interbreed with wild cats has introduced genes from other subspecies of *Felis silvestris*, including the European wildcat (*F. s. silvestris*) and the Central Asian wildcat (*F. s. ornata*).

Distribution in the United States. Throughout.

Description. Feral cats are indistinguishable from domestic house pets and display a full range of coat patterns. After many generations in a free-roaming state, they tend to revert to the "wild type" tabby color pattern, with varying degrees of white on belly and chest. Adults have a shoulder height of 8–12 in. (20–30.5 cm) and weigh 3–8 lbs. (1.4–3.6 kg). Body length is 14–24 in. (35.5–60 cm); the long, flexible tail adds an additional 8–12 in. (20–30.5 cm). These agile predators have retractable claws, sharp teeth, long whiskers, and keen hearing and eyesight, including acute night vision.

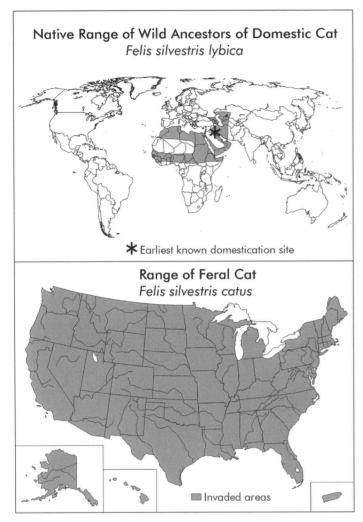

Native Range of Wild Ancestors of Domestic Cat
Felis silvestris lybica

✱ Earliest known domestication site

Range of Feral Cat
Felis silvestris catus

■ Invaded areas

Top: Cats were probably first domesticated in Mesopotamia. Their ancestors were the Near East subspecies of wildcat, *Felis silvestris lybica*. (Adapted from http://en.wikipedia.org/wiki/File:Wiki-Felis_sylvestris .png) Bottom: Feral cats are found in all 50 states and Puerto Rico.

Related or Similar Species. Free-roaming but owned domestic cats are identical in appearance, although they tend to be tame. Bobcats (*Lynx rufus*) are twice the size and have black-tipped, stump tails.

Introduction History. Domestic cats were first introduced to the mainland of the United States by European traders and colonists in the seventeenth century. They arrived in Hawai'i in the 1800s. Cats were commonly kept on sailing ships to kill vermin such as rats and mice. Before they became pampered companion animals, they were used as work animals to control rodent pests in barns and fields and were only loosely managed. Many undoubtedly escaped to a feral state. Feral cat populations continue to be supplemented by unneutered pets that wander off and produce litters in the wild. However, large numbers are the result of the deliberate abandonment of unwanted pets and litters of kittens. Often, well-intentioned people release domesticated animals into the wild rather than subject them to the likelihood of euthanasia at animal shelters. The United States, a country where cats are the most popular household pet, has an estimated 30–60 million feral cats.

Habitat. Feral cats are highly adaptable and occupy a wide variety of habitats and climate regions. They tend to congregate where food is readily available and shelter is nearby. Colonies may live in urban alleyways, near waste collection sites behind restaurants and fast food operations where food waste, as well as foraging rodents, may be abundant; in towns, where people often feed them; and on farms. They are also common on college campuses, at military bases, and in other places where the human population is transient and pet abandonment is frequent.

Diet. Midsize carnivores, feral cats prey upon small mammals, birds, reptiles, amphibians, and insects.

A. Feral cats are indistinguishable from house pets and display a full range of coat patterns. (Matt Valentine/Shutterstock.) B. After many generations in a free-roaming state, feral cats tend to revert to the "wild type" tabby pattern. (Daz/Shutterstock.)

Life History. Feral cats are very prolific. Sexual maturity is achieved by females between 7 and 10 months of age and sometimes even younger. Gestation is 63–65 days. Reproduction can occur year round. A female typically produces three litters of 4–6 kittens a year. The young are weaned in 35–40 days. Life expectancy for a feral cat is only 2–3 years, compared with 10–20 years for a cat that is a household pet.

Impacts. Cats, feral and free-roaming pets alike, are a major threat to wildlife because of the great numbers of small mammals, reptiles, and birds that they kill. The destruction of native birds often receives the most attention, but the number of small mammals such as shrews, chipmunks, and rabbits killed each year by cats is even greater and may reach more than a billion. A single cat may kill 100 or more small mammals and birds each year. While island populations of endemic species are most vulnerable to losses, the depredations occurring in suburban and urban "habitat islands" are significant.

In Florida, cats are considered a major threat to several federally endangered small mammals, including the Key Largo cotton mouse (*Peromyscus gossypinus allapaticola*), rice rat (*Oryzomys palustris natator*), Key Largo woodrat (*Neotoma floridana smalli*), Lower Keys marsh rabbit (*Sylvilagus palustris hefneri*), Choctwhatchee beach mouse (*Peromyscus polionotus allophrys*), Perdido Key beach mouse (*Peromyscus polionotus trissyllepsis*), Anastasia Island beach mouse (*Peromyscus polionotus phasma*), and Southeastern beach mouse (*Peromyscus polionotus niveiventris*). They are also known to prey upon federally listed birds such as Roseate Tern (*Sterna dougallii*), Least Tern (*Sternula antillarum*), and Florida Scrub-jay (*Aphelocoma coerulescens*) and on green sea turtles.

In Hawai'i, cat predation is one more factor pressing endangered birds—such as the Hawaiian Crow or 'Alalā (*Corvus hawaiiensis*); Hawaiian Goose or Nēnē (*Branta sandvicensis*); Palila (*Loxioides bailleui*), a Hawaiian honeycreeper; and Hawaiian Petrel or 'Ua'u (*Pterodroma sandwichensis*)—toward extinction.

Feral cats can be reservoirs for feline diseases such as feline leukemia and FIV, threatening native bobcats and mountain lions as well as domestic house cats. They also can transmit rabies to pets and humans and may carry parasites such as *Toxoplasmosis gondii*, roundworm, and hookworm. Fleas in cat colonies can spread to nearby human dwellings and workplaces.

Nominated by the IUCN as among "100 of the World's Worst" invaders, feral cats are notorious as introduced predators on islands all over the world.

Management. Feral cat control is an emotionally charged issue. Animal welfare advocates and conservationists tend to have different goals and therefore different approaches to the problem. Humane organizations are interested in the well-being of the cats; conservationists focus on halting depredation of native fauna. Trap-neuter-release (TNR) programs tend to be least controversial in most communities, but do not prevent the ecological damage a dense colony of carnivores can impose on native small animal populations. While reducing the number of breeding adults in a cat colony should over time reduce the feral population, recruitment of new members will continue as long as people abandon or fail to sterilize their pets. Cats are thought of as solitary, but they are only slightly territorial and, where food is abundant, will congregate in large numbers. Permanent removal of animals from a feral cat colony tends to open space for new arrivals. A sterilized cat released back to the colony, on the other hand, creates competition for food and shelter and may actually increase the mortality of kittens born into a feral state.

Many feral cat colonies are managed and fed by volunteers, who capture, spay, and vaccinate new strays entering the population. Kittens may be removed, tamed, and put up for adoption. Nonetheless, small mammals, birds, and reptiles will continue to be killed. As every cat owner knows, even a well-fed cat kills.

On some islands, eradication of feral colonies by trapping and euthanasia may be possible, but on the mainland, the ecological impacts of cats will continue as long as millions of pet cats are neither sterilized nor confined, and as long as people feed and care for strays.

Selected References

Driscoll, Carlos A., et al. "The Near Eastern Origin of Cat Domestication." *Science* 37: 519–23, 2007.

LaBruna, Danielle. "Domestic Cat (*Felis catus*)." Introduced Species Summary Project, Columbia University, 2001. http://www.columbia.edu/itc/cerc/danoff-burg/invasion_bio/inv_spp_summ/Felis_catus.html.

Longcore, Travis, Catherine Rich, and Lauren M. Sullivan. "Critical Assessment of Claims Regarding Management of Feral Cats by Trap-Neuter-Return." *Conservation Biology* 23: 887–94, 2009.

Masterson, J. "*Felis catus*, Feral House Cat." Smithsonian Marine Station at Fort Pierce, 2007. http://www.sms.si.edu/IRLspec/Felis_catus.htm.

Verdon, Daniel R. "Feral Cats: Problems Extend to Wildlife Species, Ecologists Say." *DVM Newsmagazine*, 2002. http://veterinarynews.dvm360.com/dvm/article/articleDetail.jsp?id=31506&sk=&date=&&pageID=1.

■| Feral Goat

Scientific name: *Capra hircus*
Synonym: *Capra aegagrus hircus*
Family: Bovidae

Native Range. Goats were first domesticated in the Zagros Mountains region of western Iran. Most island populations of feral goats in the United States likely derived from Iberian domestic breeds.

Distribution in the United States. Invasive populations of feral goats are or were found in the Channel Islands of California and all the major islands of Hawai'i. Feral goats also exist on the island of Mona off Puerto Rico. Small populations of feral goats descended from escaped or abandoned livestock are found in isolated areas of the mainland United States (e.g., in California foothill habitats), but generally are not considered a problem because their numbers are held in check by predation and hunting.

Description. San Clemente Island (California) goats are the best-described population. These animals adapted to an island environment over several hundred years in part by becoming relatively small and fine-boned. They are frequently described as deer-like and have a shoulder height of 21.5–29.5 in. (55–75 cm)—only slightly larger than dwarf breeds of domestic goats. Both males and females possess horns that spiral upward and outward. The males have heavier horns than females and also have beards. Before extensive culling and removal of goats to the mainland (see Management below),

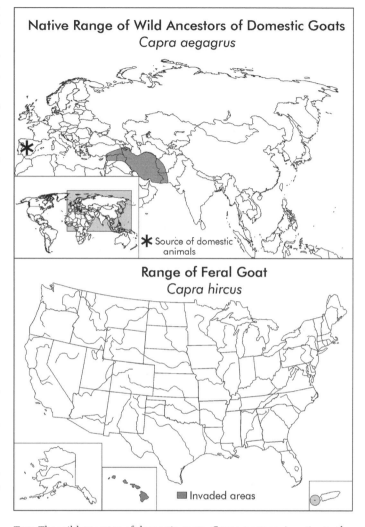

Top: The wild ancestor of domestic goats, *Capra aegagrus*, is native to the mountains of western Iran, where domestication seems to have first taken place. (Adapted from Altaileopard, http://commons.wikipedia.org/wiki/File:Capra_aegragus_map.png.) Bottom: Feral goats are notorious invaders on islands. Today they remain on the Hawaiian Islands and on Mona Island, Puerto Rico.

San Clemente Island goats displayed a variety of coat colors; today, they are mostly red or brown with black markings, especially on the head and forelegs.

Feral populations usually assemble in herds of 5–20 animals. Herds may consist of mixed age and sex groups, all-male groups, or females and their young.

Related or Similar Species. Goats are often confused with sheep. An easy way to tell them apart is to look at the tails: goats' tails are held up, while sheep tails droop downward.

Introduction History. Early Spanish and Portuguese explorers frequently dropped goats off on islands around the world to run wild and breed in order to serve as a meat supply for sailors arriving in the future. It has long been believed that the goats on the Channel Islands of California were introduced in this manner, but later supplemented by stock from

A. Feral goats on Mauna Loa, Hawai'i, in the 1970s. (National Park Service.) B. Slopes on Santa Catalina Island in the Channel Islands of California were deeply gullied and overgrown with prickly pear cactus as a consequence of overgrazing by feral goats when this photo was taken in the late 1970s. (Susan Woodward.)

California's Spanish missions and later still by farmers living on the islands. Recent genetic analysis of San Clemente goats, transplanted to that island from Santa Catalina in 1875, indicated that the island goats were not Spanish in origin, but a genetically distinct breed.

Captain James Cook brought the first goats to Hawai'i in 1778.

Habitat. Goats are hardy animals and adapt to a variety of habitat types. In California, feral goats subsist in Mediterranean scrub habitats, valley foothill hardwood, and valley foothill hardwood–conifer woodlands. They seek shelter on steep slopes and rock outcrops. In Hawai'i, they live from sea level to high alpine habitats with distinct dry seasons and prefer rocky slopes and open lava fields. A population on an island off South Carolina adapted to a hot, humid, and swampy environment.

Diet. Goats are browsing animals and thrive on brush and other coarse vegetation unpalatable to sheep and cattle. They also consume grasses.

Life History. Dominant males breed with estrous females serially. Gestation lasts 147–155 days, after which time 1–3 precocial young are born. Twinning is common. Sexual maturity in both sexes may be reached by six months and definitely by one year of age. They may breed twice a year under favorable conditions.

Impacts. Severe overgrazing by feral goats on islands can leave areas essentially devoid of vegetation. Not only are endemic plant species threatened in this manner, but the removal of food and cover endangers native island birds and other animals as well. In less severe situations, vegetation structure and species composition can be altered. Soil erosion and downslope sedimentation are other consequences of unmanaged browsing by goats. Ecosystem degradation and loss of biodiversity are potential outcomes.

Management. On both the Channel Islands and in Hawai'i, hunting was tried as a way to reduce feral goat populations without success. Reports from Hawai'i Volcanoes National Park, for example, indicate that 70,000 goats were removed between 1920 and 1970, but the goat population remained high. On San Clemente Island, owned by the U.S. Navy since 1934, systematic removal of goats began in 1972. Large numbers were driven into traps with net-wing fences and killed. Some 16,000 animals were removed between 1975 and 1979 and another 8,000 between 1979 and 1982, when an estimated 4,000 goats still roamed the island.

Goats have since been completely eradicated from Lana'i Island, Hawai'i, and San Clemente Island using a combination of techniques including aerial hunting from helicopters; specially trained goat-hunting dogs to flush out remaining survivors; and "Judas goats," sterilized goats outfitted with radio-transmitters that seek out herds to join and thus detect any goats missed by other methods.

When the navy proposed shooting goats from helicopters to eliminate them from San Clemente, the Fund for Animals, an animal welfare group, stepped in and live-trapped the remaining animals with nets and helicopters and removed some 3,000 goats to the mainland, where most were sterilized and put up for adoption. A few animals were not neutered and formed the base of a new breed of domestic goat, the San Clemente Goat. These hardy, disease- and parasite-resistant animals are used as meat animals and also as pets.

On Santa Catalina Island (California) and parts of Hawai'i, exclusion fencing has kept feral goats out of localized areas and allowed the regeneration of native plants.

Selected References

"*Capra hircus*." Introduced Species in Hawaii. Senior Seminar Biology Department, Earlham College, 2002.

Coblentz, Bruce. "*Capra hircus* (Mammal)." USA and IUCN/SSC Invasive Species Specialist Group (ISSG). ISSG Global Invasive Species Database, 2008. http://www.issg.org/database/species/ecology.asp?si=40.

Edmundson, Leslie. "San Clemente." Breeds of Livestock. Oklahoma State University, 1997. http://www.ansi.okstate.edu/breeds/goats/sanclemente/index.htm.

"San Clemente Goat." American Livestock Breeds Conservancy, n.d. http://www.albc-usa.org/cpl/sanclementegoat.html.

■|Feral Horse

Also known as: Wild horse, mustang; Banker horse
Scientific name: *Equus caballus*
Family: Equidae

Native Range. Europe. Feral horses descend from domestic horses bred in Spain and parts of northwestern Europe. The earliest domesticated horses have been traced to the steppes of Kazakhstan and Ukraine, but Iberian wild horses may have contributed to the development of European breeds.

Distribution in the United States. Populations of free-roaming horses occur in 10 western states, primarily on the public lands administered by the Bureau of Land Management. In addition, a population is maintained in Theodore Roosevelt National Park, North Dakota. Other herds are found at several sites along the Atlantic coast, including North Carolina's Outer Banks and Assateague National Seashore and Chincoteague National Wildlife Refuge in Virginia. About half all feral horses reside in Nevada.

Description. Feral horses are indistinguishable from domestic horses, although a few populations have been isolated long enough to preserve or evolve unique genetic information. Western horses generally stand 14–15 hands (4.75–5 ft. or 1.4–1.5 m); males weigh 795–860 lbs. (360–390 kg) and females 595–750 lbs. (270–340 kg). Assateague ponies and other eastern insular horses tend to be smaller, with shoulder heights seldom more than 13 hands (4.3 ft. or 1.3 m). Feral horses come in all colors and patterns, including solid colors ranging from black to cream, paints of every type, and leopards.

Of genetic significance are the few herds of pure Spanish mustangs descending from the earliest domestic horses brought to the Americas. These horses have short, straight backs and deep, narrow chests that are V-shaped when viewed from the front. The croup (rump)

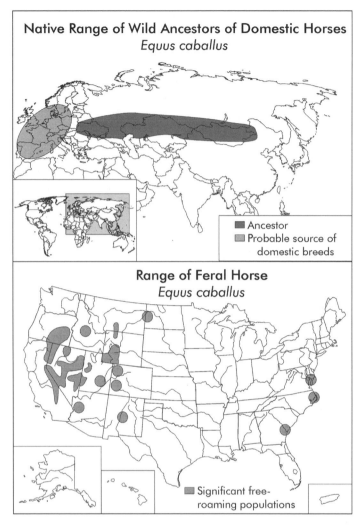

Native Range of Wild Ancestors of Domestic Horses
Equus caballus

■ Ancestor
■ Probable source of domestic breeds

Range of Feral Horse
Equus caballus

■ Significant free-roaming populations

Top: The wild ancestor of the domestic horse was native to the steppes of Eurasia, but it is now extinct in the wild. Bottom: Free-roaming feral horses still live in the deserts and grasslands of 10 western states and a few islands off the East Coast of the United States. (Based on U.S. Bureau of Land Management's herd management areas maps and data and other reported occurrences.)

is low, and the tail set low. They usually have straight to concave foreheads and convex noses. Examples include the Kiger mustangs in southeastern Oregon and the Cerbat mustangs found near Kingman, Arizona.

The feral horses of Theodore Roosevelt resemble horse types common in the nineteenth century, but rare today. They have large heads and short backs and are frequently blue or red roans with "bald" or "apron" (white) faces and white patches on their sides.

Related or Similar Species. Feral horses are generally indistinguishable from domestic horses. In western states, ranchers and Native Americans often allow their horses to run free when they are not using them, and these animals could easily be mistaken for feral horses.

Introduction History. Christopher Columbus, on his second voyage to the New World in 1493, brought the first horses to the Americas when he landed them on Hispaniola. From that Caribbean island, Spanish conquistadors introduced them to mainland North America in the early 1500s. Spanish horses spread northward through Mexico along the chain of missions set up by the Spanish. Native Americans raided the missions and settlements of the frontier, stole the horses, and became adept horsemen, warriors, and bison hunters. They exchanged and stole horses from other tribes, effectively spreading the animal throughout the continent. Some groups bred horses and developed distinctive breeds, such as the original Appaloosa. Other Spanish horses escaped from military outposts, where stallions were used as cavalry mounts.

Large herds of semidomesticated horses developed on the Great Plains, their numbers and genes added to when European settlers with breeds from Western Europe pushed westward from the Atlantic coast of the United States. Herds of wild horses from east of the Mississippi were displaced westward with settlement. French breeds moved south and west from the Detroit region and north from the New Orleans colony. The U.S. Cavalry in the late

1880s and early 1900s introduced old-style East Friesian blood to the mix. These draft horses were preferred for pulling artillery and heavy wagons; any that escaped would have joined the herds of feral horses roaming the Plains.

Settlers lost horses when wild stallions broke through fences and liberated mares. Furthermore, it was common practice to let horses roam freely on the range until needed for work on ranches. Many likely escaped captivity and added to the feral herds.

Wild horses existed in the badlands of North Dakota when Theodore Roosevelt visited in the 1880s. He recognized them as escapes from nearby ranches or Indian reservations, and many carried brands. Most horses were removed from the area in 1954, a few years after the national park was established. A few small bands eluded capture and formed the core of the current herd managed by the National Park Service.

Feral horses are generally indistinguishable from the domestic horses from which they descend. However, a few populations have been isolated long enough to preserve or develop unique genetic information. (Robert Broadhead/Shutterstock.)

The twentieth century saw the value of free-roaming horses decline as ranching and other interests grew, and many were shot or taken to slaughterhouses. Populations declined to such an extent that some feared the feral horse was headed for extinction. The mustang became celebrated by some as a romantic symbol of the Old West and its preservation demanded. A successful campaign initiated by "Wild Horse Annie" (Velma B. Johnston), who was aided by the letter-writing of schoolchildren across the country, resulted in a federal law to save the wild horse, the Wild Free-Roaming Horse and Burro Act, passed in 1971. Responsibility for managing the herds remaining on public lands fell to the Bureau of Land Management (BLM); the U.S. Forest Service gained jurisdiction over horses in national forests and grasslands. Under the protection of these agencies, feral horse populations began to grow again. In 2010, BLM estimated some 33,700 horses were on the rangelands they manage.

In their evolutionary history, horses evolved in North America and crossed the Bering Land Bridge into Eurasia during the Pleistocene Epoch. They subsequently became extinct in North America 11,000–13,000 years ago, but were domesticated in Ukraine about 5,000 years ago. Horses were reintroduced—as domestic animals—by the Spanish early in the sixteenth century. The origin of the species on this continent has led some to propose that feral horses are native, not exotic species and that they are part of the natural heritage of the West.

Habitat. In the West, feral horses generally inhabit open shrublands and woodlands in arid and semiarid regions. Optimal rangelands have grassy areas and riparian zones. On the offshore islands of the Atlantic coast, feral horses are associated with salt marshes.

Diet. On western ranges, feral horses prefer grasses, forbs, and sedges but also browse shrubs such as shadscale saltbush (*Atriplex confertifolia*), sagebrush (*Artemesia tridentata*), and rabbitbrush (*Chrysothamnus nauseosus*). Assateague ponies graze preferentially on salt-marsh cordgrass (*Spartina alterniflora*), but will eat American beach grass (*Ammophila breviligulata*) and three-square rush (*Scirpus americanus*); they also browse woody plants.

Life History. Most feral horses live in small, stable social groups of 5–15 animals. A band consists of a mature stallion and his harem, a group of 4–6 mares and their younger offspring. Sometimes a subdominant male also is part of the band. The group maintains a strict social hierarchy and often roams in an established territory. The mares will stay together even when the stallion is lost. Young stallions that have yet to acquire harems form bachelor herds. Old stallions that have lost their harems to younger males are usually the only solitary animals.

The harem forms the breeding unit. Mating may occur year round but peaks during the foaling season, which is typically early spring. Usually a mare produces a foal every other year after a gestation period of about 340 days. When a mare is ready to foal, she temporarily leaves the harem and gives birth to a single offspring. The newborn can run and swim shortly after birth; and its mother brings it into the band a few hours later. The foal will continue to nurse until its mother is about to give birth again. Both fillies and colts are driven from the band when they reach sexual maturity around 2–3 years of age. The young mares will be collected into an existing or new harem by a stallion that is not their father. Young males will form into bachelor groups until such time as they can acquire mares of their own either by successfully challenging an aging stallion or by assembling a group of mares through stealth or upon the death of another male. Mature mares rarely move to another band.

In the absence of natural predators and on good range, feral horse populations may grow rapidly, doubling in numbers in four years. Recruitment rates are especially high after major population reductions due to severe winter weather or roundups.

Impacts. It is difficult to assess the real impacts of feral horses on native ecosystems because of the high degree of emotion surrounding the issue of their role on western rangelands. Apparently from 1600 to 1850, vast herds of horses occurred throughout the Great Plains, where they had transformed life for the Plains Indians. They shared the range with bison, pronghorn, elk, and mule deer. Extermination of horses accompanied the campaign to remove Native Americans from the West and also the development of modern cattle ranching. Although free-roaming horses were captured to become cow ponies and saddle horses, cattlemen more and more regarded them as nuisances that may lure away their domestic horses and compete with cattle for forage and water.

Scientific studies reveal a dietary overlap of 80 percent or higher between feral horses and cows, both species of which are predominantly grazers. Since horses also browse, they are actually better suited to much western rangeland than are cattle; but Americans do not consume horsemeat, so the argument that horses should replace cattle is moot. With respect to impacts on other animals, comparatively little research has been conducted. Their diets do not seem to overlap significantly with deer, elk, or pronghorn, all of which are primarily browsers. Potentially, horses trample the nests of ground-dwelling birds. Reptile diversity and abundance may be reduced in areas occupied by feral horses. On well-managed ranges, plant diversity has been shown to increase with a variety of large grazing and browsing mammals, including feral horses, present. Greater abundance and diversity of large vertebrates occurred where horse droppings accumulated and soil was intermediately disturbed in California's Anza-Borrego State Park. Trampling of vegetation and compacting of soils on trails made by horses could be localized problems and could accelerate soil erosion.

Management. Feral horses are federally protected under the 1971 Wild Free-Roaming Horse and Burro Act. The law is administered by the BLM (Department of the Interior) and the U.S. Forest Service (Department of Agriculture). Shooting or poisoning feral horses in the wild became a federal crime. When BLM assumed authority for feral horses, an estimated 17,300 mustangs roamed western rangelands. In 2010, the number was estimated to be 33,700. As populations grew in the 1970s, amendments to the law in 1976, 1978, 1996, and 2004 mandated management of herds in such a way as to sustain their habitats and authorized humane removal of excess animals from the public lands. The 1976 Federal Land Policy and Management Act included feral horse management as part of BLM's and the Forest Service's multiple-use missions that require managing public lands for recreation, livestock grazing, mineral and energy production, and the conservation of natural, historical, and cultural resources. Feral horses are considered part of our national heritage.

Appropriate Management Levels (AMLs) have been determined for the 201 Herd Management Areas designated by the BLM and the 37 that it comanages with the U.S. Forest Service. Herd reductions to achieve these population sizes and to maintain them have relied largely upon roundups and selecting excess animals to be removed and put up for adoption. The roundups use helicopters and cowboys on horseback, and are expensive and dangerous to both the horses and their captors. The Adopt-A-Horse program began as a way to transition feral horses back to private ownership and domestic status. Since 1971, BLM has placed more than 225,000 horses and burros in private care. Horses that prove impossible to adopt are held in short-term holding pens and on long-term pastures. In May 2010, 10,700 feral horses (and burros) were in corrals and another 24,400 maintained on pastures in the Midwest. An estimated 33,700 were on BLM-managed range, whereas the total AML has been set at 26,582.

Fertility control is now a viable practice to reduce herd growth rates. The administration of porcine zona pellucida vaccine (PZP), an immunocontraceptive, by dart guns in the field has proved to be a safe and effective means of population control in conjunction with removal and natural mortality. Currently, booster shots are required annually.

In June 2010 the secretary of the interior proposed new solutions to wild horse management, which he deemed unsustainable for the horses, the habitat, and the taxpayer in its current form. In part, he advocated more aggressive use of fertility control measures and the establishment of new wild horse preserves across the country in order to showcase these animals to the American public and to enhance local economies through increased tourism.

Selected References

"Colonial Spanish Horse." American Livestock Breeds Conservancy, n.d. http://www.albc-usa.org/cpl/colonialspanish.html.

Sponenberg, D. P. "The Colonial Spanish Horse in the USA: History and Current Status." *Archivos de zootecnia* 41(extra): 335–48, 1992.

"Wild Horse and Burro Quick Facts." U.S. Bureau of Land Management, 2010. http://www.blm.gov/wo/st/en/prog/wild_horse_and_burro.html.

"Wild Horse: *Equus caballus*." Enature.com, 2007. http://www.enature.com/fieldguides/detail.asp?recnum=MA0169.

"Wild Horses." Theodore Roosevelt National Park, n.d. http://www.theodore.roosevelt.national-park.com/nat.htm#wil.

■ | Feral Pig

Also known as: Wild pigs, feral hog, feral swine, Pineywoods rooter, razorback; Eurasian or European wild boar, Russian wild boar
Scientific name: *Sus scrofa*
Family: Suidae

Native Range. Europe. Feral hogs are descendents of domestic breeds originally brought to the United States from Europe. Recent genetic analysis suggests that these breeds arose from wild boars native to Europe. In addition to domestic breeds, some truly wild boars from Europe were deliberately released in several states as game animals.

Distribution in the United States. Feral hogs are found primarily in the southern tier of states from California to North Carolina. Isolated populations are reported as far north as Wisconsin and New Hampshire. The greatest numbers occur in California, Florida, and Texas. Feral pigs inhabit all of the major islands of Hawai'i and occur on St. John in the Virgin Islands.

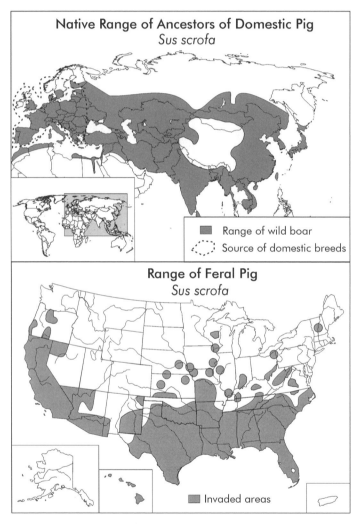

Native Range of Ancestors of Domestic Pig
Sus scrofa

■ Range of wild boar
⌐⌐⌐ Source of domestic breeds

Range of Feral Pig
Sus scrofa

■ Invaded areas

Description. Three types of free-roaming pig occur in the United States: feral hogs derived from escaped or released domestic stock, European wild boars, and hybrids between them. Most are feral livestock. These medium-size, cloven-hoofed mammals have long, pointed heads and stocky bodies. Males and females are similar in appearance. Reflecting their domestic origins, feral pigs display a variety of coat colors and patterns, from solid black, brown, white, or red, to spotted with several colors, or belted with a wide pale band across the shoulders. Their coat is coarser; bristles tend to be longer than those of domestic pigs, but shorter than those of wild boars or hybrids. The tail is straighter than a domestic pig's and never coiled. The flattened snout is flexible and elongate. Total body length is 3.6–4.9 ft. (1.1–1.5 m). The average adult has a shoulder height of about 3 ft. (1 m); males (boars) weigh on average 130 lbs. (60 kg) and females (sows) about 110 lbs. (50 kg). Maximum weight is probably about 300 lbs. (135 kg). Males have four very sharp tusks that

Top: The wild boar, ancestor of domestic pigs, is widespread in Eurasia. However, the domestic breeds that were brought to the United States and that contributed to the feral pig gene pool seem to have descended only from European populations. (Range of wild boar adapted from Altaileopard, http://en.wikipedia.org/wiki/File:Sus_scrofa_range_map.jpg. Source of domestic breeds based on Larson et al., "Worldwide Phylogeography of Wild Boar Reveals Multiple Centers of Pig Domestication." *Science* 307: 1618–21, 2005.) Bottom: In the United States, feral pigs are found primarily in southern states and on all the major islands of Hawai'i. (Adapted from "Feral/Wild Pigs: Potential Problems for Farmers and Hunters." *USDA Agricultural Bulletin* No. 799, 2005.)

continue to grow throughout the lifetime of the boar, although they usually break or wear down to lengths of approximately 5 in. (12.5 cm). The lower tusks are continually sharpened as they rub against the upper pair. Tusks are used to establish dominance among males and to defend against predators and other males. The shoulder skin of males thickens as they age into a shield, a product of both aging and fighting; it consists of tough scar tissue and cartilage. Some but not all feral piglets are striped, much like the piglets of wild boars.

European wild boars are usually light brown or black with cream-colored tips on the bristles. The ends of these long stiff hairs are often frayed. The belly hairs are lighter and the legs, ears, and tail darker than the rest of the body. Long side whiskers develop and hairs on the back of the neck give them the razorback appearance. Purebreds have longer legs and snouts and relatively larger heads than a feral pig; they have little fat. Piglets are reddish with black, lengthwise stripes.

Hybrids show characteristics of both parents. The length of the bristles is longer than in feral pigs, but shorter than in wild boars. The diameter of the bristle shaft, however, is smaller than either feral or wild pigs.

Occasionally, truly enormous pigs are captured in the wild. The infamous "Hogzilla," shot in Georgia in 2004 and estimated to have weighed some 800 lbs. (363 kg) and have had a total body length of 8 ft. (2.4 m), was determined by National Geographic researchers to be a hybrid between a domestic pig and a wild boar. Others are likely escaped or abandoned domestic pigs, which are often bred and fattened to weigh over 1,000 lbs. (454 kg).

The presence of feral pigs can be determined by a variety of signs. Hog tracks are square with rounded, splayed toes. Their rooting is often extensive; it can reach depths of 3 ft. (1 m) and may leave a field looking like it has been plowed. Wallows are depressions formed when pigs roll in mud to protect their skin from sun and insects; they often fill with water. Rubs develop where pigs scratch against trees, fence posts, and rocks to remove external parasites, dried mud, and dead hair. Telltale bristles are often left behind.

Related or Similar Species. In the southwestern states of Arizona, New Mexico, and southern Texas, there is a native pig-like animal, the collared peccary or javelina (*Tayassu tajuca*). These are not true pigs and not closely related to feral swine, but belong in a New World family of mammals, Tayassuidae. Peccaries are gray with a narrow white band around the shoulders, and smaller than a feral hog. They congregate in herds.

Introduction History. The first domesticated pigs came to North America with early Spanish explorers in the sixteenth century. Two hundred pigs arrived in Florida in 1539 with Hernando de Soto as a walking larder to accompany his expedition through the Southeast. Native Americans as well as later European settlers in the Southeast used the descendents of these hogs, often keeping them under free-range conditions. In the early nineteenth century, Spanish hogs accompanied the southern Native American groups resettled by the U.S. government in the Oklahoma Territory. (The small, wattled Choctaw hog, with its fused toes that look like a mule's hoof, is a pure Spanish breed and is still managed today as a free-range animal in Oklahoma.) Undoubtedly, many pigs escaped to live in a wild state from the time of de Soto onward.

Hernando Cortes brought pigs to Mexico, and they spread north with colonizers, probably reaching Texas some 300 years ago. Many may have been abandoned when people left Texas during its war for independence from Mexico (1834–1835). In Hawai'i, small pigs of Asian descent were originally introduced by Polynesians around 400 AD, but were replaced by European breeds about 200 years ago. English settlers brought their own breeds to

A. The coat and bristles of feral pigs tend to be longer and coarser than those of the domestic pigs from which they descend. Their piglets sometimes are reddish with lengthwise black stripes like those of wild boars. (Laurie L. Snidow/Shutterstock.) B. Some free-roaming pigs descend from wild European boars introduced as game animals. They have longer legs and snouts than a truly feral pig. (Alan Lucas/Shutterstock.)

North America beginning in the seventeenth century, and Americans developed new breeds, such as Duroc, Hereford, and Poland China. All these stock interbred in the free state to form the common, genetically mixed feral hog populations that roam the United States today.

Free-roaming feral pigs have occurred in southern states for a long time. However, range expansions into Colorado, Illinois, Indiana, Kansas, Kentucky, Missouri, Nebraska, Nevada, Ohio, Oregon, and West Virginia occurred mostly in the last decade of the twentieth century. Many of these were deliberate introductions by hunting clubs or private landowners for sport hunting. In all locations, they have been successful invaders because of their high reproductive potential, improved disease control in domestic livestock, pasture improvements related to modern livestock grazing practices, and development of watering sources for domestic stock in arid lands.

Pure European or Russian boars were first released in the late 1800s and early 1900s for sport hunting in New Hampshire, North Carolina, Missouri, Arkansas, and Tennessee. They were introduced to Texas with the same purpose in the 1930s.

Habitat. Feral pigs are habitat generalists but prefer densely vegetated moist forests and bottomlands where they can find shelter and make wallows.

Diet. Pigs are opportunistic omnivores. Their diet depends on seasonal and geographic availability and is dominated by plant matter. Pigs will consume grasses, forbs, roots and tubers, fruits, acorns, and nuts. They also eat fungi and live earthworms, mollusks, amphibians, reptiles, and birds as well as eggs, and also scavenge carrion.

Life History. Feral pigs display several characteristics that reflect ancestral domestic breeds that were purposefully selected for fast growth and early maturation. After a gestation period of 115 days, feral sows generally farrow litters of 4–6 but, under prime conditions, may give birth to 10–12 young. (Wild boars usually have smaller litters, with 3–8 piglets being typical.) Some scientists consider pigs to be the most prolific large mammal on the planet. As a rule, a 1:1 sex ratio exists at birth. Some females have two litters a year, but most have only one. Although young may be born throughout the year, peak production occurs in early spring. Piglets weigh 1–2 lbs. (0.45–0.91 kg) at birth. Young are weaned at 3–4 months. Family groups or sounders composed of 1–3 sows and their offspring may contain three generations and 50 or more individuals; they are maintained until the young mature. Females may stay in the family group or leave with their sisters to form a new group. Males always leave the family group by 16 months of age. Adult males are solitary, traveling among sounders to breed. Feral pigs are sexually mature at 6 months, but many females do not mate until 10–12 months of age. Young males may be prevented from breeding by more dominant boars. Adult stature is attained by age 3. Average lifespan is 4–5 years, with a maximum of 8 years. Hunting is the primary cause of death in adults.

Groups rather than individuals have home ranges, and they vary in size from several hundred to a few thousand acres depending upon habitat quality. Pigs tend to occupy different parts of the home range seasonally. The need to cope with heat dictates the location of summer home ranges near water and can result in nocturnal activity. Neither individuals nor sounders maintain territories within the home range.

Impacts. Due to their rooting and trampling of vegetation, feral pigs and wild boars may be the most extreme vertebrate modifier of natural ecosystems in the United States. They disturb soil horizons and compact soil, decreasing water infiltration and increasing erosion. Soil erosion, along with contamination by bacteria from fecal deposits, can increase sedimentation and pollution in waterways; pig activity has been implicated in declining freshwater mussel and aquatic insect populations in some areas. Their feeding can alter plant species composition and plant community structure. Disturbance of the soil and herb layer allows invasive plants to spread by creating habitat in which weedy exotics generally outcompete native plants. Feral pigs certainly compete with native animals for food and other resources. This is particularly true of mast-eaters such as white-tailed deer (*Odocoileus virginianus*), Wild Turkey (*Meleagris gallopavo*), and squirrels. They may also negatively affect wildlife populations through predation, habitat destruction, and the spread of parasites and disease. On the Atlantic beaches of the Southeast, feral pigs are threatening the nesting success of several endangered sea turtles, including loggerhead (*Caretta caretta*), leatherback (*Dermochelys coriacea*), hawksbill (*Eretmochelys imbricata*), green (*Chelonia mydas*), and Kemp's ridley (*Lepidochelys kempii*) turtles. In Great Smoky Mountains National Park, their destruction of the leaf litter reduces the habitat of redbacked voles (*Myodes gapperi*) and short-tailed shrews (*Blarina brevicauda*), while direct predation reduces populations of threatened red-cheek salamanders (*Plethodon jordani*) and Jones middle-toothed snail (*Mesodon jonesianus*). Tree ferns are a major part of the diet of feral pigs in Hawai'i, where they also eat other native trees and epiphytes and reduce vegetative cover. Feral hogs also facilitate the spread of the invasive strawberry guava (*Psidium cattleianum;* see Volume 2, Trees, Strawberry Guava), a plant threatening the survival of several native Hawaiian plants and animals.

Ecological and economic damage to crops, timber, and pasture can also be significant; crop depredations alone have been estimated at $1.5 billion a year in the United States. The greatest impact is to hay, small grains, corn, and peanuts, although tree fruits, cotton, vegetables, and conifer seedlings are not immune. Feral pigs are known to prey upon lambs, kids, and newborn calves, attracted to birthing areas by afterbirth. Predation on animals is difficult to document, since feral pigs also scavenge carcasses left by other animals. When they do hunt for themselves, they typically kill their prey by biting and crushing the skull. However, since the entire carcass is usually eaten, little or no evidence is left to point to the actual killer.

Feral pigs carry several parasites and disease. Among those that could affect humans are trichinosis, leptospirosis, toxoplasmosis, and brucellosis. Hunters should be careful handling live animals and carcasses, and all pig meat should be well cooked. Livestock diseases

Ossabaw Island Hog

The feral hogs on Osssabaw Island, Georgia, have been isolated long enough to develop unique characteristics but also are the closest representative, genetically, of original Spanish stock. They are a rare resource for science and for future animal breeding. These small pigs are able to store huge amounts of fat to survive lean seasons, a biochemical adaptation analogous to non-insulin-dependent diabetes in humans, which makes them an excellent model for medical research. Natural selection has also left Ossabaw Island hogs tolerant of high amounts of salt in their diet. Animals removed from the island have served as founding stock for a newly domesticated Ossabaw hog breed, which appears well adapted to sustainable or pastured pork production. Feral hogs may no longer be taken from the island because some individuals have tested positive for pseudorabies and porcine vesicular stomatitis (PSV); their future as a feral population depends on management goals of the state of Georgia.

Source: "Ossabaw Island Hog." American Livestock Breeds Conservancy, n.d. http://www.albc-usa.org/cpl/Ossabaw.html.

are also potentially harbored in feral pigs. Those of greatest concern currently are swine brucellosis (*Brucella suis*) and pseudorabies, a herpes virus not related to true rabies. Were foot-and-mouth disease to be reintroduced to the United States, feral pigs could become a reservoir for this highly infectious virus (*Aphtae epizooticae*) and make eradication difficult if not impossible. Swine brucellois is a disease of the reproductive tract that causes spontaneous abortions, stillbirths, and infertility in both male and female pigs. Pseudorabies, also known as "mad itch," causes abortion and mummified fetuses in pregnant domestic sows and can be fatal in piglets less than a month old. In other livestock, pseudorabies is an infection of the central nervous system that causes loss of appetite, staggering, and spasms, and is almost always fatal. Its early symptoms include scratching and rubbing and biting that can result in self mutilation. The virus spreads from wild pigs to domestic pigs through venereal contact.

The Invasive Species Specialist Group (ISSG) of the IUCN has nominated the feral pig as one of "100 of the World's Worst" invasive species.

Management. Management of feral hogs can be controversial when the interests of hunters conflict with those of agriculturalists and environmentalists. It is also extremely difficult to eradicate established populations because they are so prolific and so mobile.

Hunting is the best means of controlling populations. Using specially trained tracking and catch dogs or mules and dogs are popular practices. Live-trapping is also used. Regulations for hunting feral hogs, operating hunting facilities, and importing feral hogs vary from state to state.

Selected References

"Feral Pig Hunting Information." Wisconsin Department of Natural Resources, 2008. http://dnr.wi.gov/org/land/wildlife/HUNT/Pig/Pig_Hunting.htm.

IUCN/SSC Invasive Species Specialist Group (ISSG). "Sus scrofa (mammal)." ISSG Global Invasive Species Database, 2008. http://www.issg.org/database/species/ecology.asp?si=73&fr=1&sts.

Stevens, Russell. "The Feral Hog in Oklahoma." Samuel Roberts Noble Foundation, Inc., 1999. http://www.noble.org/ag/wildlife/feralhogs/.

Taylor, Rick. "The Feral Hog in Texas." Texas Parks and Wildlife, 2003. http://www.tpwd.state.tx.us/huntwild/wild/nuisance/feral_hogs/.

West, B. C., A. L. Cooper, and J. B. Armstrong. "Managing Wild Pigs: A Technical Guide." *Human-Wildlife Interactions Monograph* 1: 1–55, 2009. Berryman Institute. http://www.berryman institute.org/pdf/managing-feral-pigs.pdf.

■| House Mouse

Scientific name: *Mus musculus*
Family: Muridae

Native Range. Most sources state Asia or, more specifically, Central Asia as the place of origin of this species, but recent genetic evidence suggests that the Indian subcontinent may have been the earliest center of radiation.

Distribution in the United States. Throughout, including all Hawaiian islands, but absent from most of Alaska.

Description. These small Old World mice have relatively large round ears and prominent black eyes. The fur is gray to brown, with the undersides either a lighter shade than the back or a buffy white. The feet are a drab buff color, and the tips of the toes are white. The long grayish-brown tail, close to half the total body length, is essentially hairless; although lighter on the underside, it is not distinctly bicolor. Unlike New World mice, the incisors are not grooved. Adult head-body length is about 3 in. (65–95 mm); the tail is 3–4 in. (60–105 mm) long. Average adult weight is 0.5–0.8 oz. (17–25 g).

The presence of their droppings often alerts people to their presence. Fecal pellets are 0.25 in. (6 mm) long and have longitudinal ridges and square ends.

Related or Similar Species. New World mice that also may occupy dwellings and other buildings include deer mice (*Peromyscus maniculatus*) and white-footed mice (*P. leucopus*). Both of these rodents have white bellies, sharply demarcated from the darker back, and tails that are covered with hair and distinctly bicolor. The tail of the white-footed mouse is shorter than its head-body length. The incisors of both are grooved.

Introduction History. The house mouse has long been associated with humans. The earliest evidence of its commensal relationship with people goes back to a Neolithic site in Turkey dated at 8,000 BP. It was able to spread from settlement to settlement across Asia with expanding Neolithic populations and later along trade routes into Europe, where two subspecies, *M. m. musculus* and *M. m. domesticus*, were found by 4,000 years ago. It was likely accidentally introduced into Florida by Spanish explorers in the early 1500s and came to what is now the northern United States with French fur traders and English settlers in the early 1600s. This small rodent easily stows away in tiny spaces and in grain and other food stores on ships. Ships carrying goods to English colonists in the Pacific spread house mice to remote islands.

Habitat. House mice have had a commensal relationship with humans for at least 8,000 years. Thus, they are usually found in close association with human dwellings and other structures such as granaries, barns, and stores. They may live in stone walls, fence-rows, cultivated fields, and other areas of dense cover close to buildings during warmer seasons and retreat to buildings during winter. They do not hibernate. House mice nest in cracks and crevices in stone walls, in woodpiles, behind rafters, or in other snug places near food, rarely moving more than 50 ft. (16 m) from these secure spots to feed. In dense grass,

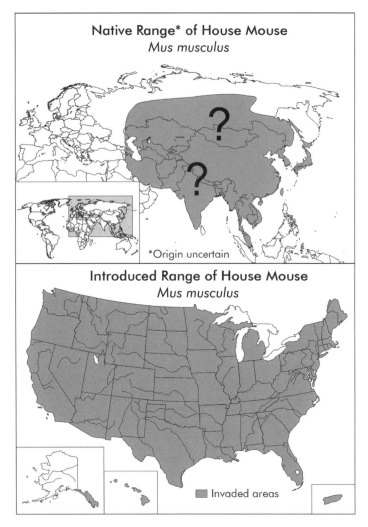

Native Range* of House Mouse
Mus musculus

*Origin uncertain

Introduced Range of House Mouse
Mus musculus

■ Invaded areas

Top: The house mouse is of Asian origin, and recent genetic analysis supports the Indian subcontinent as its likely source. However, many have long thought of Central Asia as the starting place for this rodent, which has traveled around the world with humans. Bottom: The house mouse occurs throughout the United States, except for most of Alaska.

they will make their own runways or share those made by native mice. House mice are rarely encountered in undisturbed or natural habitats.

Diet. House mice prefer cereal grains, but consume a variety of other plant material including fleshy roots, leaves, and stems. They also eat insects and sometimes meat.

In houses, they will nibble on any type of human food as well as glue, paste, and soap. They are physiologically capable of extracting much of the water they need from their food and concentrate their urine, so they may live without free water if the food supply available to them allows.

Life History. House mice have a very high reproductive potential and may, in mild climates, breed all year. Free-living populations outside of human habitations tend to have a seasonal breeding pattern and may only breed from early summer into fall. A female produces 15–150 young in a year, depending on conditions. Individual mice will make their own nests even though they live in groups and share escape holes, latrines, and

House Mice in the Laboratory

Domesticated house mice are common laboratory animals in medical and genetic research. Many inbred strains have been developed since Clarence Cook Little developed the first such population in 1909. Mutant strains have been since created by normal breeding, by insertion of foreign genes, and by gene knockout, a process that makes selected genes inoperable. Cloning mice has been possible since 1998. In 2002, sequencing the mouse genome, which has many human homologues that make it such a valuable model for research on human diseases, was finalized.

feeding areas. The nests are constructed of finely shredded paper and other soft fibrous materials. After a gestation period of 19–21 days, a female gives birth to a litter of 5 or 6 (range 2–13) blind, hairless young whose ears and eyes are closed. Young mice are fully furred at 10 days of age, open their eyes at 14 days, and are weaned at 21 days. After weaning, they leave their mother's territory. Females are sexually mature when 35 days old, males when 60 days old. A female may have 5–10 litters in a year. The average lifespan is 12–18 months.

The house mouse has relatively large rounded ears and prominent black eyes. Neither the dark back nor the upper tail is sharply demarcated from paler undersides. (Gertjan Hooijer/Shutterstock.)

Impacts. House mice have not contributed to the extinction of any species in the United States, nor do they compete with native mice. They do not carry the hantavirus that can be problematic with infestations of native mice in the western states. They also have not caused the serious health problems associated with Old World rats (see Mammals, Black Rat, and Norway Rat), although they may carry some of the same pathogens. However, they are major agricultural pests in some areas because they consume grains and other feedstuffs and contaminate food with their urine and droppings. In buildings, they gnaw and can destroy electric wiring (creating a fire hazard), insulation, woodwork, furniture, upholstery, and clothing.

Management. House mice can be controlled with poisons, fumigants, traps, and repellants. To mouse-proof a structure, all openings greater than 0.25 in. (6 mm) must be blocked. All grain, pet food, and human food should be cleaned up and stored in rodent-proof containers. Outside, removing debris and cutting and thinning dense vegetation can discourage house mice. On those islands where mice have been eradicated, an anticoagulant poison was administered.

Selected References

Ballenger, L. "*Mus musculus.*" Animal Diversity Web, University of Michigan Museum of Zoology, 1999. http://animaldiversity.ummz.umich.edu/site/accounts/information/Mus_musculus.html.

Barwell, Ezra. "*Mus musculus* (mammal)." ICUN/SSC Invasive Species Group (ISSG), ISSG Global Invasive Species Database, 2006. http://www.issg.org/database/species/ecology.asp?fr=1&si=97.

Davis, William B., and David J. Schmidly. "House Mouse." Mammals of Texas—Online version. Texas Tech University, 1997. http://www.nsrl.ttu.edu/tmot1/mus_musc.htm.

"House Mouse." Wikipedia, 2009. http://en.wikipedia.org/wiki/Mus_musculus#Mice_as_an_invasive_species.

Timm, R. M. "House Mouse." Pests of Homes, Structures, People, and Pets. Pest Notes, Publication 7483, University of California, Division of Agriculture and Natural Resources, 2006. http://www.ipm.ucdavis.edu/PMG/PESTNOTES/pn7483.html.

■|Indian Mongoose

Also known as: small Indian mongoose, small Asian mongoose
Scientific name: *Herpestes javanicus*
Family: Herpestidae

Native Range. Southwest and South Asia, from Iran through India and Myanmar (Burma) to the Malay Peninsula and Java. Populations in the West Indies and Hawai'i derive from animals originating in eastern India that were brought first to Jamaica.

Native Range of Indian Mongoose
Herpestes javanicus

Introduced Range of Indian Mongoose
Herpestes javanicus

■ Invaded areas

Distribution in the United States. Hawai'i (Hawai'i, Maui, Moloka'i, and O'ahu); Puerto Rico; U.S. Virgin Islands (St. Croix and St. John).

Description. Mongooses are small, weasel-like carnivores with slender bodies and short legs. The long tail, equal to about 40 percent of head-body length, is robust and muscular at the base and tapers gradually toward the tip. The narrow head has a pointed snout and short ears. The short, soft, brown fur has golden flecks and is paler on the undersides than on the back. Their eyes are amber. Adults have head-body lengths of 20–26 in. (500–650 mm) and weigh 0.6–2 lbs. (300–900 g). Males are larger than females and have wider heads and more robust bodies.

Introduction History. The first successful introduction of mongooses to the Caribbean region occurred in 1872, when four males and five females from Kolkatta (Calcutta) were released in Jamaica. They were intentionally brought to the islands to control rats that were destroying the sugarcane crop. From Jamaica, mongooses were

Top: The Indian mongoose is native to southern Asia from Iran to Southeast Asia. It was brought to Jamaica to control rats and from there transported to other Caribbean islands and to Hawai'i. Bottom: The Indian mongoose is well established on Puerto Rico and several of the Hawaiian Islands. (Both maps adapted from Yamada, F., and K. Sugimura. "Negative Impact of an Invasive Small Indian Mongoose *Herpestes javanicus* on Native Wildlife Species and Evaluation of a Control Project in Amami-Ohshima and Okinawa Islands, Japan." *Global Environmental Research* 8: 117–24, 2004. http://www.aires.or.jp/publication/ger/pdf/08-02-02.pdf.)

A. The Indian mongoose is a weasel-like animal with a robust tail. (Bill Hubick Photography.) B. The head is narrow with a pointed snout and amber eyes. (Bill Hubick Photography.)

deliberately taken to other Caribbean islands including, between 1877 and 1879, Puerto Rico, St. Croix, and St. John for the same rat-control purpose. In 1883, Jamaican mongooses were imported by sugar growers on the Big Island of Hawai'i. Later they were taken to Mau'i, Moloka'i, and O'ahu.

Habitat. In Puerto Rico and the U.S. Virgin Islands, mongooses are most abundant in drier habitats, but in Hawai'i, they inhabit rainforest. They may also be found near human settlements.

Diet. Mongooses feed primarily on invertebrates, but will take small vertebrates and fruits. On O'ahu and Moloka'i, cockroaches are a major part of the diet. Studies in the West Indies and on the Hawaiian islands indicate that the main mammals consumed are introduced rats and house mice. Birds and reptiles usually comprise minor parts of the diet, although consumption of the eggs of ground-nesting birds and of sea turtles is well documented. They are strictly diurnal hunters.

Life History. The time and length of the breeding season varies with latitude. In Hawai'i and the U.S. Virgin Islands, pregnant females have been trapped from February through August. Gestation lasts 49 days. Litter size ranges from one to five pups, but usually is two. At six weeks of age, young mongooses begin to hunt with their mothers, and they stay with her until they reach sexual maturity at 4–6 months. Females may produce 2–3 litters a year. Life expectancy is 4–5 years.

Impacts. Although the IUCN lists the Indian mongoose as one of the world's 100 worst invaders, its damage to the native fauna of Caribbean islands and Hawai'i may be exaggerated. In the early twentieth century in Puerto Rico, for example, mongooses were blamed for the decline of five ground-nesting birds (Key West Quail-dove [*Geotrygon chrysia*], Bridled Quail-dove [*G. mystacea*], Black Rail [*Laterallus jamaicensis*], Short-eared Owl [*Asio flammeus*], and Puerto Rican Nightjar [*Caprimulgis moctitherus*]). The nightjar was believed to have gone extinct, but was rediscovered in 1961 in areas with and without mongoose populations. With the exception of the Black Rail, populations of all these birds have rebounded, suggesting a balance may have developed between the native birds and exotic predator. By the 1980s, the Bridled Quail-dove had shifted to nesting in trees and was common on St. Croix.

In Hawai'i also, the mongoose is blamed for the reduction or extirpation of several endemic birds, but documentation of predation is difficult and rare. Among the birds listed as threatened by mongooses are eight on the federal list of endangered species: the Hawaiian Goose or Nēnē (*Branta sandvicensis*), the Hawaiian Crow or 'Alalā (*Corvus hawaiiaensis*), the Hawaiian Duck or Koloa (*Anas wyvilliana*), the Hawaiian Coot or Alae ke'koke'o (*Fulica alai*), the Hawaiian Stilt or A'eo (*Himantopus mexicanus knudseni*), the Hawaiian Gallinule or Alae 'ula (*Gallinula chloropus sandvicensis*), the Hawaiian Petrel or 'Ua'u (*Pterodroma sandwichensis*), and the Newell Shearwater or 'A'o (*Puffinus auricularis newelli*).

Contrary to popular stories, mongooses are not responsible for black rats nesting in roofs and trees; this is the rats' natural behavior. Mongooses, however, do seem to have shifted the relative abundance of introduced rats in Hawai'i in favor of black rats to the detriment of Norway rats and Polynesian rats (*Rattus exulans*), both ground-nesters. A similar pattern has been noted on St. Croix, where there are more black rats than Norway rats in mongoose habitat. In Puerto Rico, Norway rats exist only in areas free of mongooses, but black rats co-occur with mongooses.

Mongooses were not introduced to St. Croix or Nevis until well after the disappearance of two native snakes, the Saint Croix racer (*Alsophis sanctaecrucis*) and orange-bellied racer (*A. rufiventris*), so blaming mongooses for the reptiles' extinctions is erroneous. However, they are implicated in the loss of the Saint Croix ground lizard (*Ameiva polops*) and in reduced populations of lizards and amphibians elsewhere. Mongooses are known to prey upon the eggs and hatchlings of four sea turtle species on Caribbean islands: the hawksbill sea turtle (*Eretmochelys imbricata*), leatherback turtle (*Dermochelys coriacea*), green sea turtle (*Chelonia mydas*), and loggerhead sea turtles (*Caretta caretta*).

Since mongooses do prey on reptiles and birds, they represent one more danger to already threatened species on islands and are considered a major deterrent to the recovery or reestablishment of island endemics.

Management. Mongooses are often naturalized members of ecosystems into which they were introduced and may not need management. Trapping is the most common tool used in sensitive areas such as nesting grounds of seabirds and sea turtles. However, trapping is expensive and labor intensive, and has only temporary results, so it is not generally employed. The use of diphacinone, an anticoagulant, has had positive results in trials. Prevention of introduction into areas where mongooses do not already exist is the preferred course of action. In Hawai'i, state law forbids any person to introduce, keep, or breed mongooses without permits from the Hawai'i Department of Agriculture, which will not issue them for mongoose-free islands.

Selected References

Hays, Warren S. T., and Sheila Conant. "Biology and Impacts of Pacific Island Invasive Species. 1. A Worldwide Review of Effects of the Small Indian Mongoose, *Herpestes javanicus* (Carnivora: Herpestidae)." *Pacific Science* 61 (1): 3–16, 2007.

"Mongoose (*Herpestes javanicus*)." Hawai'i Invasive Species Partnership, 2008. http://www .hawaiiinvasivespecies.org/pests/mongoose.html.

Roy, Sugoto. "*Herpestes javanicus* (Mammal)." IUCN/SSC Invasive Species Specialist Group, ISSG Global Invasive Species Database, 2006. http://www.issg.org/database/species/ecology.asp?si=86.

■|Norway Rat

Also known as: Brown rat, common rat, sewer rat, wharf rat
Scientific name: *Rattus norvegicus*
Family: Muridae

Native Range. Northern China and Mongolia, perhaps originally living along stream banks, but for millennia living in close association with human settlement. (This rodent did not originate in Norway, as its name mistakenly implies. Indeed, it was unknown in Europe until the medieval period and did not make its way into Western Europe until the early 1700s.)

Distribution in the United States. Found throughout the United States.

Description. Norway rats have coarse brown or dark-gray fur dorsally with underparts a lighter shade of the general body color. The tail is hairless and scaly, and shorter than the head-body length. The ears are prominent but relatively short and bald. Adult head-body length averages about 10 in. (25 cm), total length about 15.5 in. (39 cm). Adult body weight ranges from 0.5 to 1.0 lb. (200–400 g); males are larger than females.

The presence of these nocturnal animals is often indicated by their droppings, the effects of gnawing and rubbing, and sounds such as scratching and squeaking in the walls. Droppings occur in runways, feeding areas, and near sites of shelter; single pellets can reach 0.75 in. (2 cm) in length and 0.25 in. (0.6 cm) in diameter.

Related or Similar Species. The smaller black rat (*Rattus rattus*), another invasive species (see Mammals, Black Rat), has a tail that is longer than its head-body length. Native woodrats (*Neotoma* spp.) have white undersides. The black rat is much rarer in the United States than the Norway, having been largely replaced by it after the Norway rat arrived in this country.

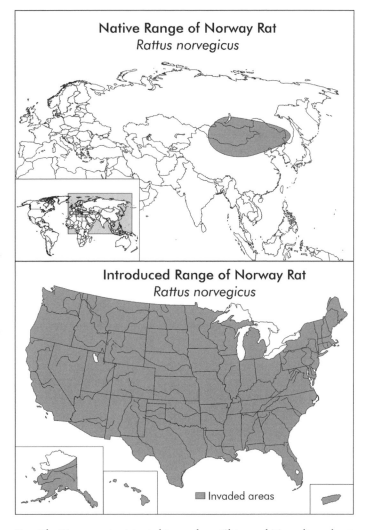

Top: The Norway rat originated in northern China and Mongolia and not Norway, as its common and scientific names imply. Bottom: The Norway rat arrived in the United States at the time of the American Revolution and largely replaced alien black rats. Today it is found in all 50 states and Puerto Rico.

Norway rats have coarse brown or gray fur on the back; the underparts are lighter. The hairless scaly tail is shorter than head-body length. (S. Cooper Digital/Shutterstock.)

Introduction History. The early history of the rat as a human commensal is still to be unraveled, but it appears to have spread along routes of human migration and trade out of northern China and Mongolia during the Middle Ages. Stowaways on ships, Norway rats were present in England by 1730 and soon were reported in other European countries, reaching Spain by 1800. The first rats showed up in North America prior to or during the American Revolution (1770s), according to some accounts, arriving in grain stores brought in with Hessian soldiers fighting with the British against the American colonists. Rats then moved from port to port as accidental hitchhikers on ships. On land, their prodigious ability to reproduce, their adaptability to a wide range of habitat and food types, their propensity to live with humans, and their ability to climb, swim, jump, and dig allowed them to expand their range and invade farmstead, village, town, and city. When overcrowding occurs, they will undertake mass migrations to new areas.

The rats on Rat Island, Alaska, finally eradicated in 2009, derived from the wreck of a Japanese ship in 1780.

Habitat. The Norway rat occurs in close association with humans, occupying such man-made environments as cities, towns, farms, garbage dumps, and sewers. They may also be found in disturbed and human-modified habitats such as marshes, vacant lots, and open fields and may dig complex systems of burrows in the banks of streams and canals. They are excellent swimmers. Away from water, they are ground dwellers, and in buildings, they prefer cellars and lower floors, although they can climb. They may also burrow under foundations and sidewalks as well as roads and railroad tracks.

Diet. True omnivores, Norway rats eat everything humans eat and more. Experimental studies show that, if available, they will select a well-balanced, nutritious diet of fresh foods including cereals, meat, fish, nuts, and fruits. Household garbage usually provides adequate moisture and limits their need for drinking water. They forage mainly at night, led by their excellent sense of smell. Cautious about new food, they avoid unfamiliar items at first and then test small amounts to see if it will make them sick. They can detect very low levels of contaminants, making poisoning a problematic strategy for rat control.

Life History. Norway rats construct nests below ground and line them with leaves, twigs, shredded paper, or other fibrous materials. Each female gives birth to a litter of about eight pups after a gestation period lasting 22–24 days. Several females may use the same communal nest, and all females care for all of the young. The small, hairless pups do not open their eyes for 14–17 days and will not be weaned until 3–4 weeks old, at which time they leave the nest. Females come into estrous about 18 hours after giving birth and will mate.

Good Rats

Rats are often associated with filth and disease, but domestic descendents of the Norway rat have had a positive influence on humanity. They are commonly used in research and are familiar to undergraduates as the albino laboratory rat that runs through mazes and is subjected to a variety of behavioral experiments in psychology classes. Domestication has also produced the fancy or pet rat, a clean and intelligent companion animal that comes in a variety of colors and markings.

Most American laboratory rats trace their origins to the Wistar Institute in Philadelphia, where in 1906 an albino strain was developed by Helen Dean King for use in biological research. Later strains include the Sprague-Dawley rat, a multipurpose research animal prized for its docility; the black-hooded Long Evans rat used in behavioral and obesity studies; Zucker rats, lean and obese forms of which are used in the understanding the genetics of obesity; and hairless rats that serve as models for research on immune-deficiency diseases and genetic kidney diseases.

Domestic rats may go back to the eighteenth and nineteenth centuries, when rat-catchers were employed throughout Europe to trap rats. Many of the captives ended up as the bait in a bloodsport in which terriers competed (and people made wagers) to see how fast they could kill all the rats in a pit. Some naturally occurring albinos or oddly colored and marked rodents were apparently kept and the tamest bred and sold as pets. Rat fancy first became a formal hobby in England when the National Mouse Club accepted rats at an exhibition in 1901. In 1976 the National Fancy Rat Society was established in England, and in 1983 the American Fancy Rat and Mouse Association was founded in California.

Sources: "Fancy Rat." Wikipedia. http://en.wikipedia.org/wiki/Fancy_rat.

Hanson, Anne. "History of the Norway rat (*Rattus norvegicus*)" Rat behavior and biology. 2003, 2004. http://www.ratbehavior.org/history.htm.

"Laboratory rat." n.d. Wikipedia. http://en.wikipedia.org/wiki/Laboratory_rat.

Breeding occurs year round, although mating peaks in spring and fall in the temperate climates of the United States. An individual female may breed as many as seven times and produce some 60 offspring a year. Young males reach reproductive maturity at three months of age, females at four months. Normal lifespan for a wild rat is two years.

Impacts. Worldwide, especially on islands, Norway rats have been implicated in the extinction or range reduction of native species through both competition and predation. On Rat Island in the western Aleutians of Alaska, for example, rat predation on eggs and chicks led to the extirpation of such burrow-nesting seabirds as Cassin's Auklet (*Ptychorampus aleuticus*), Tufted Puffin (*Fratercula cirrhata*), and Storm Petrels (*Oceanodroma* spp.), and likely contributed to population losses among several other ground- and crevice-nesting shorebirds. In most of the United States, they are mainly an agricultural pest and urban menace. Rats consume crops and contaminate stored food with their feces and urine. They eat bird eggs and kill young poultry. They gnaw and dig their way into buildings, where they can cause structural damage to walls and floors, destroy insulation as they nest in and burrow in walls, and gnaw electrical wires to cause short circuits and fires. Their burrows may undermine roads, bridges, canals, and levees.

The fleas and lice on rats carry infectious bacterial diseases to both human and livestock, including murine typhus and bubonic plague, although the latter has been more closely associated with black rats throughout history. Flies, mosquitoes, and ticks transmit tularemia from rats to humans; their ticks also carry spotted fever. Contact with the urine of rats in contaminated soil, water, or plants can transmit the bacterium *Leptospira interrogans*, the cause of infectious jaundice or leptospirosis, to livestock and humans. Food poisoning (salmonellosis) from contaminated feed is particularly dangerous to horses. Contaminated food may also harbor the nematodes that cause trichinosis. Bites can transmit rat bite fever. Norway rats are responsible for millions of dollars of damage to crops and buildings in the United States every year.

Management. On islands, poisoning rats with anticoagulants has been successful in eradicating populations. Less slow-acting poisons are quickly learned and avoided by these highly intelligent rodents. Other effective control methods include rat-proofing buildings by blocking entry points, storing food in rat-proof containers, and removing potential shelter sites such as trash heaps, overgrown vegetation, and woodpiles.

Selected References

Hanson, Anne. "History of the Norway Rat (*Rattus norvegicus*)." Rat Behavior and Biology, 2004. http://www.ratbehavior.org/history.htm.

ICUN/SSC Invasive Species Group (ISSG). "*Rattus norvegicus* (mammal)." ISSG Global Invasive Species Database, 2006. http://www.issg.org/database/species/ecology.asp?si=159.

Myers, P., and D. Armitage. "*Rattus norvegicus*." Animal Diversity Web, University of Michigan Museum of Zoology, 2004. http://animaldiversity.ummz.umich.edu/site/accounts/information/Rattus _norvegicus.html.

"Norway Rat, *Rattus norvegicus*." eNature.com, 2007. http://www.enature.com/fieldguides/detail.asp ?recnum=MA0095.

"Norway Rats." Internet Center for Wildlife Damage Management, Cornell University, Clemson University, University of Nebraska–Lincoln, and Utah State University, 2005. http://icwdm.org/handbook/rodents/NorwayRats.asp.

■| Nutria

Also known as: Coypu, swamp beaver
Scientific name: *Myocastor coypus*
Family: Myocastoridae

Native Range. Subtropical South America: Argentina, southern Brazil, Bolivia, Chile, Uruguay, and Paraguay. These large, semiaquatic rodents are found chiefly in the lowlands, but may range to elevations of 4,000 ft. (1,190 m) in the Andes.

Distribution in the United States. Alabama, Arkansas, Colorado, Delaware, Florida, Georgia, Idaho, Louisiana, Maryland, Mississippi, New Mexico, North Carolina, Oklahoma, Oregon, Tennessee, Texas, Virginia, and Washington.

Description. This is a fairly large, stocky rodent with a large, nearly triangular head and short neck. The eyes and ears are small and positioned high on the head. The nostrils and mouth can be sealed to prevent the intake of water when it dives or feeds underwater. The prominent front teeth range in color from yellow to dark orange. The tail, which comprises roughly one-third of total body length, is round and scaly with a sparse cover of bristles. Four toes on their hind feet are webbed, but the outermost, fifth toe is free. The much smaller, black forefeet have four unwebbed toes that provide dexterity in digging and manipulating food items. The 4–5 pairs of teats on the female are located high on the flank so that young may nurse while their mother is in the water or lying on her abdomen. The fur that made them so valuable in the past is actually a dense, velvety slate-gray undercoat concealed beneath long, coarse, but glossy yellowish-brown guard hairs. White hairs cover the chin. Head and body length ranges from 20 to 25 in.; tail length from 10–17 in. (25–43 cm). Nutria may weigh as much as 15–20 lbs. (6.8–9.0 kg); males are larger than females.

Nutria droppings are cylindrical in shape and as much as 3 in. (7.6 cm) long. Fine, length-wise grooves distinguish them from the fecal pellets of other animals. Droppings will be seen floating on the water, along the shore, and at feeding sites.

Nutria are nocturnal and most commonly seen at twilight, usually when they are swimming. A distinguishing factor, then, is the narrow tail snaking behind them or arched out of the water.

Related or Similar Species. Nutria are similar to native beaver (*Castor canadensis*) and muskrat (*Ondatra zibethica*), but intermediate between the

Top: Nutria are native to the wetlands of South America, south of the Tropic of Capricorn. (Adapted from map at http://en.wikipedia.org/wiki/File:Nutria-SouthAmerica.gif.) Bottom: Nutria being raised on fur farms escaped captivity and established populations in numerous states, where they damage native marsh communities and can become agricultural pests. (Adapted from Fuller, Pam. "*Myocastor coypus.*" USGS Noningidenous Aquatic Species Database, Gainesville, FL, 2005. http://nas.er.usgs.gov/queries/factsheet.aspx?speciesID=1089.)

two in size and appearance. The nutria's head resembles that of a beaver, but its tail is more like that of a muskrat. Beavers have flat, paddle-shaped tails, whereas muskrat tails are slender and flattened from side to side. Nutria are twice as large as muskrats and about a third the size of beavers. The tracks of nutria can be confused with those of beavers, which have five webbed toes. Nutria slides, slick muddy trails into the water, are much narrower than those of beavers. Nutria make flattened, circular platforms of coarse emergent vegetation 3–6 ft. (1–2 m) in diameter, where they feed, loaf, and groom, and sometimes give birth. Multiple runways radiate out from these platforms, which could be mistaken for muskrat houses.

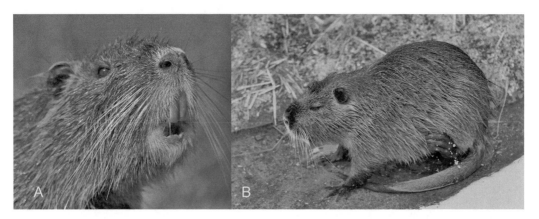

A. The nutria's prominent teeth range in color from yellow to dark orange. (Sters/Shutterstock.) B. Nutria are large, stocky rodents. Their small eyes and ears are set high on their large heads. (Bodil 1955/Shutterstock.)

Introduction History. The first nutria were deliberately introduced into the United States from South America in 1899 in an effort to establish new ventures in fur farming. The first attempt was made in Elizabeth Lake, California, but the animals failed to reproduce. Between 1899 and 1940, nutria ranches were started in Washington, Oregon, Michigan, New Mexico, Louisiana, Ohio, and Utah. The boom years for production were in the 1930s, after which time the industry collapsed due to low fur prices during World War II, competition from beaver pelts, and poor reproduction. Some ranchers released their nutria, while others simply failed to round up any that escaped during storms and floods, or because of poor confinement structures. Since these animals had all been bred on ranches, the free-roaming animals now are considered feral.

For a time, nutria were viewed as weed cutters to control noxious vegetation or as ways to increase the take of fur trappers. Both state and federal agencies, as well as individuals, introduced nutria into Alabama, Arkansas, Georgia, Kentucky, Maryland, Mississippi, Oklahoma, Louisiana, and Texas.

California's small feral population was eradicated in the 1970s. Nutria were eradicated or simply failed to survive in Idaho, Indiana, Kansas, Kentucky, Michigan, Minnesota, Missouri, Montana, Nebraska, Ohio, and Utah.

Maryland received nutria in 1943 as part of a federal program to establish an ill-fated experimental fur station in Blackwater National Wildlife Refuge. The first feral animals were reported in 1952, after the fur station proved unprofitable. Early attempts to eradicate nutria failed, and the population exploded and expanded their range into Delaware. Only recently have they finally been eradicated from Blackwater NWR, but small populations remain in Maryland. Elsewhere in the Chesapeake Bay region, a small population in Virginia is believed to have moved in from North Carolina.

In the early 1930s, nutria were introduced into Louisiana near New Orleans, where they were quickly trapped out. They were introduced again in 1938, this time for fur farming, and many escaped captivity during a 1940 hurricane. In the next 60 years, feral populations dispersed into West Texas (1941), Arkansas (early 1960s) and Mississippi, reaching Tennessee in 1996.

In Oregon and Washington, nutria were first imported in the late 1930s for fur farming. By the 1950s, some 600 farms had been established in the two states. Storms and flooding damaged holding pens and allowed animals to escape. Viable feral populations continue to exist in the Pacific Northwest. In Washington State, after a series of warm winters, nutria seem to be expanding their distribution area.

Small feral populations survive in Colorado, Florida, New Mexico, North Carolina, Oklahoma, and Texas.

Habitat. Nutria remain close to permanent water and may be associated with fresh and brackish marshes, swamps, and the shores of rivers, bayous, lakes, and drainage canals. They shelter in burrows that they excavate themselves, but they will also occupy the burrows of other species. Their burrows may be simple tunnels or a complex network of passages and chambers up to 50 ft. (15 m) long.

Diet. Nutria are herbivores and consume a wide variety of terrestrial and aquatic plants, including the bark of trees. In the United States, favored foods include cordgrasses (*Spartina* spp.), bulrushes (*Scirpus* spp), spikerushes (*Eleocharis* spp.), flatsedges (*Cyperus* spp.) pickerel-weeds (*Pontederia* spp.), arrowheads (*Sagittaria* spp.), and cattails (*Typha* spp.). They also eat baldcypress (*Taxodium distichum*) seedlings, duckweeds, and waterhyacinth (*Eichornia crassipes*; see Volume 2, Aquatic Plants, Waterhyacinth), an invasive plant species. Often nutria will cut a plant off at the waterline and carry it to a feeding platform that they have constructed in shallow water. Nutria also dig beneath the marsh surface and feed on the root mat.

Life History. Nutria construct nests of coarse plant material in dens within their burrows. They reproduce throughout the year, producing two or litters annually. The female will come into estrus every 24–26 days and stay in heat 1–4 days. Gestation is about 130 days. Typically, 4–6 precocial young are born, although the original number of embryos may have been much higher. Prenatal embryo losses are common, especially in cold weather and among poorly nourished females; miscarriage rates are high. Estrus begins soon after the female miscarries or gives birth. She will mate with several males each time she comes into estrus.

Young are born with full coats of fur and open eyes. Within a day, they swim and feed on plants. They will be weaned in 5–8 weeks. Nutria are sexually mature in 4–6 months. Adults usually live in pairs, but large numbers may congregate in favorable habitat. Individuals occupy only a small area throughout their lives, rarely traveling more than 600 ft. (180 m) from their dens each day. Wild nutria rarely live more than three years. Young may be taken by coyotes, foxes, raccoons, alligators, and owls; humans are the only predators of adults.

Impacts. The damage nutria do to native ecosystems is related to both their feeding and their burrowing. Animals feeding directly on the root mat of marsh plants produce what is known as an "eat out." The mat binding the marsh substrate together is weakened and soil is washed away, pockmarking the marsh with holes. In Louisiana some "eat outs" are 500 ac. (202 ha) in size.

Deep swimming channels come to fragment the marsh and, in coastal areas, allow the influx of salt water. The channels or "runs" create more edge habitat exposed to erosion by wind and water. Productive wetlands can be turned into barren mudflats or open water, destroying habitat for marshland invertebrates, fish, and birds. Nutria feeding on the rhizomes of sea oat on the barrier islands of Mississippi has exposed the sand dunes to erosion.

Large burrows made by nutria weaken river banks, levees, dams, and the sides of drainage canals and can lead to cave-ins and erosion during heavy rains.

In man-made environments, nutria may girdle orchard trees as well as ornamental trees and shrubs. In the 1950s, expanding populations in Louisiana became pests in rice and sugar cane fields.

Nutria has been nominated by the IUCN as one of "100 of the World's Worst" invasive species.

Management. Trapping and shooting have been the most effective ways of controlling feral populations of nutria, and high fur prices once made these viable strategies. With the fall of prices, Louisiana established a nutria control program that included a bounty system to encourage the harvest of 400,000 nutria a year from coastal areas. Part of that program involves promoting nutria as a source of lowfat, high-protein meat.

The legal status of nutria varies from state to state, and influences control methods. In Delaware, they are considered furbearers and regulated as such. In Maryland, they are listed as "unprotected" and can be controlled as wildlife species when they cause damage to ecological or economic resources. In Virginia, nutria are "nuisance" species; there is open season all year for trapping them. In Washington State, nutria are classified as "prohibited," meaning they may not be released into the wild and may not be transported without a permit from the Washington Department of Fish and Wildlife.

Selected References

Bertolino, S. "*Myocastor coypus* (Mammal)." IUCN/SSC Invasive Species Specialist Group, ISSG Global Invasive Species Database, 2005. http://www.invasivespecies.net/database/species/ecology.asp?si=99&fr=1&sts=.

"Exotic Aquatics of the Gulf Coast: Nutria (*Myocastor coypus*)." La Mer, Louisiana Sea Grant, n.d. http://www.lamer.lsu.edu/invasivespecies/nutria/index.html.

"Invasive Species in the Chesapeake Watershed." Chesapeake Bay Program and Maryland Sea Grant, 2002. http://www.mdsg.umd.edu/issues/restoration/non-natives/workshop/nutria.html.

Link, Russell. "Living with Wildlife: Nutria." Washington Department of Fish and Wildlife, 2006. http://wdfw.wa.gov/wlm/living/nutria.pdf.

"Worldwide Distribution, Spread of, and Efforts to Eradicate the Nutria (*Myocastor coypus*)." National Wetlands Research Center, U.S. Geological Survey, 2008. http://www.nwrc.usgs.gov/special/nutria/namerica.htm.

■ State-by-State Occurrences of Invasive Microorganisms, Fungi, and Animals

Includes only the species featured in the *Encyclopedia*.

Alabama

Microorganisms: West Nile virus
Fungi: Chestnut blight fungus, chytrid frog fungus, Dutch elm disease
Invertebrates
        Cnidarian: Australian spotted jellyfish
        Mollusks: Asian clam, zebra mussel
        Arachnids: Honeybee tracheal mite, varroa mite
        Insects: Argentine ant, Asian tiger mosquito, common bed bug, Formosan subterranean termite, Japanese beetle, multicolored Asian lady beetle, red imported fire ant
Vertebrates
        Fish: Brown trout, rainbow trout
        Birds: Cattle Egret, Eurasian Collared-Dove, European Starling, House Finch, House Sparrow, Monk Parakeet, Rock Pigeon
        Mammals: Black rat, feral cat, feral pig, house mouse, Norway rat, nutria

Alaska

Microorganisms
Fungi: Chytrid frog fungus
Invertebrates
        Tunicates: Chain tunicate
        Insect: Common bed bug
Vertebrates
        Fish: Rainbow trout
        Birds: Cattle Egret, European Starling, Rock Pigeon
        Mammals: Feral cat, house mouse, Norway rat

Arizona

Microorganisms: West Nile virus
Fungi: Chytrid frog fungus
Invertebrates
        Annelid worms: European earthworms
        Mollusks: Asian clam, Chinese mystery snail, New Zealand mud snail, quagga mussel
        Arachnids: Honeybee tracheal mite, varroa mite
        Insect: Africanized honey bee, Argentine ant, Common bed bug, Formosan subterranean termite, multicolored Asian lady beetle
Vertebrates
        Fish: Bighead carp, brown trout, gizzard shad, mosquitofish, rainbow trout
        Amphibians: African clawed frog, American bullfrog
        Birds: Cattle Egret, European Starling, House Sparrow, Rock Pigeon
        Mammals: Feral burro, feral cat, feral horse, feral pig, house mouse, Norway rat

(Continued)

Arkansas

Microorganisms: West Nile virus
Fungi: Chytrid frog fungus, Dutch elm disease
Invertebrates
       Annelid worms: European earthworms
       Mollusks: Asian clam, zebra mussel
       Arachnids: Honeybee tracheal mite, varroa mite
       Insects: Africanized honey bee, Argentine ant, Asian tiger mosquito, common bed bug, Formosan subterranean termite, Japanese beetle, multicolored Asian lady beetle, red imported fire ant
Vertebrates
       Fish: Bighead carp, brown trout, grass carp, rainbow trout
       Birds: Cattle Egret, Eurasian Collared-Dove, European Starling, House Finch, House Sparrow, Rock Pigeon
       Mammals: Feral cat, feral pig, house mouse, Norway rat, nutria

California

Microorganisms: Lyme disease bacterium, West Nile virus
Fungi: Chytrid frog fungus, sudden oak death, white pine blister rust
Invertebrates
       Tunicates: Chain tunicate, colonial tunicate
       Annelid worms: European earthworms
       Mollusks: Asian clam, Chinese mystery snail, golden apple snail, New Zealand mud snail, quagga mussel, zebra mussel
       Crustaceans: Chinese mitten crab, green crab
       Arachnids: Honeybee tracheal mite, varroa mite
       Insects: Africanized honey bee, Argentine ant, Brown marmorated stink bug, common bed bug, Formosan subterranean termite, glassy-winged sharpshooter, multicolored Asian lady beetle, red imported fire ant
Vertebrates
       Fish: Bighead carp, brown trout, mosquitofish, rainbow trout, spotted tilapia
       Amphibians: African clawed frog, American bullfrog
       Birds: Cattle Egret, Eurasian Collared-Dove, European Starling, House Sparrow, Rock Pigeon
       Mammals: Black rat, feral burro, feral cat, feral goat, feral horse, feral pig, house mouse, Norway rat

Colorado

Microorganisms: West Nile virus
Fungi: Chytrid frog fungus, white pine blister rust
Invertebrates
       Annelid worms: European earthworms
       Mollusks: Asian clam, Chinese mystery snail, New Zealand mud snail, quagga mussel, zebra mussel
       Arachnids: Honeybee tracheal mite, varroa mite
       Insects: Common bed bug, Japanese beetle, multicolored Asian lady beetle
Vertebrates
       Fish: Alewife, bighead carp, brown trout, gizzard shad, mosquitofish, rainbow trout
       Amphibians: American bullfrog:
       Birds: Cattle Egret, Eurasian Collared-Dove, European Starling, House Sparrow, Rock Pigeon
       Mammals: Feral cat, feral horse, feral pig, house mouse, Norway rat, nutria

Connecticut

Microorganisms: Lyme disease bacterium, West Nile virus
Fungi: Bat white-nose syndrome, chestnut blight fungus, chytrid frog fungus, Dutch elm disease, white pine blister rust
Invertebrates
        Bryozoan: Lacy crust bryozoan
        Tunicates: Chain tunicate, colonial tunicate
        Annelid worms: European earthworms
        Mollusks: Asian clam, Chinese mystery snail, common periwinkle, zebra mussel
        Crustaceans: Green crab, rusty crayfish
        Arachnids: Honeybee tracheal mite, varroa mite
        Insects: Common bed bug, gypsy moth, hemlock woolly adelgid, multicolored Asian lady beetle
Vertebrates
        Fish: Brown trout, mosquitofish, rainbow trout
        Birds: Cattle Egret, European Starling, House Finch, House Sparrow, Monk Parakeet, Mute Swan, Rock Pigeon
        Mammals: Feral cat, house mouse, Norway rat

Delaware

Microorganisms: Lyme disease bacterium, West Nile virus
Fungi: Bat white-nose syndrome, chestnut blight fungus, chytrid frog fungus, Dutch elm disease
Invertebrates
        Tunicates: Chain tunicate
        Annelid worms: European earthworms
        Mollusks: Asian clam, common periwinkle
        Crustaceans: Green crab
        Arachnids: Honeybee tracheal mite, varroa mite
        Insects: Asian tiger mosquito, brown marmorated stink bug, common bed bug, gypsy moth, hemlock woolly adelgid, Japanese beetle, multicolored Asian lady beetle
Vertebrates
        Fish: Brown trout, rainbow trout
        Birds: Cattle Egret, European Starling, House Finch, House Sparrow, Monk Parakeet, Mute Swan, Rock Pigeon
        Mammals: Feral cat, house mouse, Norway rat, nutria

Florida

Microorganisms: West Nile virus
Fungi: Chytrid frog fungus, Dutch elm disease
Invertebrates
        Cnidarian: Australian spotted jellyfish
        Mollusks: Asian clam, Asian green mussel, golden apple snail
        Arachnids: Honeybee tracheal mite, varroa mite
        Insects: Africanized honey bee, Argentine ant, Asian tiger mosquito, common bed bug, Formosan subterranean termite, multicolored Asian lady beetle, red imported fire ant
Vertebrates
        Fish: Asian swamp eel, bighead carp, brown trout, lionfish, mosquitofish, rainbow trout, spotted tilapia, walking catfish
        Amphibians: Coqui (noninvasive), Cuban treefrog
        Reptiles: Brown anole, Burmese python, green iguana, Nile monitor
        Birds: Cattle Egret, Common Myna, Eurasian Collared-Dove, European Starling, House Finch, House Sparrow, Monk Parakeet, Rock Pigeon
        Mammals: Black rat, feral cat, feral pig, house mouse, Norway rat, nutria

(Continued)

Georgia

Microorganisms: West Nile virus
Fungi: Chestnut blight fungus, chytrid frog fungus, Dutch elm disease
Invertebrates
> Annelid worms: European earthworms
> Mollusks: Asian clam, Asian green mussel
> Arachnids: Honeybee tracheal mite, varroa mite
> Insects: Argentine ant, Asian tiger mosquito, common bed bug, Formosan subterranean termite, hemlock woolly adelgid, Japanese beetle, multicolored Asian lady beetle, red imported fire ant

Vertebrates
> Fish: Alewife, Asian swamp eel, brown trout, lionfish, rainbow trout
> Reptiles: Brown anole
> Birds: Cattle Egret, Eurasian Collared-Dove, European Starling, House Finch, House Sparrow, Rock Pigeon
> Mammals: Black rat, feral cat, feral horse, feral pig, house mouse, Norway rat, nutria

Hawai'i

Microorganisms: Avian malaria
Fungi: Chytrid frog fungus
Invertebrates
> Cnidarian: Australian spotted jellyfish
> Mollusks: Asian clam, giant African snail, golden apple snail, zebra mussel
> Insects: Argentine ant, Asian tiger mosquito, common bed bug, Formosan subterranean termite

Vertebrates
> Fish: Asian swamp eel, bighead carp, brown trout, mosquitofish, rainbow trout
> Amphibians: American bullfrog, coqui
> Reptiles: Green iguana
> Birds: Cattle Egret, Common Myna, European Starling, House Finch, House Sparrow, Japanese White-eye, Rock Pigeon
> Mammals: Black rat, feral cat, feral goat, feral pig, house mouse, Indian mongoose, Norway rat

Idaho

Microorganisms: West Nile virus
Fungi: Chytrid frog fungus, white pine blister rust
Invertebrates
> Annelid worms: European earthworms
> Mollusks: Asian clam, New Zealand mud snail
> Arachnids: Honeybee tracheal mite, varroa mite
> Insects: Common bed bug, multicolored Asian lady beetle

Vertebrates
> Fish: Brown trout, mosquitofish, rainbow trout
> Birds: Cattle Egret, European Starling, House Sparrow, Rock Pigeon
> Mammals: Feral cat, house mouse, Norway rat, nutria

Illinois

Microorganisms: Lyme disease bacterium, West Nile virus
Fungi: Chestnut blight fungus, chytrid frog fungus, Dutch elm disease, white pine blister rust
Invertebrates
> Annelid worms: European earthworms
> Mollusks: Asian clam, Chinese mystery snail, quagga mussel, zebra mussel

        Arachnids: Honeybee tracheal mite, varroa mite

        Insects: Argentine ant, Asian longhorned beetle, Asian tiger mosquito, common bed bug, emerald ash borer, gypsy moth, Japanese beetle, multicolored Asian lady beetle

Vertebrates

        Fish: Alewife, brown trout, gizzard shad, mosquitofish, rainbow trout, round goby, sea lamprey, silver carp

        Birds: Cattle Egret, Eurasian Collared-Dove, European Starling, House Finch, House Sparrow, Monk Parakeet, Rock Pigeon

        Mammals: Feral cat, feral pig, house mouse, Norway rat

Indiana

Microorganisms: Lyme disease bacterium, West Nile virus

Fungi: Chestnut blight fungus, chytrid frog fungus, Dutch elm disease, white pine blister rust

Invertebrates

        Annelid worms: European earthworms

        Mollusks: Asian clam, Chinese mystery snail, zebra mussel

        Arachnids: Honeybee tracheal mite, varroa mite

        Insects: Asian tiger mosquito, common bed bug, emerald ash borer, gypsy moth, Japanese beetle, multicolored Asian lady beetle

Vertebrates

        Fish: Alewife, bighead carp, brown trout, gizzard shad, mosquitofish, rainbow trout

        Birds: Cattle Egret, European Starling, House Finch, House Sparrow, Rock Pigeon

        Mammals: Feral cat, feral pig, house mouse, Norway rat

Iowa

Microorganisms: Lyme disease bacterium, West Nile virus

Fungi: Chestnut blight fungus, chytrid frog fungus, Dutch elm disease, white pine blister rust

Invertebrates

        Annelid worms: European earthworms

        Mollusks: Asian clam, Chinese mystery snail, quagga mussel, zebra mussel

        Crustaceans: Rusty crayfish

        Arachnids: Honeybee tracheal mite, varroa mite

        Insects: Common bed bug, Japanese beetle, multicolored Asian lady beetle

Vertebrates

        Fish: Bighead carp, brown trout, mosquitofish, rainbow trout

        Amphibians: American bullfrog (in DeSoto NWR)

        Birds: Cattle Egret, Eurasian Collared-Dove, European Starling, House Finch, House Sparrow, Rock Pigeon

        Mammals: Feral cat, feral pig, house mouse, Norway rat

Kansas

Microorganisms: West Nile virus

Fungi: Chytrid frog fungus, Dutch elm disease

Invertebrates

        Mollusks: Asian clam, zebra mussel

        Arachnids: Honeybee tracheal mite, varroa mite

        Insects: Asian tiger mosquito, common bed bug, Japanese beetle, multicolored Asian lady beetle

Vertebrates

        Fish: Bighead carp, brown trout, gizzard shad, mosquitofish, rainbow trout

        Birds: Cattle Egret, Eurasian Collared-Dove, European Starling, House Finch, House Sparrow, Rock Pigeon

        Mammals: Feral cat, feral pig, house mouse, Norway rat

(Continued)

Kentucky

Microorganisms: West Nile virus
Fungi: Chestnut blight fungus, chytrid frog fungus, Dutch elm disease, white pine blister rust
Invertebrates
 Annelid worms: European earthworms
 Mollusks: Asian clam, quagga mussel, zebra mussel
 Arachnids: Honeybee tracheal mite, varroa mite
 Insects: Africanized honey bee, Asian tiger mosquito, common bed bug, emerald ash
   borer, hemlock woolly adelgid, Japanese beetle, multicolored Asian lady beetle
Vertebrates
 Fish: Alewife, bighead carp, brown trout, gizzard shad, grass carp, mosquitofish, rainbow
   trout
 Birds: Cattle Egret, Eurasian Collared-Dove, European Starling, House Finch, House
   Sparrow, Rock Pigeon
 Mammals: Feral cat, feral pig, house mouse, Norway rat

Louisiana

Microorganisms: West Nile virus
Fungi: Chytrid frog fungus, Dutch elm disease
Invertebrates
 Cnidarian: Australian spotted jellyfish
 Annelid worms: European earthworms
 Mollusks: Zebra mussel
 Arachnids: Honeybee tracheal mite, varroa mite
 Insects: Africanized honey bee, Argentine ant, Asian tiger mosquito, common bed bug,
   Formosan subterranean termite, multicolored Asian lady beetle, red imported fire ant
Vertebrates
 Fish: Bighead carp, grass carp, rainbow trout, silver carp
 Birds: Cattle Egret, Eurasian Collared-Dove, European Starling, House Finch, House
   Sparrow, Monk Parakeet, Rock Pigeon
 Mammals: Black rat, feral cat, feral pig, house mouse, Norway rat, nutria

Maine

Microorganisms: West Nile virus
Fungi: Chestnut blight fungus, chytrid frog fungus, Dutch elm disease, white pine
  blister rust
Invertebrates
 Bryozoan: Lacy crust bryozoan
 Tunicates: Chain tunicate, colonial tunicate
 Annelid worms: European earthworms
 Mollusks: Asian clam, Chinese mystery snail, common periwinkle
 Crustaceans: Green crab, rusty crayfish
 Arachnids: Honeybee tracheal mite, varroa mite
 Insects: Common bed bug, gypsy moth, hemlock woolly adelgid, multicolored Asian lady
   beetle
Vertebrates
 Fish: Alewife, brown trout, rainbow trout
 Birds: Cattle Egret, European Starling, House Finch, House Sparrow, Mute Swan, Rock
   Pigeon
 Mammals: Feral cat, house mouse, Norway rat

Maryland

Microorganisms: Lyme disease bacterium, West Nile virus

Fungi: Bat white-nose syndrome, chestnut blight fungus, chytrid frog fungus, Dutch elm disease, white pine blister rust
Invertebrates

    Tunicates: Chain tunicate
    Annelid worms: European earthworms
    Mollusks: Asian clam
    Crustaceans: Green crab, rusty crayfish
    Arachnids: Honeybee tracheal mite, varroa mite
    Insects: Argentine ant, Asian tiger mosquito, brown marmorated stink bug, common bed bug, emerald ash borer, gypsy moth, hemlock woolly adelgid, Japanese beetle, multicolored Asian lady beetle

Vertebrates

    Fish: Alewife, brown trout, mosquitofish, northern snakehead, rainbow trout
    Birds: Cattle Egret, European Starling, House Finch, House Sparrow, Mute Swan, Rock Pigeon
    Mammals: Feral cat, house mouse, Norway rat, nutria

Massachusetts

Microorganisms: Lyme disease bacterium, West Nile virus
Fungi: Bat white-nose syndrome, chestnut blight fungus, chytrid frog fungus, Dutch elm disease, white pine blister rust
Invertebrates

    Bryozoan: Lacy crust bryozoan
    Tunicates: Chain tunicate, colonial tunicate
    Annelid worms: European earthworms
    Mollusks: Asian clam, Chinese mystery snail, common periwinkle, zebra mussel
    Crustaceans: Green crab, rusty crayfish
    Arachnids: Honeybee tracheal mite, varroa mite
    Insects: Asian longhorned beetle, common bed bug, gypsy moth, hemlock woolly adelgid, Japanese beetle, multicolored Asian lady beetle

Vertebrates

    Fish: Alewife, brown trout, mosquitofish, rainbow trout
    Amphibians: American bullfrog (on Nantucket, Martha's Vineyard, and Cape Cod)
    Birds: Cattle Egret, European Starling, House Finch, House Sparrow, Mute Swan, Rock Pigeon
    Mammals: Feral cat, house mouse, Norway rat

Michigan

Microorganisms: Lyme disease bacterium, West Nile virus
Fungi: Chestnut blight fungus, chytrid frog fungus, Dutch elm disease, white pine blister rust
Invertebrates

    Annelid worms: European earthworms
    Mollusks: Asian clam, Chinese mystery snail, New Zealand mud snail, quagga mussel, zebra mussel
    Crustaceans: Rusty crayfish, spiny water flea
    Arachnids: Honeybee tracheal mite, varroa mite
    Insects: Common bed bug, emerald ash borer, gypsy moth, Japanese beetle, multicolored Asian lady beetle

Vertebrates

    Fish: Alewife, brown trout, gizzard shad, mosquitofish, rainbow trout, round goby, sea lamprey
    Birds: Cattle Egret, European Starling, House Finch, House Sparrow, Mute Swan, Rock Pigeon
    Mammals: Feral cat, house mouse, Norway rat

(Continued)

Minnesota

Microorganisms: Lyme disease bacterium, West Nile virus
Fungi: Chytrid frog fungus, Dutch elm disease, white pine blister rust
Invertebrates
> Annelid worms: European earthworms
> Mollusks: Asian clam, Chinese mystery snail, New Zealand mud snail, quagga mussel, zebra mussel
> Crustaceans: Rusty crayfish, spiny water flea
> Arachnids: Honeybee tracheal mite, varroa mite
> Insects: Asian tiger mosquito, common bed bug, emerald ash borer, Japanese beetle, multicolored Asian lady beetle

Vertebrates
> Fish: Alewife, brown trout, mosquitofish, rainbow trout, round goby, sea lamprey
> Birds: Cattle Egret, European Starling, House Finch, House Sparrow, Rock Pigeon
> Mammals: Feral cat, house mouse, Norway rat

Mississippi

Microorganisms: West Nile virus
Fungi: Chestnut blight fungus, chytrid frog fungus, Dutch elm disease
Invertebrates
> Cnidarian: Australian spotted jellyfish
> Annelid worms: European earthworms
> Mollusks: Asian clam, zebra mussel
> Arachnids: Honeybee tracheal mite, varroa mite
> Insects: Africanized honey bee, Argentine ant, Asian tiger mosquito, common bed bug, Formosan subterranean termite, Japanese beetle, multicolored Asian lady beetle, red imported fire ant

Vertebrates
> Fish: Bighead carp, rainbow trout
> Birds: Cattle Egret, Eurasian Collared-Dove, European Starling, House Finch, House Sparrow, Rock Pigeon
> Mammals: Black rat, feral cat, feral pig, house mouse, Norway rat, nutria

Missouri

Microorganisms: West Nile virus
Fungi: Bat white-nose syndrome, chestnut blight fungus, chytrid frog fungus, Dutch elm disease
Invertebrates
> Annelid worms: European earthworms
> Mollusks: Asian clam, Chinese mystery snail, quagga mussel, zebra mussel
> Crustaceans: Rusty crayfish
> Arachnids: Honeybee tracheal mite, varroa mite
> Insects: Argentine ant, Asian tiger mosquito, common bed bug, emerald ash borer, Japanese beetle, multicolored Asian lady beetle

Vertebrates
> Fish: Brown trout, grass carp, mosquitofish, rainbow trout
> Birds: Cattle Egret, Eurasian Collared-Dove, European Starling, House Finch, House Sparrow, Rock Pigeon
> Mammals: Feral cat, feral pig, house mouse, Norway rat

Montana

Microorganisms: West Nile virus
Fungi: Chytrid frog fungus, Dutch elm disease, white pine blister rust

Invertebrates
        Annelid worms: European earthworms
        Mollusks: New Zealand mud snail
        Arachnids: Honeybee tracheal mite, varroa mite
        Insects: Common bed bug, multicolored Asian lady beetle
Vertebrates
        Fish: Brown trout, mosquitofish, rainbow trout
        Birds: Cattle Egret, Eurasian Collared-Dove, European Starling, House Sparrow, Rock
            Pigeon
        Mammals: Feral cat, feral horse, house mouse, Norway rat

Nebraska

Microorganisms: West Nile virus
Fungi: Chytrid frog fungus, Dutch elm disease
Invertebrates
        Mollusks: Asian clam, Chinese mystery snail, zebra mussel
        Arachnids: Honeybee tracheal mite, varroa mite
        Insects: Asian tiger mosquito, common bed bug, Japanese beetle, multicolored Asian lady
            beetle
Vertebrates
        Fish: Alewife, bighead carp, brown trout, gizzard shad, mosquitofish, rainbow trout
        Birds: Cattle Egret, Eurasian Collared-Dove, European Starling, House Finch, House
            Sparrow, Rock Pigeon
        Mammals: Feral cat, feral pig, house mouse, Norway rat

Nevada

Microorganisms: West Nile virus
Fungi: chytrid frog fungus, white pine blister rust
Invertebrates
        Annelid worms: European earthworms
        Mollusks: Asian clam, New Zealand mud snail, quagga mussel
        Arachnids: Honeybee tracheal mite, varroa mite
        Insects: Africanized honey bee, Argentine ant, common bed bug, multicolored Asian lady
            beetle
Vertebrates
        Fish: Brown trout, mosquitofish, rainbow trout, spotted tilapia
        Amphibians: American bullfrog
        Birds: Cattle Egret, European Starling, House Sparrow, Rock Pigeon
        Mammals: Feral burro, feral cat, feral horse, feral pig, house mouse, Norway rat

New Hampshire

Microorganisms: Lyme disease bacterium, West Nile virus
Fungi: Bat white-nose syndrome, chestnut blight fungus, chytrid frog fungus, Dutch elm disease,
   white pine blister rust
Invertebrates
        Bryozoan: Lacy crust bryozoan
        Tunicates: Chain tunicate, colonial tunicate
        Annelid worms: European earthworms
        Mollusks: Chinese mystery snail
        Crustaceans: Green crab, rusty crayfish
        Arachnids: Honeybee tracheal mite, varroa mite
        Insects: Common bed bug, gypsy moth, hemlock woolly adelgid, Japanese beetle

(Continued)

Vertebrates
>> Fish: Alewife, brown trout, rainbow trout
>> Birds: Cattle Egret, European Starling, House Finch, House Sparrow, Rock Pigeon
>> Mammals: Feral cat, feral pig, house mouse, Norway rat

New Jersey

Microorganisms: Lyme disease bacterium, West Nile virus
Fungi: Bat white-nose syndrome, chestnut blight fungus, chytrid frog fungus, Dutch elm disease, white pine blister rust
Invertebrates
>> Tunicates: Chain tunicate, colonial tunicate
>> Annelid worms: European earthworms
>> Mollusks: Asian clam, Chinese mystery snail, common periwinkle
>> Crustaceans: Green crab, rusty crayfish
>> Arachnids: Honeybee tracheal mite, varroa mite
>> Insects: Asian longhorned beetle, Asian tiger mosquito, brown marmorated stink bug, common bed bug, gypsy moth, hemlock woolly adelgid, Japanese beetle, multicolored Asian lady beetle
Vertebrates
>> Fish: Asian swamp eel, brown trout, mosquitofish, rainbow trout
>> Amphibians: American bullfrog (Cape May NWR)
>> Birds: Cattle Egret, European Starling, House Finch, House Sparrow, Monk Parakeet, Mute Swan, Rock Pigeon
>> Mammals: Feral cat, house mouse, Norway rat

New Mexico

Microorganisms: West Nile virus
Fungi: Chytrid frog fungus, white pine blister rust
Invertebrates
>> Annelid worms: European earthworms
>> Mollusks: Asian clam
>> Arachnids: Honeybee tracheal mite, varroa mite
>> Insects: Africanized honey bee, common bed bug, Formosan subterranean termite, multicolored Asian lady beetle, red imported fire ant
Vertebrates
>> Fish: Brown trout, rainbow trout
>> Birds: Cattle Egret, Eurasian Collared-Dove, European Starling, House Sparrow, Rock Pigeon
>> Mammals: Feral cat, feral horse, house mouse, Norway rat, nutria

New York

Microorganisms: Lyme disease bacterium, West Nile virus
Fungi: Bat white-nose syndrome, chestnut blight fungus, chytrid frog fungus, Dutch elm disease, white pine blister rust
Invertebrates
>> Tunicates: Chain tunicate, colonial tunicate
>> Annelid worms: European earthworms
>> Mollusks: Asian clam, Chinese mystery snail, common periwinkle, quagga mussel, zebra mussel
>> Crustaceans: Green crab, rusty crayfish, spiny water flea
>> Arachnids: Honeybee tracheal mite, varroa mite
>> Insects: Asian longhorned beetle, brown marmorated stink bug, common bed bug, emerald ash borer, gypsy moth, hemlock woolly adelgid, Japanese beetle, multicolored Asian lady beetle

Vertebrates
>> Fish: Alewife, brown trout, mosquitofish, rainbow trout, round goby
>> Birds: Cattle Egret, European Starling, House Finch, House Sparrow, Monk Parakeet, Mute Swan, Rock Pigeon
>> Mammals: Feral cat, house mouse, Norway rat

North Carolina

Microorganisms: West Nile virus
Fungi: Chestnut blight fungus, chytrid frog fungus, Dutch elm disease, white pine blister rust
Invertebrates
>> Annelid worms: European earthworms
>> Mollusks: Asian clam
>> Arachnids: Honeybee tracheal mite, varroa mite
>> Insects: Argentine ant, Asian tiger mosquito, common bed bug, Formosan subterranean termite, gypsy moth, hemlock woolly adelgid, Japanese beetle, multicolored Asian lady beetle, red imported fire ant
Vertebrates
>> Fish: Alewife, brown trout, lionfish, rainbow trout
>> Birds: Cattle Egret, Eurasian Collared-Dove, European Starling, House Finch, House Sparrow, Mute Swan, Rock Pigeon
>> Mammals: Feral cat, feral horse, feral pig, house mouse, Norway rat, nutria

North Dakota

Microorganisms: West Nile virus
Fungi: Chytrid frog fungus, Dutch elm disease
Invertebrates
>> Annelid worms: European earthworms
>> Arachnids: Honeybee tracheal mite, varroa mite
>> Insects: Common bed bug, multicolored Asian lady beetle
Vertebrates
>> Fish: Brown trout, rainbow trout
>> Birds: Cattle Egret, European Starling, House Sparrow, Rock Pigeon
>> Mammals: Black rat, feral cat, feral horse, house mouse, Norway rat

Ohio

Microorganisms: West Nile virus
Fungi: Chestnut blight fungus, chytrid frog fungus, Dutch elm disease, white pine blister rust
Invertebrates
>> Annelid worms: European earthworms
>> Mollusks: Asian clam, Chinese mystery snail, quagga mussel, zebra mussel
>> Crustaceans: Spiny water flea
>> Arachnids: Honeybee tracheal mite, varroa mite
>> Insects: Asian tiger mosquito, common bed bug, emerald ash borer, gypsy moth, Japanese beetle, multicolored Asian lady beetle
Vertebrates
>> Fish: Alewife, brown trout, mosquitofish, rainbow trout, round goby, sea lamprey
>> Birds: Cattle Egret, European Starling, House Sparrow, Rock Pigeon
>> Mammals: Feral cat, feral pig, house mouse, Norway rat

Oklahoma

Microorganisms: West Nile virus
Fungi: Chytrid frog fungus, Dutch elm disease

(Continued)

Invertebrates
> Crustaceans: Zebra mussel
> Arachnids: Honeybee tracheal mite, varroa mite
> Insect: Africanized honey bee, Argentine ant, Asian tiger mosquito, multicolored Asian lady beetle, red imported fire ant

Vertebrates
> Fish: Bighead carp, brown trout, rainbow trout
> Birds: Cattle Egret, Eurasian Collared-Dove, European Starling, House Finch, House Sparrow, Rock Pigeon
> Mammals: Feral cat, feral pig, house mouse, Norway rat, nutria

Oregon

Microorganisms: West Nile virus
Fungi: Chytrid frog fungus, sudden oak death, white pine blister rust
Invertebrates
> Tunicates: Chain tunicate
> Annelid worms: European earthworms
> Mollusks: Asian clam, Chinese mystery snail, New Zealand mud snail
> Crustaceans: Green crab
> Arachnids: Honeybee tracheal mite, varroa mite
> Insects: Argentine ant, brown marmorated stink bug, common bed bug, multicolored Asian lady beetle

Vertebrates
> Fish: Brown trout, mosquitofish, rainbow trout
> Amphibians: American bullfrog
> Birds: Cattle Egret, European Starling, House Sparrow, Monk Parakeet, Rock Pigeon
> Mammals: Black rat, feral burro, feral cat, feral horse, feral pig, house mouse, Norway rat, nutria

Pennsylvania

Microorganisms: Lyme disease bacterium, West Nile virus
Fungi: Bat white-nose syndrome, chestnut blight fungus, chytrid frog fungus, Dutch elm disease, white pine blister rust
Invertebrates
> Annelid worms: European earthworms
> Mollusks: Asian clam, Chinese mystery snail, quagga mussel, zebra mussel
> Crustaceans: Rusty crayfish, spiny water flea
> Arachnids: Honeybee tracheal mite, varroa mite
> Insects: Asian tiger mosquito, brown marmorated stink bug, common bed bug, Japanese beetle, multicolored Asian lady beetle

Vertebrates
> Fish: Alewife, brown trout, gizzard shad, mosquitofish, northern snakehead, rainbow trout
> Birds: Cattle Egret, European Starling, House Finch, House Sparrow, Rock Pigeon
> Mammals: Feral cat, house mouse, Norway rat

Rhode Island

Microorganisms: Lyme disease bacterium, West Nile virus
Fungi: Chestnut blight fungus, chytrid frog fungus, Dutch elm disease, white pine blister rust
Invertebrates
> Bryozoan: Lacy crust bryozoan
> Tunicates: Chain tunicate, colonial tunicate
> Annelid worms: European earthworms
> Mollusks: Chinese mystery snail, common periwinkle
> Crustaceans: Green crab

        Arachnids: Honeybee tracheal mite, varroa mite

        Insects: Common bed bug, gypsy moth, hemlock woolly adelgid, Japanese beetle, multicolored Asian lady beetle.

Vertebrates

        Fish: Brown trout, rainbow trout

        Birds: Cattle Egret, European Starling, House Finch, House Sparrow, Monk Parakeet, Mute Swan, Rock Pigeon

        Mammals: Feral cat, house mouse, Norway rat

South Carolina

Microorganisms: West Nile virus

Fungi: Chestnut blight fungus, chytrid frog fungus, Dutch elm disease

Invertebrates

        Annelid worms: European earthworms

        Mollusks: Asian clam, Asian green mussel

        Arachnids: Honeybee tracheal mite, varroa mite

        Insects: Argentine ant, Asian tiger mosquito, common bed bug, Formosan subterranean termite, hemlock woolly adelgid, Japanese beetle, multicolored Asian lady beetle, red imported fire ant

Vertebrates

        Fish: Alewife, brown trout, lionfish, rainbow trout

        Birds: Cattle Egret, Eurasian Collared-Dove, European Starling, House Finch, House Sparrow, Mute Swan, Rock Pigeon

        Mammals: Black rat, feral cat, feral pig, house mouse, Norway rat

South Dakota

Microorganisms: West Nile virus

Fungi: Chytrid frog fungus, Dutch elm disease, white pine blister rust

Invertebrates

        Mollusks: Zebra mussel

        Arachnids: Honeybee tracheal mite, varroa mite

        Insects: Common bed bug, multicolored Asian lady beetle

Vertebrates

        Fish: Bighead carp, brown trout, rainbow trout

        Birds: Cattle Egret, European Starling, House Sparrow, Rock Pigeon

        Mammals: Feral burro, feral cat, house mouse, Norway rat

Tennessee

Microorganisms: West Nile virus

Fungi: Bat white-nose syndrome, chestnut blight fungus, chytrid frog fungus, Dutch elm disease, white pine blister rust

Invertebrates

        Annelid worms: European earthworms

        Mollusks: Asian clam, zebra mussel

        Arachnids: Honeybee tracheal mite, varroa mite

        Insects: Argentine ant, Asian tiger mosquito, brown marmorated stink bug, common bed bug, Formosan subterranean termite, hemlock woolly adelgid, Japanese beetle, multicolored Asian lady beetle, red imported fire ant

Vertebrates

        Fish: Alewife, bighead carp, brown trout, grass carp, mosquitofish, rainbow trout

        Birds: Cattle Egret, Eurasian Collared-Dove, European Starling, House Finch, House Sparrow, Rock Pigeon

        Mammals: Feral cat, feral pig, house mouse, Norway rat, nutria

(Continued)

Texas

Microorganisms: West Nile virus
Fungi: Chestnut blight fungus, chytrid frog fungus, Dutch elm disease
Invertebrates
        Mollusks: Asian clam, Chinese mystery snail, golden apple snail, zebra mussel
        Arachnids: Honeybee tracheal mite, varroa mite
        Insects: Africanized honey bee, Argentine ant, Asian tiger mosquito, common bed bug,
           Formosan subterranean termite, multicolored Asian lady beetle, red imported fire ant
Vertebrates
        Fish: Bighead carp, brown trout, grass carp, mosquitofish, rainbow trout
        Reptiles: Green iguana
        Birds: Cattle Egret, Eurasian Collared-Dove, European Starling, House Sparrow, Monk
           Parakeet, Rock Pigeon
        Mammals: Black rat, feral cat, feral pig, house mouse, Norway rat, nutria

Utah

Microorganisms: West Nile virus
Fungi: Chytrid frog fungus
Invertebrates
        Annelid worms: European earthworms
        Mollusks: Asian clam, New Zealand mud snail, zebra mussel
        Arachnids: Honeybee tracheal mite, varroa mite
        Insects: Africanized honey bee, common bed bug, multicolored Asian lady beetle
Vertebrates
        Fish: Brown trout, gizzard shad, mosquitofish, rainbow trout
        Amphibians: American bullfrog
        Birds: Cattle Egret, European Starling, House Sparrow, Rock Pigeon
        Mammals: Feral burro, feral cat, feral horse, house mouse, Norway rat

Vermont

Microorganisms: Lyme disease bacterium, West Nile virus
Fungi: Bat white-nose syndrome, chestnut blight fungus, chytrid frog fungus, Dutch elm disease,
   white pine blister rust
Invertebrates
        Annelid worms: European earthworms
        Mollusks: Asian clam, Chinese mystery snail, zebra mussel
        Arachnids: Honeybee tracheal mite, varroa mite
        Insects: Common bed bug, gypsy moth, hemlock woolly adelgid, Japanese beetle,
           multicolored Asian lady beetle
Vertebrates
        Fish: Alewife, brown trout, rainbow trout
        Birds: Cattle Egret, European Starling, House Finch, House Sparrow, Rock Pigeon
        Mammals: Feral cat, house mouse, Norway rat

Virginia

Microorganisms: Lyme disease bacterium, West Nile virus
Fungi: Bat white-nose syndrome, chestnut blight fungus, chytrid frog fungus, Dutch elm disease,
   white pine blister rust
Invertebrates
        Tunicates: Chain tunicate
        Annelid worms: European earthworms
        Mollusks: Asian clam, veined rapa whelk, zebra mussel
        Crustaceans: Green crab

> Arachnids: Honeybee tracheal mite, varroa mite
> Insects: Asian tiger mosquito, brown marmorated stink bug, common bed bug, emerald ash borer, Formosan subterranean termite, gypsy moth, hemlock woolly adelgid, Japanese beetle, multicolored Asian lady beetle, red imported fire ant

Vertebrates

> Fish: Alewife, bighead carp, brown trout, mosquitofish, northern snakehead, rainbow trout
> Birds: Cattle Egret, European Starling, House Finch, House Sparrow, Monk Parakeet, Mute Swan, Rock Pigeon
> Mammals: Black rat, feral cat, feral horse, feral pig, house mouse, Norway rat, nutria

Washington

Microorganisms: West Nile virus
Fungi: Chytrid frog fungus, white pine blister rust
Invertebrates

> Tunicates: Chain tunicate, colonial tunicate
> Annelid worms: European earthworms
> Mollusks: Asian clam, Chinese mystery snail, New Zealand mud snail
> Crustaceans: Green crab
> Arachnids: Honeybee tracheal mite, varroa mite
> Insects: Common bed bug, multicolored Asian lady beetle

Vertebrates

> Fish: Brown trout, rainbow trout
> Amphibians: American bullfrog
> Birds: Cattle Egret, European Starling, House Sparrow, Rock Pigeon
> Mammals: Black rat, feral cat, house mouse, Norway rat, nutria

West Virginia

Microorganisms: West Nile virus
Fungi: Bat white-nose syndrome, chestnut blight fungus, chytrid frog fungus, Dutch elm disease, white pine blister rust
Invertebrates

> Annelid worms: European earthworms
> Mollusks: Asian clam, zebra mussel
> Arachnids: Honeybee tracheal mite, varroa mite
> Insects: Asian tiger mosquito, brown marmorated stink bug, common bed bug, emerald ash borer, gypsy moth, hemlock woolly adelgid, Japanese beetle, multicolored Asian lady beetle

Vertebrates

> Fish: Alewife, bighead carp, brown trout, mosquitofish, rainbow trout
> Birds: Cattle Egret, European Starling, House Finch, House Sparrow, Rock Pigeon
> Mammals: Feral cat, feral pig, house mouse, Norway rat

Wisconsin

Microorganisms: Lyme disease bacterium, West Nile virus
Fungi: Chytrid frog fungus, Dutch elm disease, white pine blister rust
Invertebrates

> Annelid worms: European earthworms
> Mollusks: Asian clam, Chinese mystery snail, New Zealand mud snail, zebra mussel
> Crustaceans: Rusty crayfish, spiny water flea
> Arachnids: Honeybee tracheal mite, varroa mite
> Insects: Common bed bug, emerald ash borer, gypsy moth, Japanese beetle, multicolored Asian lady beetle

(Continued)

Vertebrates
        Fish: Alewife, brown trout, mosquitofish, rainbow trout, sea lamprey
        Birds: Cattle Egret, European Starling, House Finch, House Sparrow, Rock Pigeon
        Mammals: Feral cat, feral pig, house mouse, Norway rat

Wyoming

Microorganisms: West Nile virus
Fungi: Chytrid frog fungus, Dutch elm disease, white pine blister rust
Invertebrates
        Annelid worms: European earthworms
        Mollusks: New Zealand mud snail
        Arachnids: Honeybee tracheal mite, varroa mite
        Insects: Common bed bug, multicolored Asian lady beetle
Vertebrates
        Fish: Brown trout, mosquitofish, rainbow trout
        Birds: Cattle Egret, European Starling, House Sparrow, Rock Pigeon
        Mammals: Feral cat, feral horse, house mouse, Norway rat

Commonwealth of Puerto Rico

Fungi: Chytrid frog fungus
Invertebrates
        Cnidarian: Australian spotted jellyfish
        Mollusks: Asian clam
        Insects: Red imported fire ant
Vertebrates
        Fish: Brown trout, gizzard shad, mosquitofish, rainbow trout?
        Amphibians: Cuban treefrog
        Reptiles: Burmese python, green iguana
        Birds: Cattle Egret, European Starling, House Sparrow, Monk Parakeet, Rock Pigeon
        Mammals: Feral cat, feral goat, house mouse, Indian mongoose, Norway rat

■ Glossary

Achene. A small, dry, hard one-seed fruit.

Adventive. Refers to an introduced species that has arrived in a new habitat or environment without the aid of humans and that has not established a self-replacing population.

Aeciospore. A fungal spore produced in an aecium. Each spore has two nuclei and is part of a chain of spores.

Aecium. The cuplike fruiting body of some rust fungi.

Aerenchyma. Pithy respiratory tissue, common in stems of some aquatic plant species.

Agnathan. Member of the class Agnatha, the jawless fish.

Alate. Winged reproductive adult of a social insect, such as ants and termites.

Alien (species). A nonnative species. A species found beyond its normal range limits. *Synonyms*: exotic, nonindigenous.

Allee effect. The consequences of low population density when the presence of too few individuals greatly reduces reproductive success.

Alleleopathy. Condition in which one plant or species exudes chemicals that prevent the growth of other plants in the immediate vicinity.

Altricial. Refers to recently hatched birds or other newborn animals that have closed eyes and little or no down or fur, and that are unable to leave the nest and therefore must depend upon the parents for food.

Anadromous. Refers to fish that spend most of their lives in salt water but ascend freshwater streams to spawn.

Anecic. Refers to deep-burrowing earthworms that inhabit the lower layers of the soil.

Annelid worm. Any member of the phylum Annelida, the segmented worms.

Annual. A plant that germinates from seed, matures, and dies in one season.

Apical (snail). The tip of a spiraling shell.

Apomictic. Refers to a flower than does not require pollination to produce seed.

Aquatic. Refers to a plant growing primarily or entirely in water, either rooted or free-floating.

Aril. The fleshy coating around a seed.

Arthropod. Member of the phylum Arthropoda, invertebrates with exoskeletons, segmented bodies, and jointed appendages. The phylum includes arachnids, insects, and crustaceans.

Ascospore. A type of spore bearing a single copy of each chromosome formed by sexual reproduction in fungi in the Division/Phylum Ascomycetes.

Asexual reproduction. The multiplication of individuals without the fusion of gametes. Can occur in fungi and animals through cell splitting, budding, cloning, or sporation. In plants, formation of new plants without the transfer of pollen. In some plants, new individuals can be generated vegetatively from parts of the parent plant.

Auricle. Earlike appendage at the base of some leaves, which clasps the stem.

Awn. A bristle-shaped appendage on a grass.

Axis. The central line of any organ, such as a stem.

Barbel. Whiskerlike tactile organ in catfish and carp that houses taste glands and helps them to find food in murky water.

Basidiospore. A spore bearing a single copy of each chromosome and found in fungi of the Division/Phylum Basidiomycetes.

Beak (bivalve). The highest raised part of each valve, which is generally pointed and located near the hinge.

Benthos. A collective term referring to organisms living on the seabed.

Bergmann's Rule. An ecogeographic pattern wherein the higher the latitude or colder the climate, the larger the body size of warmed-blooded animals compared to close relatives living at lower latitudes and/or in warmer climates.

Biennial. A plant that lives for two years, usually flowering and setting seed in the second year.

Bilabiate. Refers to a corolla, two-lipped.

Biodiversity. The total variation and variability of life found in genes, species, communities, ecosystems, and landscapes.

Biogeography. The science that studies the distribution patterns of species and the processes that determine those patterns.

Biotype. A subset of a species with a particular set of genetic features.

Bivalve. A mollusk, such as a clam or mussel, that has its body covered by two rigid shells joined by a hinge.

Blade. The portion of the leaf that extends from the leaf sheath, flat, folded, or with rolled margins.

Bolt. Rapid growth of flower stalk.

Bract. A small scale-like leaf, usually associated with a flower.

Bulbil. Small bulb, usually growing from leaf axils.

Byssal threads. Filaments that some mollusks produce and use to fasten themselves to hard surfaces.

Calyx. The leaf-like sepals that enclose the petals of a flower.

Canker. A localized area of dead tissue on the trunk or branch of a woody plant.

Cardinal teeth (bivalve). Ridges and grooves on the inner surfaces of both valves of a bivalve near the front end of the hinge that help hold the shells in alignment.

Carton. Material made of undigested cellulose, mud, and termite saliva.

Catadromous. Refers to fish that spend most of their lives in freshwater but migrate to the sea to breed.

Chasmogamous. Refers to flowers that open to allow cross-pollination.

Chitin. A strong, semitransparent, horny substance that is the main material composing the exoskeletons of arthropods and the internal structures of certain other invertebrates.

Chlamydospore. Large, thick-walled resting spore of several kinds of certain fungi. It is the part of the life cycle that allows survival during unfavorable conditions, such as excessive drought or heat.

Cilia. Hairlike structures used by some cells to move themselves or to move food particles.

Clambering. Refers to shrubs or vines with stems that climb onto and over other plants.

Cleistogamous. Refers to flowers that do not open and are self-pollinated as buds.

Clitellum. Thickened, saddlelike section of the body of earthworms that secretes a viscous fluid in which the worm's eggs are deposited.

Colubrid. Any snake of the large and poorly defined group of nonvenomous snakes placed in the family Colubridae.

Columella (snail). The central structural spine of coiled snail and whelk shells.

Community (ecological). All species living in the same area or a subset of them, such as the bird community or the plant community.

Compound. Refers to leaves that are divided into leaflets.

Conidia. Asexual, nonmotile spores of a fungus such as chestnut blight.

Contact. Refers to herbicides that kill only the plant portions contacted.

Coppicing. Refers to trees that sprout many shoots from a cut stump.

Corm. A type of bulb.

Corolla. The petals of a flower.

Corona. A distinct circular growth between the corolla and the stamens, especially in the milkweed family.

Culm. Stem of a grass, usually hollow.

Cuticle (insect). The exoskeleton, composed mostly of chitin.

Cuticle (plant). A protective waxy coating produced by the outermost cells of a leaf or other aerial part of a plant.

Cyme. A broad and flat-topped determinate flower cluster, with central flowers opening first.

DBH (Diameter at breast height). Standard way of expressing the diameter of a living tree. There is no universal standard, however, as to what breast height is. In the United States, DBH is usually measured at a height of 1.4 m (about 4.5 ft.) above ground.

Decumbent. Refers to a stem that is reclining or lying on the ground, but with the tip upright.

Dehiscent. Refers to a seed capsule that opens, sometimes explosively.

Determinate. Refers to when a branch or stem ceases to grow after flowering.

Detritus. Organic debris composed of parts of plants, the remains of animals, and waste products that accumulates on the ground or moves into water bodies from surrounding terrestrial areas.

Diapause. A suspension of development in response to adverse environmental conditions.

Dioecious. Refers to male and female flowers being on different plants.

Disjunct. A distribution pattern in which parts of the range are noncontiguous, i.e., separated geographically.

Drupe. A fleshy, one-seeded fruit. Druplet is one part of a berry fruit.

Ecology. The interrelationships among organisms and the nonliving aspects of the environment in which they live; the science that studies such interrelationships.

Ecosystem. The totality of living and nonliving elements in a given area that function as a unit to cycle nutrients and maintain a flow of energy.

Ecotype. A population that is adapted to a particular environment and displays characteristics that set it apart from related populations but that has not evolved into a distinct species.

Emergent. Refers to aquatic plants that grow primarily above the water surface.

Entire. Refers to leaf margins that are smooth, not toothed or serrated.

Epiphyte. A plant that physically lives on another but obtains no nutrients from the host. Bromeliads and tropical orchids are frequently epiphytic. The leaves of some epiphytic bromeliads fuse to form a tank in which water collects, creating prime breeding grounds for mosquitoes and some treefrogs.

Erect. Growing upright, not sprawling or trailing.

Established (species). A species not native to a geographic area and with a self-replacing population.

Exotic (species). Any nonnative species. *Synonyms:* alien, nonnative, nonindigenous.

Fasciated. Abnormal growth of a plant part, such as an inflorescence, causing it to be twisted or incurved. Also called crested.

Floret. The individual flower of a grass, comprised of two bracts, the lemma and the palea, and the pistil and stamens. Also the individual flower of a composite.

Follicle. A dry, dehiscent fruit or seed pod that splits open on the front.

Forb. A broad-leaved, green-stemmed, nonwoody plant. One type of herb.

Fouling (organism). Any organism that accumulates on solid surfaces in an aquatic environment and impedes the normal mechanical functioning of the equipment or host on which it resides.

Frass. Fine, powdery material that wood-eating insects produce as waste after digesting plant matter.

Fruiting body. Multicellular structure of fungi that carries spore-forming bodies. When the sexual stages of the life cycle are aerial, they are usually visible to the naked eye.

Gamete. A mature sexual reproductive cell, either sperm, pollen, or egg, that fuses with another cell during fertilization to form a new organism.

Genotype. The total complement of genes in an individual or an entire species.

Gill rakes. Bony or cartilaginous, finger-like projections off the gill arch of fish that allow filter-feeders to retain food particles and keep solids from entering the gill cavity. Also called gill rakers.

Glabrous. Smooth; not rough, fuzzy, or hairy.

Glaucous. Covered with a bloom, a whitish substance that rubs off.

Gloger's Rule. An ecogeographic pattern in which warm-blooded animals in humid environments tend to have darker pigments in skin, feathers, or hair than close relatives living in drier environments.

Glume. Bract on a grass that does not have associated flowers.

Gonopodium. An anal fin that on some male live-bearing fish has been modified to allow passage of sperm and internal fertilization.

Graminoid. Herbaceous plant that includes grasses, reeds, rushes, and sedges.

Granivorous. Feeding on seeds.

Gravid. Refers to a female carrying eggs or developing young; pregnant.

Habitat. The place in which an organism lives and the physical attributes of that place.

Halophyte. Plants adapted to salty conditions.

Hammock. Slightly raised tree islands surrounded by other vegetation, usually sawgrass, in the Everglades.

Harborage (insects). Shelter or refuge.

Haustoria. The root-like, absorbing organs of a parasitic plant.

Herb. A plant with no persistent woody stem above ground. Herbaceous.

Hibernaculum. The wintering place that shelters hibernating bats.

Holoparasite. A parasitic plant that can obtain nutrients and water in no way other than from a host plant.

Host specific. Refers to a biological control that affects only the intended plant.

Hyphae. Long, branching threads that are the main vegetative structural feature of many fungi. *See also* Mycelium.

Indehiscent. Refers to a fruit that does not split open at maturity to release seeds.

Indeterminate. Refers to when a branch or stem continues to grow after flowering.

Inflorescence. Flower stalk and how the flowers are arranged.

Injurious wildlife. Species of wild mammals, wild birds, fish, mollusks, crustaceans, amphibians, and reptiles listed under the auspices of the Lacey Act that the secretary of the interior has determined may be harmful to the health and welfare of humans; the interests of agriculture, horticulture, or forestry; and the welfare and survival of wildlife resources of the United States. Such species require a permit in order to be imported or transported between states.

Instar. A developmental stage in larval insects that begins and ends with a molt until the individual is sexually mature. Often, but not always, morphological changes occur from one instar to the next.

Introduced species. A species that has been transported, either deliberately or unintentionally, by humans to a location it had not previously occupied.

Introduction. The transport and release into a free-living state of a nonnative species.

Invasive species. (1) A nonnative species that is currently spreading rapidly or that has done so in the past; (2) "An alien species whose introduction does or is likely to cause economic or environmental harm or harm to human health" (Executive Order 13112).

Involucre. Whorl of small leaves or bracts beneath a flower or flower cluster, especially thistles.

Irruption. A sudden, rapid increase in numbers in an animal population, usually accompanied by the migration of many individuals.

Keratin. A fibrous structural material composed of protein in skin, hair, nails, feathers, and beaks of vertebrates.

Lateral line. A sense organ in aquatic organisms such as fish and amphibians that is used to detect movement and vibration. Commonly visible as a faint line running lengthwise down each side.

Lateral teeth (bivalve). Elongated, interlocking projections along the hinge line of a shell that prevent the two valves from sliding against each other when the shell is closed.

Lemma. The lower of the two bracts that enclose the flower in a grass.

Lenticel. Corky cells in the bark of trees that allow air to penetrate into the interior.

Ligule. Ring on the inside of a grass leaf where the blade meets the sheath.

Lore (bird). The area between the eye and the bill on the side of the head

Macroalgae. Large, multicellular algae. Seaweed.

Macrophyte. Any plant large enough to be visible to the naked eye.

Margins. The edges of leaves or leaflets.

Membranous. Thin, parchment-like texture.

Meristem. The growing point of a plant.

Monocarpic. Refers to a plant in which the growing point of the plant becomes the flowering stem, and the plant dies after flowering.

Monoecious. Refers to both female and male flowers on the same plant.

Monospecific. Consisting of one species.

Monotypic. Consisting of one genotype or ecotype, such as a clone.

Mucilaginous. Soft, moist, sticky, or gel-like.

Mycelium. A mass of branching, filamentous hyphae through which a fungus absorbs nutrients and decomposes plant material.

Naïve. Previously unexposed to a pathogen and therefore having no natural immunity.

Native. In this encyclopedia, describes species, habitats, or ecosystems known to have existed in North America prior to European colonization. Considered by many to be the natural elements of a continent's biodiversity that would occur even if humans had not settled the region.

Native transplant. A species that is native to the country or region in question but has been transported beyond its natural range limits.

Naturalized (species). Refers to a nonindigenous species that is able to sustain itself reproductively in the wild outside of cultivation, and has become a functioning member of a native ecosystem.

Nitrogen fixer. A plant that, with the help of certain soil bacteria that form nodules on its roots, can utilize atmospheric nitrogen.

Node. Joint in a stem, usually where the leaves grow.

Nonindigenous (species). A species that is not native to the place in which it now occurs.

Nonnative (species). An alien, exotic, or nonindigenous species.

Noxious (weed). A plant specified by law as being especially undesirable, troublesome, and difficult to control.

Nuisance (species). According to the Nonindigenous Nuisance Aquatic Prevention and Control Act of 1990, an alien aquatic species that "threatens the diversity or abundance of native species or the ecological stability of infested waters, or commercial, agricultural or recreational activities dependent upon such waters."

Nymph (insect). The immature form of insects that undergo a gradual and incomplete metamorphosis before reaching the adult stage. A nymph resembles the adult form and never enters a pupal stage. It becomes an adult after the final molt.

Opercle. A bony plate that supports the gill covers of fishes, especially the most posterior one.

Organelle. Any of the distinct structures within a cell that performs a specific and vital function.

Outcompete. When a plant or animal displaces another plant or animal by being a better competitor for some resource.

Palate (plant). A bulge in the lower lip of a figwort (Scrophulariaceae) flower that closes off the throat.

Palea. The upper of the two bracts that enclose the flower in a grass.

Palpus. A jointed organ for touching or tasting attached to a mouthpart in arthropods.

Panicle. A loose, irregularly compound inflorescence with flowers on pedicels.

Pantropical. Found throughout the tropics.

Pappus. Appendage to a flower in the sunflower family (Asteraceaceae), such as a thistle, which may remain attached to the fruit; may be bristled, plume-like, or scaly.

Paradioecious. Refers to male and female flowers occurring on separate plants, but any individual can develop flowers of either gender.

Parietal callus. In some snails, a thickened deposit on the margin of the aperture and the wall of the body whorl closest to the central spine (columella). It is often smooth and glossy and may be adorned with raised ribs or wrinkles.

Parthenogenesis. A form of asexual reproduction in which growth and development of embryos occurs without fertilization of the ovum by sperm.

Pathway. The means by which a species arrives at a new region.

Pedicel. Stalk that supports an individual flower or fruit.

Peduncle. Stalk that supports a flower cluster.

Perennial. A plant that lives for more than one season, although aerial parts may die back.

Perigynium. The inflated sac, which encloses the ovary in *Carex* species.

Petiole. A leaf stalk.

pH. The measure of acidity or alkalinity of soil or water. Using a logarithmic scale, it describes the amount of hydrogen ions in the solution.

Pharyngeal teeth (fish). Teeth in the throat located at the back of a fish's head.

Pheromone. Chemicals released by an organism into its environment to communicate with other members of its own species. Some pheromones are alarm signals, while others attract individuals to food or to a mate.

Phloem. Plant tissues that conduct foods made in the leaves to all other parts of the plant.

Photoperiod. The number of hours of daylight.

Phreatophyte. Refers to plants with roots that extend into the water table.

Phytoplankter. A tiny, usually microscopic plant that is part of the plankton.

Pinna. The primary division, or branch, of a pinnate leaf. Leaflets are on the pinna. Plural, pinnae.

Pinnate. Describes compound leaves that have pairs of leaflets on either side of a stalk. Evenly pinnate leaves have an even number of paired leaflets and terminate in a pair. Oddly pinnate leaves have an uneven number of leaflets and terminate in a single leaflet.

Pinnatifid. Describes leaves that resemble pinnately compound leaves, but with lobes that do not reach the midrib of the leaf.

Plankton. A collective term referring to all organisms that drift in open water unable to move under their own power against tides and currents.

Pollard. Describes the method of severely pruning tree limbs back to the trunk or to a main branch.

Polychaete. A member of the Polychaete class of annelid worms characterized by having bristles on each body segment. Also called bristle worms and lugworms.

Postemergent. Describes a herbicide that affects growing plants.

Precocial. Referring to hatchlings or newborns that are born with their eyes open, fully feathered, or furred, and that leave the nest a short time after birth or hatching,

Preemergent. Describes a herbicide that prevents seeds from germinating.

Pronotum. In insects, the upper surface of the first segment of the thorax.

Propagule. In animals, the minimum number of individuals of a species capable of colonizing a new area. This may be fertilized eggs, a mated female, a single male and a single female, or a whole group of organisms, depending upon the biological and behavioral requirements of the species. In plants, a propagule is whatever structure functions to reproduce the species, such as a seed, spore, stem, or root cutting.

Protist. A microorganism that is either single-celled or multicellular, but lacking specialized tissues, and has the genetic information carried in a cell nucleus.

Pubescence. Describes plant parts covered with soft, fine hairs.

Pupa. In the development of those insects that undergo complete metamorphosis, the life stage that immediately precedes the adult stage. Some pupae remain inside the exoskeleton of the final larval instar, but others are encased in a cocoon or chrysalis.

Pustule. A blister-like spot.

Pycnia. A flask-shaped or conical fruiting body of a rust fungus that develops below the epidermis of the host and bears pycniospores.

Pycniospore. A spore produced in a pycnia of a rust fungus. It fuses with a hypha of the opposite mating type to produce the sexual generation.

Raceme. Inflorescence with all the individual flowers on a single axis.

Rachilla. A small or secondary axis or rachis, especially the axis that bears florets in sedges and grasses.

Rachis. An axis bearing flowers or leaflets.

Recurved. Bent backward.

Rhizoid (fungi). A structure that functions like a root to anchor the fungus and absorb nutrients. Rhizoid also releases enzymes that break down organic matter.

Rhizome. A root structure below the soil surface, distinguished from a root by having nodes; can grow shoots that produce new plants. Also called a rootstock.

Rootcrown. Top portion of a root, often containing dormant buds that sprout.

Rosette. Arrangement of leaves radiating from a central point.

Ruderal. Waste places.

Samara. An indehiscent winged fruit.

Saprotroph. Any organism that gains energy and nutrients from dead organic material.

Savanna. Grassland with scattered trees.

Scrambling. Sprawling or climbing over other plants.

Sebaceous gland. A gland in the skin that secretes an oily substance to lubricate the skin and hair.

Semievergreen. Remaining green only in warm climates or sheltered locations.

Senesce. To grow old, turn brown. Senescence.

Sepal. A leaf-like bract that encloses a flower or flower bud. Sepals form the calyx.

Sessile (plant). Refers to flowers or leaves attached directly to stems, without pedicels or petioles.

Settle (mollusks and crustaceans). The process by which larvae leave the plankton stage of their life and attach to a substrate.

Sexual reproduction. The formation of new individuals from the union of two gametes, an ovum and a sperm. In the higher plants it takes places with the transfer of pollen from a male flower to a female flower.

Sheath. A leaf structure that surrounds and encloses a grass stem.

Shrub. Woody perennial smaller than a tree, usually with several stems.

Silicle. A small seed pod in the Mustard family.

Silique. A seed pod in the Mustard family.

Simple. Refers to leaves that are not divided or compound.

Species. A group of individuals of the same kind that can interbreed and produce viable offspring.

Spike. A simple inflorescence with sessile flowers on a single axis. Also branch of a grass inflorescence.

Spikelet. Secondary spike, especially a grass structure that includes glumes and florets, the cluster of grass flowers.

Spirochete. A bacterium of the phylum Spirochaete, distinguished by its spirally twisted form.

Sporangium. Structure in which spores are formed; spore case. Plural, sporangia.

Sporation. Spore formation.

Stellate. Refers to plant hairs, star-shaped.

Stipule. The basal appendage of a petiole.

Stolon. Root stem on the soil surface that roots at nodes; may produce new plants from sprouts; also called runners.

Stromata. The connective tissue framework or support of cells or organisms.

Subdioecious. Male and female flowers usually restricted to separate plants.

Submergent. Refers to aquatic plants, or parts, that grow completely underwater.

Submersed. Refers to aquatic plants that grow primarily underwater. Flowering parts may be at or slightly above the water surface. May be free-floating or rooted.

Subshrub. A shrub in which the upper branches die back during the unfavorable seasons.

Substrate. Substance in which plants are rooted; can be soil, sand, alluvium, mud, or rock.

Surfactant. Substance added to a herbicide to help the chemicals adhere to the foliage.

Systemic. Refers to herbicides that are absorbed into plant tissues and translocated throughout the plant.

Talus. Cone- or fan-shaped slope of loose rocks at the base of a cliff.

Telium. The pimplelike cluster of spore cases that is produced by rust fungi.

Ternate. In sets of three.

Thorax (insects). The central of three main segments of an insect's body: the segment between the head and the abdomen.

Tree. A woody plant with one main trunk.

Turion. A scaly, young shoot or sucker on a root or tuber.

Two-ranked. Referring to alternate arrangement of leaves; leaves are on opposite sides of the stem, in the same geometric plane.

Umbel. Often flat-topped inflorescence resembling an umbrella, with individual pedicels rising from a common point.

Uredinia. A reddish, pimplelike structure on the tissue of a plant infected by a rust fungus.

Vegetative reproduction. Formation of new plants from pieces of the parent plant, such as stems, leaves, rhizomes, and stolons. Also called asexual reproduction.

Vent (reptile). Cloaca. The common cavity into which the intestinal, genital, and urinary tracts end.

Vine. Plant whose stem requires support; can be trailing on the ground or climbing by twining, tendrils, or other means.

Whorl. Arrangement of leaves in a circle around the stem, three or more leaves at one node.

Xylem. Tissue that conducts water and dissolved minerals from the roots to all other parts of a plant, provides mechanical support, and forms the wood of trees and shrubs.

Zooanthellae. Protozoans that live symbiotically in some jellyfishes as well as corals and other marine organisms.

Zooid. One of the individual organisms composing a colonial animal, such as a bryozoan.

Zooplankter. Any animal, single-celled or multicelled, that is part of the plankton.

Zoospore. An asexual spore produced by some fungi that can move around by using a tail-like appendage (flagellum).

Zygomorphic. In plants, irregular corollas that can be equally divided into mirror-image halves in only one plane, such as pea or orchid flowers.

■ Index

Page numbers in boldfaced type refer to a main entry in the encyclopedia and "t" indicates table.

■ About the Authors

SUSAN L. WOODWARD received her PhD in Geography—with a specialization in biogeography—from the University of California, Los Angeles in 1976. Her doctoral research included three years along the Lower Colorado River studying feral burros, considered by some then and now to be an invasive species. When her work began, the burro (along with the feral horse) had just been placed under the jurisdiction of the U.S. Bureau of Land Management (BLM), which had a federal mandate to manage this living symbol of the Old West. The results of her field work provided the BLM with some of its earliest baseline data on burro population biology and ecology.

Dr. Woodward taught biogeography, physical geography, and human ecology for 22 years at Radford University in Virginia, before retiring in 2006. She is the author of *Biomes of Earth* (2003) and served as general editor and author of three volumes for *Greenwood Guides to Biomes of the World* (2009).

JOYCE A. QUINN retired from California State University, Fresno as professor emerita after 21 years of teaching a variety of courses in physical geography and mapping techniques. She earned an MA from the University of Colorado and a PhD from Arizona State University, both in Geography, specializing in the effect of climate and soils on the distribution of plants. She has traveled extensively throughout North America, Latin America, Europe, northern and southern Africa, Uzbekistan, Nepal, China, Southeast Asia, Micronesia, and elsewhere. She is a member of the Cactus and Succulent Society of America and the California Invasive Plant Council and is the author of two volumes of *Greenwood Guides to Biomes of the World* (2009).